Wireless Networks

Series Editor

Xuemin Sherman Shen, University of Waterloo, Waterloo, ON, Canada

The purpose of Springer's Wireless Networks book series is to establish the state of the art and set the course for future research and development in wireless communication networks. The scope of this series includes not only all aspects of wireless networks (including cellular networks, WiFi, sensor networks, and vehicular networks), but related areas such as cloud computing and big data. The series serves as a central source of references for wireless networks research and development. It aims to publish thorough and cohesive overviews on specific topics in wireless networks, as well as works that are larger in scope than survey articles and that contain more detailed background information. The series also provides coverage of advanced and timely topics worthy of monographs, contributed volumes, textbooks and handbooks.

Rui Chen • Wen-Xuan Long • Nan Cheng

Circular Array-Based Radio Frequency OAM Communications

Springer

Rui Chen
Xidian University
Xian, China

Wen-Xuan Long
Xidian University
Xian, China

Nan Cheng
Xidian University
Xian, China

ISSN 2366-1186 ISSN 2366-1445 (electronic)
Wireless Networks
ISBN 978-3-031-37588-0 ISBN 978-3-031-37589-7 (eBook)
https://doi.org/10.1007/978-3-031-37589-7

This Springer imprint is published by the registered company Springer Nature Switzerland AG
The registered company address is: Gewerbestrasse 11, 6330 Cham, Switzerland

Paper in this product is recyclable.

Preface

The RF OAM communication plays an important role in the next-generation wireless communication network, and it is key to improving the spectral efficiency and energy efficiency of networks. To support the practical application of RF OAM communications, the comprehensive and in-depth knowledge of the modeling and transmission theory of RF OAM communications are essential.

This book aims to fill this gap in the existing literature by adequately presenting and developing a comprehensive knowledge of the modeling and transmission theory of RF OAM communications. The studied topics contain the UCA-based RF point-to-point OAM communication system, the UCA-based RF point-to-multipoint OAM wireless backhaul network, the HCCL structure for long-distance OAM transmissions, and the UCA-based joint OAM RadCom scheme. For the UCA-based RF point-to-point OAM communication system, a 2-D ESPRIT-based distance and AoA estimation method will be first introduced, followed by an OAM reception scheme including the hybrid mechanical and electronic beam steering with the estimated AoA and the amplitude detection with the estimated distance. At last, the proposed RF OAM communication scheme will be extended to the RF OAM-MIMO system equipped with UCCAs. For the UCA-based RF point-to-multipoint OAM wireless backhaul network, an OAM-based multi-user distance and AoA estimation method will be first introduced, followed by a multi-user OAM preprocessing scheme. At last, the proposed methods will be extended to the downlink multi-user OAM-MIMO wireless backhaul system equipped with UCCAs. Moreover, a novel HCCL structure will be introduced to realize long-distance OAM transmission. When the HCCL structure is applied in the RF OAM communication, the divergence angles of OAM beams are significantly reduced, and the received SNR is expected to be significantly improved. For the UCA-based joint OAM RadCom scheme, an OAM-based 3-D super-resolution position estimation and rotation velocity detection method will be first introduced, and then the PCRB of the OAM-based estimates and the transmission rate of the integrated system will be derived and analyzed. To achieve the best performance trade-off, the transmitted integrated OAM beams will be optimized by means of an exhaustive search method. Finally, this book will discuss future research directions to inspire further investigation in the field of RF OAM communications from different perspectives.

This book aims to adequately present a comprehensive knowledge of the modeling and transmission theory of RF OAM communications. Meanwhile, future research directions to inspire further investigation in the field of RF OAM communications from different perspectives will be elaborated for the first time.

This book is dominantly written for researchers and professionals whose focus is wireless communications, while advanced-level students majoring in electrical engineering and/or computer science can also find the valuable content and suggestion.

Xian, China

Rui Chen
Wen-Xuan Long
Nan Cheng

Contents

List of Figures

List of Tables

Chapter 1
Overview of OAM Technology in Communications

1.1 OAM Waves and Their Potential

The angular momentum of the electromagnetic field associated with a volume V can be expressed as [98]

$$\mathbf{J} = \int \varepsilon_0 \mathbf{r} \times \mathrm{Re}\{\mathbf{E} \times \mathbf{B}^*\} dV, \tag{1.1}$$

where ε_0 is the dielectric constant in vacuum, $\mathbf{r}$ is the radius vector of a point in the electric field, $\mathbf{E}$ and $\mathbf{B}$ are the electric field intensity and magnetic flux density vector, respectively, $\mathrm{Re}\{\cdot\}$ represents the real part operation, and $(\cdot)^*$ represents the conjugate operator. The angular momentum $\mathbf{J}$ can be decomposed into a component $\mathbf{S}$ (SAM) associated with polarization and a component $\mathbf{L}$ (OAM) related to the spatial distribution of electromagnetic waves, i.e.,

$$\mathbf{J} = \mathbf{S} + \mathbf{L}, \tag{1.2}$$

where

$$\begin{aligned} \mathbf{S} &= \varepsilon_0 \int \mathrm{Re}\left\{\mathbf{E}^* \times \mathbf{A}\right\} dV, \\ \mathbf{L} &= \varepsilon_0 \int \mathrm{Re}\left\{i\mathbf{E}^* \left(\hat{\mathbf{L}} \cdot \mathbf{A}\right)\right\} dV. \end{aligned} \tag{1.3}$$

In (1.3), $\hat{\mathbf{L}} = -i\,(\mathbf{r} \times \nabla)$ is the OAM operator, and $\mathbf{A}$ is the vector potential.

It has been demonstrated that vortex waves having helical phase structure can carry OAM [2]. Due to the phase singularity, the wavefront of electromagnetic waves will be twisted during the propagation, forming vortex waves. The vortex waves are characterized by helical phase fronts, as shown in Fig. 1.1. The amplitude

R. Chen et al., *Circular Array-Based Radio Frequency OAM Communications*, Wireless Networks, https://doi.org/10.1007/978-3-031-37589-7_1

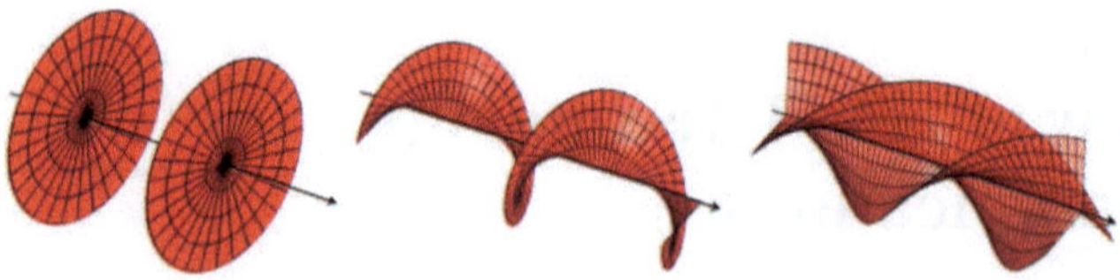

Fig. 1.1 The helical phase fronts of OAM waves: **(a)** $\ell = 0$, **(b)** $\ell = 1$, and **(c)** $\ell = 3$

of the vortex wave field is null along the axis center so that there is a "dark" core at the location of the phase singularity. The helical phase structure is described by the phase term $\exp(i\ell\theta)$, where θ is the transverse azimuthal angle and ℓ is known as the *topological charge* or *OAM mode* and can be any real number. The magnitude and sign of topological charge ℓ determine the number and chirality of the wave torsions in one wavelength, respectively. When an OAM beam has integer desired topological charges, the OAM modes are called *pure*. In practice, it might happen that some of the energy of one pure mode leaks into other modes. Mode purity is measured as the ratio of the power of the desired mode and the overall OAM power.

There are several special OAM waves that can be well described by their radial intensity distributions, among which Laguerre-Gaussian (LG) beams are probably the most widely known. LG beams are paraxial solutions of the wave equation in cylindrical coordinates and homogeneous media (e.g., free space). In cylindrical coordinates, the complex amplitude distribution of a LG beam propagating along the z-axis is given by [2, 160]

$$
\begin{aligned}
LG_{p,\ell}(r,\theta,z) = & \sqrt{\frac{2p!}{\pi w^2(z)(p+|\ell|)!}} \left[\frac{r\sqrt{2}}{w(z)}\right]^{|\ell|} L_p^{|\ell|}\left(\frac{2r^2}{w^2(z)}\right) \\
& \exp\left[\frac{-r^2}{w^2(z)}\right] \exp\left[\frac{ikr^2 z}{2(z^2+z_R^2)}\right] \exp(i\ell\theta) \\
& \exp\left[-i(2p+|\ell|+1)\tan^{-1}\left(\frac{z}{z_R}\right)\right],
\end{aligned}
\tag{1.4}
$$

where $w(z) = w_0[(z^2+z_R^2)/z_R^2]^{1/2}$ is the radius of the beam, w_0 is the waist radius, z_R is the Rayleigh range, k is the wave number, $L_p^{|\ell|}(x)$ is the associated Laguerre polynomial, and $(2p+|\ell|+1)\tan^{-1}(z/z_R)$ is the Gouy phase. It can be seen from (1.4) that LG beams are characterized by two indexes: the azimuthal index ℓ and the radial index p. The former is the topological charge, which characterizes the beam OAM, and the latter is related to the number of radial nodes on the cross section of the beam intensity. Therefore, the azimuth and radial wavefronts of LG beams can be described by the ℓ and p indexes.

The complex amplitude distribution of Bessel beams, another special OAM wave described by their radial intensity distributions, is

$$B(r, \theta, z) = J_\ell(k_r r) \exp(ik_z z) \exp(i\ell\theta), \tag{1.5}$$

where $J_\ell(k_r r)$ is the ℓ-order Bessel function of the first kind and k_r and k_z are radial and axial wave numbers, respectively.

OAM beams have many attractive properties, but probably the most important one is the orthogonality between modes. Considering two OAM waves with topological charges ℓ_1 and ℓ_2, it can be shown that the two OAM modes are orthogonal through the inner product

$$\int_0^{2\pi} e^{i\ell_1\theta} \left(e^{i\ell_2\theta}\right)^* d\theta = \begin{cases} 0 & \ell_1 \neq \ell_2 \\ 2\pi & \ell_1 = \ell_2 \end{cases}. \tag{1.6}$$

Thanks to the inherent orthogonality between OAM modes, OAM waves with different ℓ can be used as a separate set of data transmission channels adding a new dimension independent of time, frequency, and polarization. Accordingly, this new multiplexing dimension can be combined with other existing multiplexing strategies to improve system capacity.

Moreover, OAM has certain application potential in particle manipulation and imaging. Due to the presence of the "dark" core, the pattern of OAM beams has a specific *doughnut shape*. This particular shape allows OAM beams to capture particles or drive the particles to rotate around the beam axis, with great potential in biological and medical applications. Besides, this ring beam with helical phase structure is expected to obtain resolutions beyond the Rayleigh limit for imaging systems and provide new instruments for accurate radar targets imaging.

1.2 Generation of OAM Waves

Since Allen et al. in 1992 proved that the optical vortex with a spiral wavefront carries OAM, the research on OAM has gradually deepened and broadened. In this section, we discuss various state-of-the-art methods for generating OAM in the fields of light, radio, and sound waves. Because of the large number of methods discussed, Fig. 1.2 shows a diagram that summarizes a taxonomy of the techniques presented in this section.

1.2.1 Optical OAM

In optics, a vortex beam can either be obtained as the direct output of a laser cavity or can be generated by feeding a Gaussian beam into a converter, such as a cylindrical lens, a spiral phase plate, a phase hologram, a metamaterial, or a q-plate.

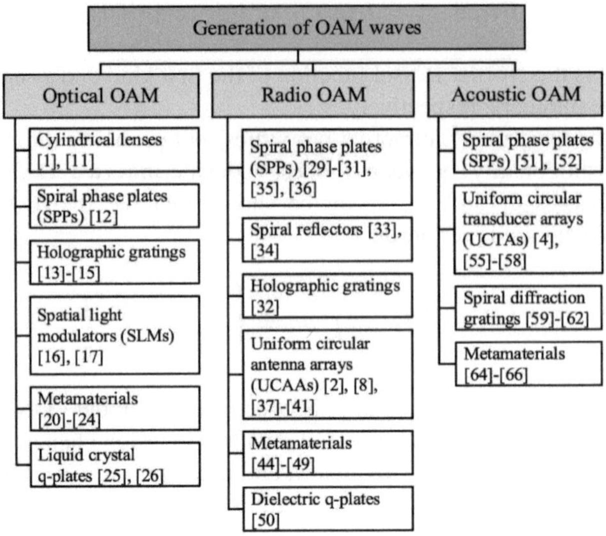

Fig. 1.2 Taxonomy diagram of generation of OAM waves

Even before Allen showed that the vortex beam carries OAM, research has been focused on vortex beams and on how to generate them. In [135], the resonance phenomenon is exploited to generate a vortex from an optical cavity. The laser cavity normally produces a mixture of multiple modes including the basic mode with $\ell = 0$. By placing a component inside the laser cavity, such as a spot-defect mirror [66], the laser cavity can be forced to resonate on a specific OAM mode.

In [2, 8], *cylindrical lenses* are used to transform a Hermite-Gaussian (HG) laser beam into a helically phased LG beam. An example of a second-order HG mode decomposition and LG mode synthesis is shown in Fig. 1.3a. The 45° angle HG mode can be decomposed into a series of HG modes, and this series of HG modes can be rephased to obtain the LG mode. This rephasing can be achieved by changing the Gouy phase shift in the HG mode. Two cylindrical lens mode converters, the $\pi/2$-converter and the π-converter, are proposed in [8], as shown in Fig. 1.3b. The functions of the mode converters are similar to the polarization of birefringence $\lambda/4$ and $\lambda/2$ plates. The $\pi/2$-converter can convert a HG mode with indices m, n, oriented at 45° angle to the lens axis, into a LG mode with topological charge $\ell = m - n$ and number of radial nodes in the intensity distribution $p = \min(m, n)$. The π-converter changes the index of the input mode, that is, $HG_{m,n}$ turns into $HG_{n,m}$, or $LG_{m,n}$ turns into $LG_{n,m}$, which has an azimuthal dependence of the opposite sign. Cylindrical lens mode converters have high conversion efficiency and generate OAM with high purity, but they require high construction precision and, at the same time, have poor flexibility because they require a very precise incident field angle.

The *spiral phase plate* (SPP) is another way to implement the vortex beam [9]. As shown in Fig. 1.4, the SPP is designed like a rotating step so that its thickness increases as the azimuth angle increases but is uniform radially. The step height is

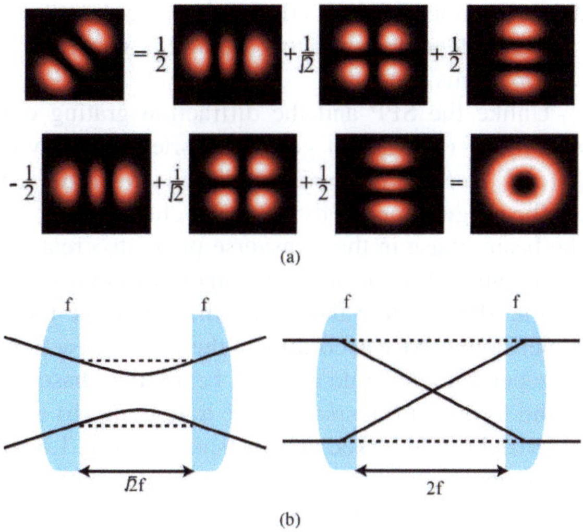

Fig. 1.3 (**a**) An example of second-order HG mode decomposition and LG mode synthesis: HG_{02} at 45° = $\frac{1}{2}HG_{02} + \frac{1}{\sqrt{2}}HG_{11} + \frac{1}{2}HG_{20}$ and $-\frac{1}{2}HG_{02} + \frac{i}{\sqrt{2}}HG_{11} + \frac{1}{2}HG_{20} = LG_{02}$. (**b**) Schematic diagram of $\pi/2$-converter and π-converter. Both converters consist of two identical cylindrical lenses of focal length f [160]

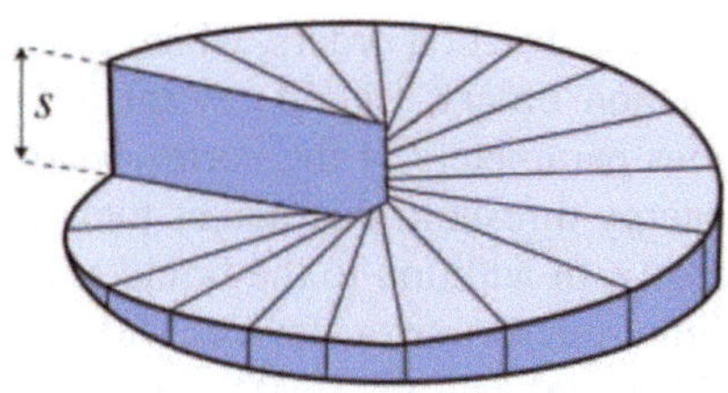

Fig. 1.4 Schematic diagram of SPP with step height s [160]

expressed as $s = \ell\lambda\theta/[2\pi(n-n_0)]$, where n is the refractive index of the transparent dielectric material of the plate, n_0 is the refractive index of the external medium, ℓ is the topological charge, λ is the wavelength of the incident light, and θ is the spatial azimuth. When a plane wave of Gaussian light passes through the phase plate, the light beam experiences a different phase in the azimuth direction due to the spiral thickness of the phase plate and is converted into a helically phased beam with topological charge ℓ. The SPP has high conversion efficiency and can be used for high-power laser beams but has also the limit that it can generate a single mode only, and it has very tight precision requirements.

Instead of producing a complex refractive optical element to generate a vortex wave, one can employ a computer-generated hologram to design a diffractive optical element whose transmittance function is related to the helical phase $\exp(i\ell\theta)$. *Holographic gratings*, such as Fresnel spiral diffraction gratings and ℓ-fold forked diffraction gratings [6, 7, 53], are generated by using a photolithographic process to record an interference pattern on a photoresist-coated substrate. When Gaussian light is incident on the ℓ-fold forked diffraction grating, a vortex beam with the topological charge of $n\ell$ can be obtained at the nth diffraction order. Holographic diffraction gratings are relatively simple and fast to produce and provide good wavefront flatness and high efficiency for a single polarization plane. However, as the number of diffraction orders increases, the quality of the vortex beam is seriously degraded, and this method is generally only used to generate low-order vortex beams.

Unlike the SPP and the diffraction grating described above, the *spatial light modulator* (SLM) can generate vortex beams with different topological charges [31, 94]. A SLM is a pixelated liquid crystal device whose liquid crystal molecules can be programmed to dynamically change the incident beam parameters, including the beam phase in the transverse plane, to create a vortex beam. In detail, a phase hologram with a transmittance function of $\exp(i\ell\theta)$ is digitally generated and loaded on the SLM so that the SLM will determine the phase of each point in the two-dimensional space according to the value of each pixel of the input phase hologram. In general, to generate a vortex beam, the phase hologram loaded onto the SLM is in the form of (a) a spiral phase hologram, (b) a ℓ-fold forked hologram, and (c) a binarized ℓ-fold forked grating (shown in Fig. 1.5a–c, respectively). The spiral phase hologram produces a vortex beam, which exits in the direction perpendicular to the hologram plane, with only a single topological charge. The forked hologram can control the exit direction of the vortex beam, while the binarized forked grating has multiple diffraction orders, which produce different topological charges at different diffraction orders. Liquid crystal SLMs are very flexible since they can control various parameters of the vortex beam, but they are relatively expensive and have an energy threshold that makes their use impossible with high-power laser beams. Recently, in addition to the liquid crystal SLM, a diffractive optical element using a digital micro-mirror device (DMD) has been introduced [95, 97]. In contrast to liquid crystal devices, DMDs cost less and are faster, but their diffraction efficiency is lower.

Transformation optics is a recently developed discipline that employs complex artificial materials, called *metamaterials*, to make transformations in optical space. Generation of optical OAM is one of the fields where the findings of transformation optics can be applied: the metamaterials are planar ultrathin optical components,

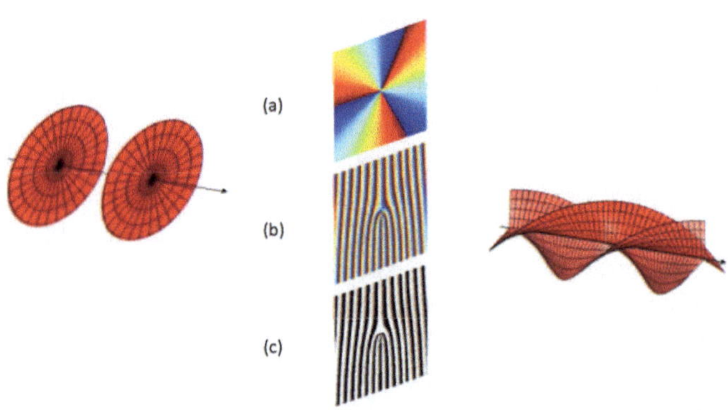

Fig. 1.5 Phase holograms with $\ell = 3$ in the form of (**a**) a spiral phase hologram, (**b**) an ℓ-fold forked hologram, and (**c**) a binarized ℓ-fold forked grating

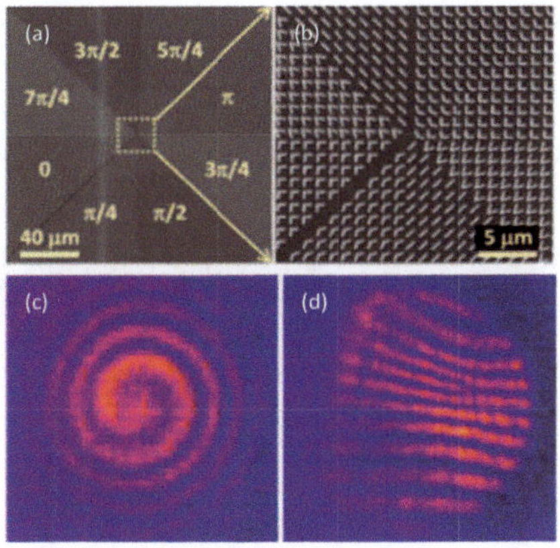

Fig. 1.6 (**a**) A metamaterial composed by a sub-wavelength V-shaped antenna array. The entire structure consists of eight regions, and the phase response difference between adjacent regions is $\pi/4$. (**b**) Partially enlarged view of the center part of (**a**). (**c**) Measured pattern with a spiral fringe created by the interference of the vortex beam with $\ell = 1$ and a co-propagating Gaussian beam. (**d**) Measured pattern with a dislocated fringe created by the interference of the vortex beam with $\ell = 1$ and a Gaussian beam when the two are tilted with respect to each other [162]

usually composed of sub-wavelength constitutional units, such as V-shaped antennas [47, 162], L-shaped antennas [68], rectangular apertures [175], and rectangular split-ring resonators [139]. Rather than relying on the continuous transitions of conventional optics, these planar ultrathin optical components operate by forcing a sudden change of beam phase, amplitude, or polarization at the interface. Thus, optical OAM is obtained by controlling the geometrical parameters (shape, size, direction, etc.) of the metamaterial to manipulate the phases of different azimuths and change the spatial phase of the incident light. As shown in Fig. 1.6, the metamaterial structure is composed by a sub-wavelength V-shaped antenna array, and the beam phase can be controlled by adjusting the angle between the two arms of the V-shaped antenna. The entire surface is divided into eight regions, and the phase response difference between the adjacent regions is $\pi/4$. This object generates a helical phase shift relative to the front of the incident beam producing a vortex with the topological charge of $\ell = 1$. The biggest advantage of the use of metamaterials is their small size that allows for easy integration. Unfortunately, only a vortex beam with a particular topological charge can be generated by a single device.

The *q-plate* [92, 93] is a liquid crystal panel that has a uniform birefringence phase retardation δ and a transverse optical axis pattern with a nonzero topological charge. The angle between the optical axis orientation and the x-axis in xy plane can be expressed as $\alpha(r, \varphi) = q\varphi + \alpha_0$, where α_0 is an initial optical axis orientation. Q-plate patterns with different values of q and α_0 are shown in Fig. 1.7. When q-plate is optimally tuned, i.e., $\delta = \pi$, a light beam incident on it is modified to have a topological charge variation $\Delta\ell = \pm 2q$. For example, a circularly polarized Gaussian passing through a tuned q-plate with $q = 1$ has a helical phase front

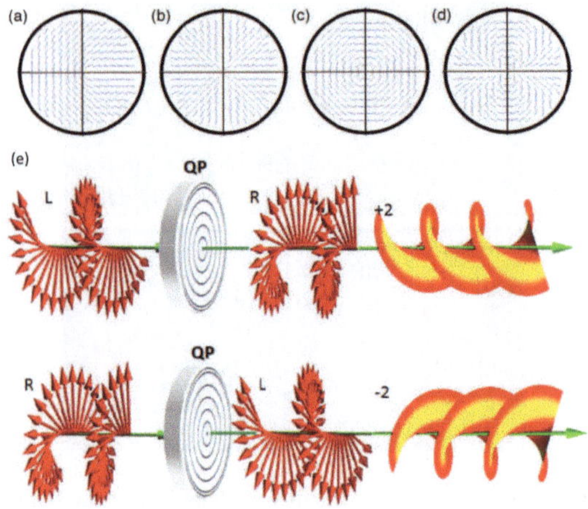

Fig. 1.7 Four examples of q-plate patterns with **(a)** $(q, \alpha_0) = (1/2, 0)$, **(b)** $(q, \alpha_0) = (1, 0)$, **(c)** $(q, \alpha_0) = (1, \pi/2)$, and **(d)** $(q, \alpha_0) = (2, 0)$. The segments indicate the optical axis orientation in the transverse plane. **(e)** Pictorial illustration of the optical action of a tuned q-plate with $q = 1$ on an input circularly polarized plane beam. A left-circular (or right-circular) polarized Gaussian beam passing through a tuned q-plate with $q = 1$ turns into a helically phased beam with $\ell = +2$ (or $\ell = -2$) and right-circular (or left-circular) polarization [93]

with topological charges $\ell = \pm 2$, where the sign depends on the chirality of the input polarized light, as shown in Fig. 1.7e. The characteristic of q-plates is that they generate OAM as a result of optical spin-to-orbital angular momentum conversion, exploiting the law of conservation of angular momentum, and for $q = 1$, the maximum topological charge that can be generated is $\ell = 2$. The q-plate is essentially a Pancharatnam-Berry phase optical component that controls the wavefront shape by transitioning the polarization state. In addition to the q-plate, the computer-generated sub-wavelength dielectric grating [13, 14] and the L-shaped antenna array metamaterial proposed in [68] are also Pancharatnam-Berry optical devices. The q-plate is capable of producing a pure OAM mode but, being made of liquid crystal, has also an energy threshold and thus cannot be used in conjunction with high-power laser beams.

Table 1.1 compares the different methods of generating optical OAM with respect to several parameters: *cost*; *speed*; *conversion efficiency*, the ratio of outgoing power to incident power; *OAM mode*, the capacity of generating pure integer modes, multiple modes, i.e., several different pure modes, or composite modes OAM; *flexibility*, the capacity of controlling the OAM wavefront parameters such as the beam direction or the size of OAM beam ring; *working frequency*; *processing difficulty*, the difficulty in manufacturing the generator; *system complexity*, the level of difficulty of integrating the generator in a transmission system; and *market readiness*, the availability on the market of the technology as a product. As can be seen, the lesson learned is that each method has advantages and disadvantages. Taking the cylindrical lens as an example, it has high mode conversion efficiency

Table 1.1 Comparison of optical OAM generation methods

Features	Cylindrical lenses	SPPs	Holographic gratings	SLMs	Metamaterials	Liquid crystal q-plates
Cost	Normal	Low	Low	High	Low	High
Speed	Normal	Normal	Fast	Normal	Normal	Fast
Conversion efficiency	High	High	Low	Normal	Relatively high	Relatively high
OAM mode	Multiple modes; pure mode	Single mode; non-pure mode	Multiple mode	Multiple modes; compos-ite mode	Single mode	Single mode; pure mode
Flexibility	Low	Low	Low	High	Low	High
Working frequency	All	One point	All	All	A range	All
Processing difficulty	High	High	High	Low	High	Low
System complexity	High	Low	Low	Low	Low	Low
Market readiness	Yes	Yes	Yes	Yes	No	Yes

and generates OAM modes with high purity, but processing difficulty and system complexity are high. The OAM generation method should be chosen according to practical considerations. Flexibility is the biggest advantage of SLMs. If different OAM modes or superimposed modes are required, SLMs should be the first choice. Besides, SPPs perform well in cost, conversion efficiency, and structural complexity. If mode purity and processing difficulty are not considered, SPPs are more recommended in a single OAM mode application due to the low cost and the simple structure.

In optical OAM communication systems, SLMs loaded with phase holograms are widely used to generate OAM, and SPPs are second only to SLMs. In general, cylindrical lenses require a special incident field that limits their application. At the same time, low conversion efficiency makes the application of holographic gratings much less practical than SLMs and SPPs. Metamaterials are low cost, small in size, and high in efficiency, so they are more suitable for small integrated vortex beam generators. The q-plate is a spin-orbital angular momentum conversion device and as such is not recommended yet for OAM communications because of the low value of the maximum topological charge and is more suitable for particle manipulation.

1.2.2 Radio OAM

OAM is a fundamental characteristic of electromagnetic waves at all frequencies. It is not limited to the optical band and can be generated also in the radio band [131].

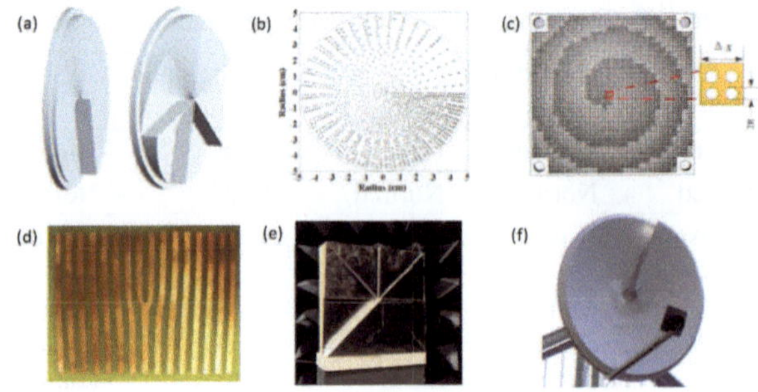

Fig. 1.8 (**a**) Single-stage and multi-stage SPPs [182], planar SPPs with varying borehole densities (**b**) [10], and radius (**c**) [28], (**d**) forked gratings [87], (**e**) stepped spiral reflecting surface [128], and (**f**) spiral parabolic antenna [129]

In analogy with the generation of optical OAM, *SPPs*, whose thickness increases with the increase of the azimuth angle, have been the first method used to generate OAM [133, 153] at high radio frequencies. To generate a high-order vortex beam, the spiral surface is usually replaced by a stepped surface to form a stepped SPP [182] as shown in Fig. 1.8a. Recently, also, planar SPPs [10, 28] have been proposed for generating radio OAM. Unlike conventional SPPs, which use thickness to change the beam phase, planar phase plates adjust the wavefront of the electromagnetic wave by controlling the change in dielectric constant using different borehole densities [10] or radius [28]. Although these devices have a planar structure, like ordinary phase plates, the requirements in terms of structural precision are very high, and the plate is difficult to manufacture.

To create OAM, the stepped spiral surface can be also built as a reflector illuminated by a conventional antenna [128]. The *spiral reflector* causes the reflected electromagnetic wave to have a wave path difference at different positions of the cross section, so that the desired electromagnetic vortex is created. As in Fig. 1.8e, the reflecting stepped spiral surface is divided into N discrete regions, each of which introduces a discontinuous phase difference $2\pi/N$. The topological charge of radio OAM of stepped surfaces is $\ell = 2\mu_0(N+1)/(\lambda N)$, where N is the number of regions into which the reflection surface is partitioned, μ_0 is the total surface spacing, and λ is the wavelength of the incident electromagnetic wave. When N tends to infinity, the stepped spiral reflector becomes a continuous form, such as the spiral parabolic antenna used in the 2012 OAM wireless communication experiment [129]. The spiral reflector is simple and easy to implement. However, it can only respond to specific frequencies and the OAM mode produced by this method is difficult to determine.

Holographic gratings designed by means of computer-generated holograms are also a method of generating radio OAM in analogy with the generation method at optical frequencies [87].

Nevertheless, a *uniform circular antenna array* (UCAA), as shown in Fig. 1.9, is the most commonly used antenna for generating electromagnetic vortex [98, 131]. The UCAA is composed by N elements, uniformly distributed on the circumference. Each array element is controlled by an input signal of the same amplitude but different phase, and the phase difference between adjacent array elements is

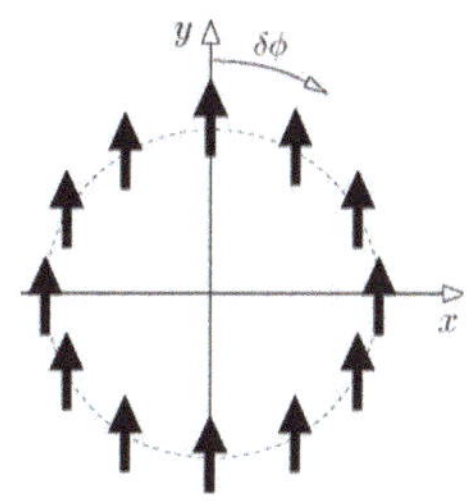

Fig. 1.9 Schematic diagram of a UCAA with N elements uniformly distributed on the circumference, where the azimuth difference between adjacent elements is $\delta\phi = 2\pi/N$ [98]

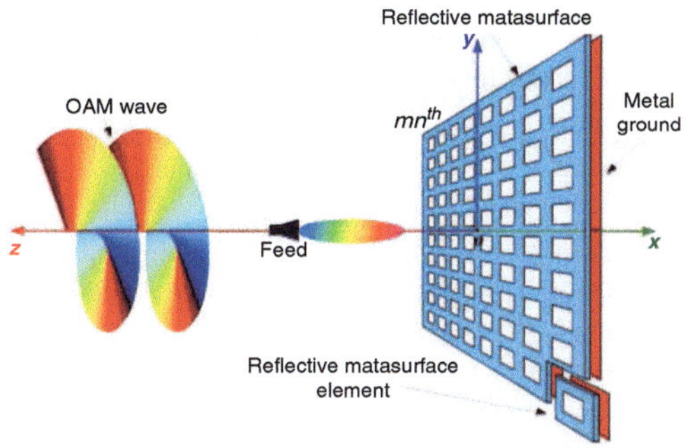

Fig. 1.10 Pictorial illustration of generating OAM with a reflective metamaterial [163]

$\Delta\varphi = 2\pi\ell/N$, where ℓ is the OAM mode, so that the phase difference over the whole array is $2\pi\ell$. Theoretically, a UCAA with N array elements can produce a distortion-free vortex wave with $|\ell| < N/2$. By precisely controlling the amplitude and the phase of the excitation signals, UCAAs have the potential to generate multiplexed radio OAM. In practice, UCAAs can be used in different forms of antennas, such as dipole antennas [98, 131], Vivaldi antennas [32], horn antennas [5], Yagi antennas [42], and microstrip patch antennas [4, 124]. In addition to UCAAs, radio OAM generation with imperfect uniform circular arrays (IUCAs) has also been discussed in [84]. In [130], a circular time-switched array (TSA) is employed to generate vortex electromagnetic waves. This TSA method uses high-speed RF switches to activate the array elements sequentially, enabling the simultaneous generation of all OAM modes at the harmonic frequencies of the TSA switching cycle. The mode of the electromagnetic vortex generated by the UCAA can be controlled flexibly, but this usually requires a complicated feed network.

Similar to the generation method at optical frequencies, *metamaterials*, which can be classified as reflective metamaterials [21, 163–165, 173] and transmissive metamaterials [70], are also used to generate vortex electromagnetic waves. The schematic diagram of generating OAM using a reflective metamaterial is shown in Fig. 1.10, where the incident wave at radio frequencies is reflected and converted into a vortex wave.

In addition to metamaterials, also *q-plates*, made by carving a series of concentric rings on a dielectric plate, can be used to generate a vortex wave at radio frequencies [86].

Table 1.2 Comparison of radio OAM generation methods

Features	SPPs	Holographic gratings	Spiral reflectors	UCAAs	Metamaterials	Dielectric q-plates
Cost	Low	Low	Low	High	Low	Low
Speed	Normal	Fast	Normal	Normal	Fast	Normal
Conversion efficiency	High	Low	Normal	Normal	Relatively high	/(Not discussed)
OAM mode	Single mode; non-pure mode	Multiple mode	Single mode; non-pure mode	Multiple modes; composite mode	Single mode	Single mode; pure mode
Flexibility	Low	Low	Low	High	Low	Low
Working frequency	One point	All	One point	All	One point	One point
Processing difficulty	High	High	High	Low	High	High
System complexity	Low	Low	Low	High	Low	Low
Market readiness	No	No	No	Yes	No	No

Table 1.2 compares the most common methods of generating vortex waves at radio frequencies. Normally, because their size depends on the signal wavelength, SPPs and holographic gratings are suitable for the generations of vortex waves at high frequencies such as millimeter-waves, while spiral reflectors and UCAAs are more suitable for OAM generation at lower frequencies. It is well-known that massive multiple-input multiple-output (MIMO) technology is going to be one of the key technologies for 5G mobile networks. Therefore, UCAAs have attracted a wide attention, and it is currently under study the introduction of OAM into 5G or beyond 5G networks. In addition, thanks to their capability of controlling the wavefront, radar imaging systems frequently use UCAAs to generate OAM waves. The low cost, high conversion efficiency, and simple structure of SPPs make them recommended for radio OAM communication. Since the spiral parabolic antenna can focus the OAM wave while it is being generated, it is more suitable for long-distance transmissions. Moreover, SPPs, spiral parabolic antennas, and other single transmit antennas are expected to be combined with MIMO technology to further increase capacity. Consistently with what happens with optical OAM, low conversion efficiency makes holographic gratings less useful, and metamaterials are more suitable for miniaturized integrated circuits due to the small size and low cost. Also in this scenario, q-plates are not particularly recommended for radio OAM communications.

Terahertz (THz) usually refers to the electromagnetic waves with the frequency of 0.1–10 THz, which is between the far-infrared (FIR) and microwave range [142]. This frequency region is of great significance in the increasingly rare spectrum resources. Its rich spectrum can provide a foundation for the sixth generation (6G), virtual reality, autonomous vehicles, and other emerging technologies [19, 30, 174]. Moreover, as a transition region from the macro-electronics to micro-photonics, THz waves have a series of special properties. Firstly, THz waves have good penetrability for many materials such as wood, paper, and plastic [37], and thus, it can efficiently complement X-ray imaging and ultrasonic imaging technology. Secondly, the photon energy of THz waves is lower than the energy of most chemical bond; therefore, it will not damage the chemical structure of the tested material [20, 51]. Lastly, many material macromolecules have obvious absorption and resonance in the THz region; hence, THz spectral imaging technology can distinguish the material composition and structure of object [52, 91, 104, 126].

Since the THz is between the radio and the optical frequency, the generation of THz-OAM can be inspired from those in both radio and optical region, yet it also has its own requirements. The uniform circular array is the most commonly used method for generating radio OAM beams owing to the simultaneous generation of multiple OAM beams and the flexibility of various antenna elements [22, 26, 27]. However, limited by the accuracy of the commercial THz phase shifter, the structure is not optimal for the THz band temporarily. Using spatial light modulator (SLM) is also very often for generating optical OAM beams [120]; however, the THz SLM is absent commercially. Currently, the main methods of generating the THz OAM beams are spiral phase plate-based method, metasurface-based method, holographic method, and Q-plate-based method.

1.2.3 Acoustic OAM

OAM has found applications in audio bands as well [55]. Because of their simple structure and high conversion efficiency, SPPs are widely used to generate optical vortices and radio vortices. Similarly, *SPPs* have also been employed to generate acoustic vortices. The acoustic SPP in [146] has thickness h that varies with azimuth, $h = \ell\theta/[2\pi f(c_0^{-1} - c^{-1})]$, where c_0 is the sound speed in the surrounding medium, c is the sound speed in the phase plate, and f is the frequency of the incident wave. Another absorbing SPP using optoacoustic technology was proposed in 2004 to generate OAM in the ultrasonic band [50]. The difference is that the optoacoustic conversion efficiency of this SPP is very low, and this device requires high-energy short-pulse excitation. In addition to passive SPPs, active sound sources with spiral thickness [34, 54] are used to generate acoustic vortices. The dimensions of those spiral structures are generally dozens of wavelengths, and their large volume limits the application in low-frequency sound waves.

In analogy with antenna arrays in radio OAM, transducer arrays especially *uniform circular transducer arrays* (UCTAs) are also widely used to generate acoustic vortices carrying OAM. A simple four-panel transducer has first been proposed to produce acoustic vortices [55]. Subsequently, hexagonal arrays [90] and UCTAs [79, 136, 159] have also been studied: UCTAs use fewer transducers than hexagonal arrays and, by adjusting the phase difference between adjacent

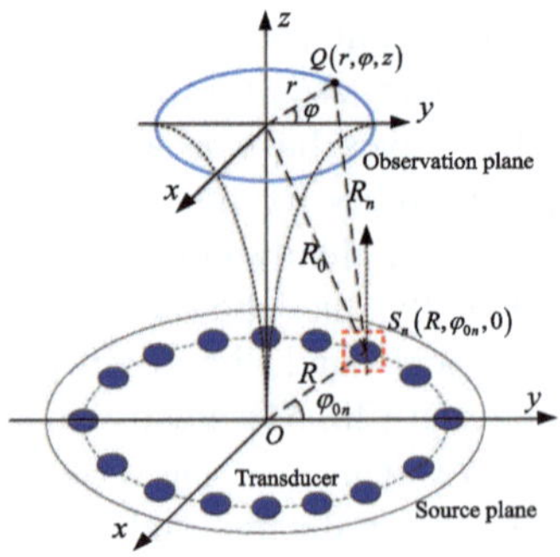

Fig. 1.11 Schematic of an acoustic vortex generation system using a UCTA composed by N uniformly distributed elements. The positions of the n-th sound source S_n and the observation point Q are $(R, \varphi_{0n}, 0)$ and (r, φ, z), respectively [79]

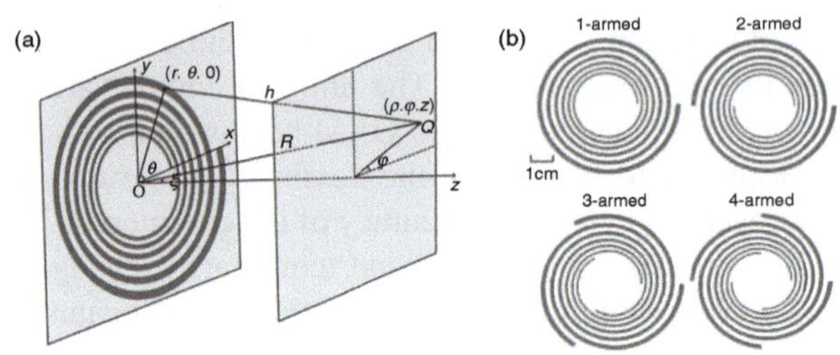

Fig. 1.12 (**a**) The multi-arm coiling slits using logarithmic spiral gratings for generating the stable acoustic vortices. (**b**) Schematic diagram of the one-armed, two-armed, three-armed, and four-armed coiling slits [61]

transducers, are capable of flexibly generating acoustic vortices with different OAM modes. Figure 1.11 shows a UCTA composed by N uniformly spaced elements: by driving each transducer with a sine wave with frequency f and phase difference $\Delta\varphi = 2\pi\ell/N$, the array can generate an acoustic vortex with $|\ell| \leq \lfloor (N-1)/2 \rfloor$, where $\lfloor x \rfloor$ indicates the greatest integer less than or equal to x. By precisely controlling the amplitude and the phase of the excitation signals, UCTAs can generate multiplexed acoustic vortices. It should be noted that each transducer can be regarded as a point source in low-frequency acoustic fields, but not at high frequencies. In high-frequency acoustic fields, the directivity of the sound source needs to be considered because the size of the transducer is larger than the wavelength of the sound wave. Like UCAAs, UCTAs also require complex control circuits, and the complexity grows with the number of transducers.

Another method for generating acoustic vortices is to employ sub-wavelength *spiral diffraction gratings* depicted on rigid materials, which, due to the diffraction effect, can generate acoustic vortices in the paraxial region. Higher-mode-acoustic OAM can be produced by increasing the number of arms of spiral gratings. Using a grating with m arms, an acoustic vortex with the topological charge of $\ell = mn$ is obtained at the nth diffraction order. Spiral gratings can be designed as Archimedes spiral type [63, 140], Fresnel spiral type [64], and logarithmic spiral type [61] shown in Fig. 1.12. The first two types can only be used for a specific frequency, while the logarithmic spiral type is for a wider frequency range. In polar coordinates, when $m = 1$ the two concentric logarithmic spirals are denoted as $r_1 = a_1 \exp(b\theta)$ and $r_2 = a_2 \exp(b\theta)$, respectively, where θ is the angular coordinate; r_1 and r_2 are the radial coordinates; a_1 and a_2 are the inner radii of the two spirals; b varies with the

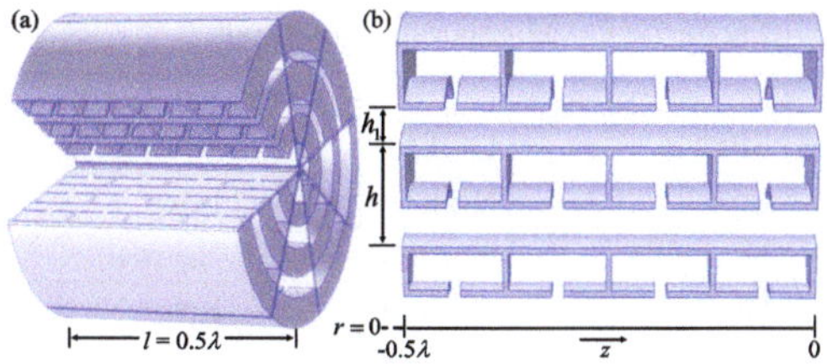

Fig. 1.13 (**a**) Schematic of the assembled planar acoustic resonance layer consisting of eight fanlike sections. The thickness of the layer is 0.5λ. (**b**) An individual section consisting of three rows of resonators with fixed height h in the radial r direction, sided by pipes of varying height h_1 to produce the needed effective wave number [152]

number of arms, determining the growth rate of the spiral; and the width of the spiral grating is $d = r_2 - r_1$. As long as the wavelength of the incident acoustic wave is greater than $2d_{min}$ and less than $2d_{max}$, effective diffraction can occur to obtain the desired acoustic OAM. Since the rigid material blocks a large amount of incident power, such a passive spiral diffraction grating has a low efficiency. Recently, an active spiral grating fabricated using a cellular ferroelectret film has been proposed to improve conversion efficiency [100].

In addition to optical OAM and radio OAM, *metamaterial* can also be used to generate acoustic OAM [36, 152, 161]. Usually, to generate acoustic OAM, a metasurface with sub-wavelength thickness is equally divided into N regions. By changing structural parameters in each region, the phase of the outgoing acoustic wave can be adjusted to be spirally distributed along the azimuthal direction. A planar metamaterial structure for converting planar acoustic waves into vortex acoustic waves using acoustic resonance is shown in Fig. 1.13. The metamaterial is divided into eight fanlike sections. Each individual section is configured to be composed of three rows of resonators in the radial direction. Each row consists of four Helmholtz cavities and a straight pipe. The phase of the outgoing sound wave can be expressed as $\phi_{out} = \phi_{in} + k^{(eff)}l$, where $k^{(eff)}$ is the equivalent wave number and varies with the ratio of the heights of the pipe and resonator h_1/h and l is the structural thickness. It can be shown that when $k^{(eff)}$ depends on azimuth change, the output phase also changes with azimuth. The cascade of four Helmholtz cavities in each row can flexibly control $k^{(eff)}$ from $-k$ to k over the whole azimuth, i.e., achieving an exit phase from 0 to 2π, while the combination of the Helmholtz cavity and the straight pipe can get a very high transmittance. The phase difference between adjacent sections of the metamaterial is $\pi/4$, so that planar acoustic waves are converted into vortex sound waves with the topological charge of $\ell = 1$. This metamaterial structure has the advantages of being highly efficient with a small size and a planar shape. However, the proposed structure can only be used for the generation of acoustic vortices at specific frequencies in the air.

Table 1.3 compares the generation methods of acoustic OAM. Due to the dimension of the structure, SPPs are more suitable for high-frequency sound fields and UCTAs are more suitable for low-frequency sound fields. Because of their

Table 1.3 Comparison of acoustic OAM generation methods

Features	SPPs	UCTAs	Spiral diffraction gratings	Metamaterials
Cost	Low	High	Low	Low
Speed	Normal	Normal	Normal	Normal
Conversion efficiency	High	Normal	Low	Relatively high
OAM mode	Single mode; non-pure mode	Multiple modes; composite mode	Multiple modes	Single mode
Flexibility	Low	High	Low	Low
Working frequency	One point	All	A range	One point
Processing difficulty	High	Low	High	High
System complexity	Low	High	Low	Low
Market readiness	No	Yes	No	No

flexibility, UCTAs are more common in acoustic communications and particle manipulation. Both spiral diffraction gratings and metamaterials are characterized by sub-wavelength dimensions. However, passive diffraction gratings have low efficiency because a large amount of incident sound field energy is blocked and lost and active diffraction gratings are more expensive. Therefore, metamaterials with small size, low cost, and high efficiency have attracted a lot of attention recently.

1.2.4 Summary and Open Challenges

Table 1.4 summarizes the most important methods for generating OAM. Optical OAM generation is a field that has been developed relatively earlier with respect to radio OAM and acoustic OAM generation. Moreover, many methods, such as SPPs and metamaterials, are proposed by analogy with optical OAM for radio and acoustic bands. Nevertheless, based on their respective characteristics, specific methods have been proposed to generate radio OAM and acoustic OAM, such as UCAAs and UCTAs.

The methods of generating optical, radio, and acoustic OAM have been compared in their respective subsections. In general, SLMs are relatively optimal in optical OAM generation. SPPs and UCAAs have also proven to be good at generating radio OAM in practice. For acoustic OAM generation, UCTAs are flexible in both air and water. However, the inherent size of the transducer makes UCTAs not suitable for many miniaturized applications. The metamaterials with small size proposed so far can only generate OAM in the air. Nevertheless, underwater acoustic communication is a research area of great interest, and so, finding a new material or structure with low cost, small size, and high efficiency to generate a stable acoustic vortex field underwater is an important research direction. In addition, the

Table 1.4 Summary of OAM generation methods

Methods	Introduction and features	Remarks
Cylindrical lenses	HG modes without OAM can be converted to LG modes with OAM. High conversion efficiency; high OAM mode purity; a special incident field is required	Optical OAM
SPPs	With spiral thickness $h = \ell\lambda\theta/[2\pi(n - n_0)]$ in optics and electromagnetics and $h = \ell\theta/[2\pi f(c_0^{-1} - c^{-1})]$ in acoustics. Simple structure; high precision; high conversion efficiency; specific working frequency	Optical OAM Radio OAM Acoustic OAM
Holographic gratings	Interference patterns of plane waves and vortex waves, such as ℓ-fold forked diffraction gratings, whose transmittance function is related to $\exp(i\ell\theta)$ Simple and fast; low efficiency	Optical OAM Radio OAM
SLMs	A pixelated liquid crystal device loaded with a phase hologram whose transmittance function related to $\exp(i\ell\theta)$ High cost; flexible control	Optical OAM
UCAAs	N elements uniformly distributed on the circumference with the excitation phase difference $2\pi\ell/N$ can generate radio OAM with $\lvert\ell\rvert < N/2$ Flexible control; a complicated feed network is required	Radio OAM
UCTAs	Similar to UCAAs	Acoustic OAM
Spiral reflectors	The reflection surfaces are divided into N discrete regions and can generate OAM with $\ell = 2\mu_0(N+1)/(\lambda N)$. Simple structure; beam convergence effect; it is difficult to determine topological charge values; specific working frequency	Radio OAM
Spiral diffraction gratings	Archimedes spiral type, Fresnel spiral type, or logarithmic spiral type gratings depicted on sub-wavelength rigid materials. Low efficiency; the topological charge is changed by controlling the number of arms of the grating	Acoustic OAM
Metamaterials	It is composed of sub-wavelength constitutional units and can flexibly manipulate wavefront phase, amplitude, and polarization. Low cost; small size; easy integration; high conversion efficiency; specific working frequency	Optical OAM Radio OAM Acoustic OAM
Q-plates	A uniform birefringent liquid crystal or dielectric plate that can generate OAM as a result of optical spin-to-orbital angular momentum conversion. High conversion efficiency; fast response; high OAM mode purity; simple chiral control of OAM beams	Optical OAM Radio OAM

capacity of focusing the beam for long distance transmissions is essential. How to generate convergent OAM beams is a challenging topic in optical, radio, and acoustic research.

1.3 Detection of OAM Waves

In this section, we discuss the various methods for detecting OAM waves. At the receiver of a communication system, the different OAM modes can be separated easily by exploiting the orthogonality of the helical phase fronts. A variety of methods for detecting OAM have been proposed for light and radio waves. In the acoustic frequency range, OAM detection is still in its early days. A taxonomy diagram of the detection methods discussed here is shown in Fig. 1.14.

1.3.1 Optical OAM

Being the detection of optical OAM the inverse process of generating OAM, OAM modes can be detected by using an *inverse SPP* designed for the specific mode to be detected [143]. Moreover, a vortex beam with the topological charge of ℓ is converted into planar Gaussian light after being irradiated through a *holographic grating* with a transmittance function related to the anti-helical phase factor $\exp(-i\ell\theta)$ [89]. Using spatial mode filters, Gaussian light can be separated from other OAM beams, detecting only one specific OAM mode at the time. For multiplexed OAM modes, a series of holographic gratings are required to sequentially detect each single OAM mode, or, alternatively, beam splitter can be combined for parallel detection. In [48] and [46], a two-dimensional forked holographic grating,

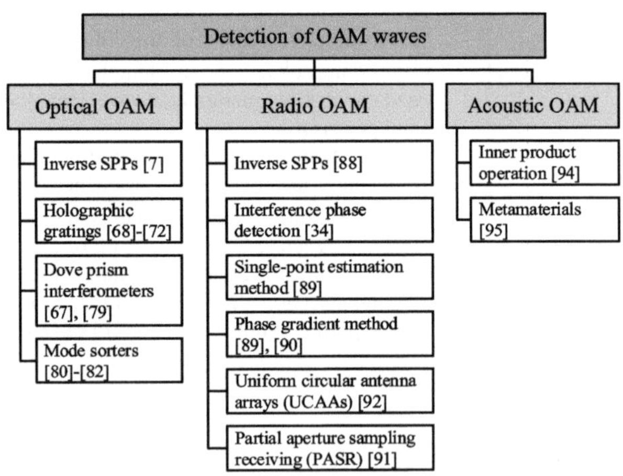

Fig. 1.14 Taxonomy diagram of detection of OAM waves

which is the superposition of a vertical grating and a horizontal grating, is designed to simultaneously detect 8 OAM modes. Two-dimensional Dammann gratings, combining the characteristics of conventional gratings and Dammann gratings, can achieve equal energy distribution on the designed diffraction order, expanding the detection range of the grating [43, 168]. At present, Dammann holographic gratings are widely used at the receiving end of optical communication systems since they can perform parallel detection on multiple OAM states. These holographic gratings require high precision and are normally loaded onto SLMs or recorded using advanced grating fabrication techniques. At the same time, their conversion efficiency is not higher than $1/N$, where N is the total number of detected modes. Recently, [65] has proposed a novel phase hologram, designed by modifying the Lin's algorithm [82], which can be used to achieve almost complete concentration of incident energy to a specified target OAM mode that can be detected more effectively than with conventional forked gratings.

A vortex wave with the topological charge of ℓ can interfere with a plane wave to obtain ℓ spiral stripes or ℓ-fold forked stripes. As such, there are many methods for detecting optical OAM using interference, like the Young's double-slit interference [125, 179], the multipoint interference [11], and the annular aperture interference [29]. However, with these methods, it is difficult to correctly detect the intended OAM mode because of the complexity of the received interference pattern. A simpler method is to use a Mach-Zehnder interferometer with a Dove prism placed in each arm to detect single-photon OAM [74], called *Dove prism interferometer*, as shown in Fig. 1.15. The angle of rotation between the two Dove prisms is $\alpha/2$, and the rotation angle of the passing beam is α. Correspondingly, the phase difference between the two arms is $\ell\alpha$. If $\alpha/2 = \pi/2$, by correctly adjusting the path length of the interferometer, all beams with even ℓ have constructive interference at one output port and cancellation at the other output port, while for all beams with odd ℓ, it is just the opposite. Therefore, such an interferometer can achieve OAM mode odd-ℓ and even-ℓ separation at the single-photon level. As shown in [75], the improved interferometer system is capable of detecting a single-photon OAM and SAM. By cascading multiple interferometers as shown in Fig. 1.16, it is possible to generalize this architecture to detect any number of OAM modes, and the theoretical efficiency is close to 100%. In principle, detecting N OAM modes

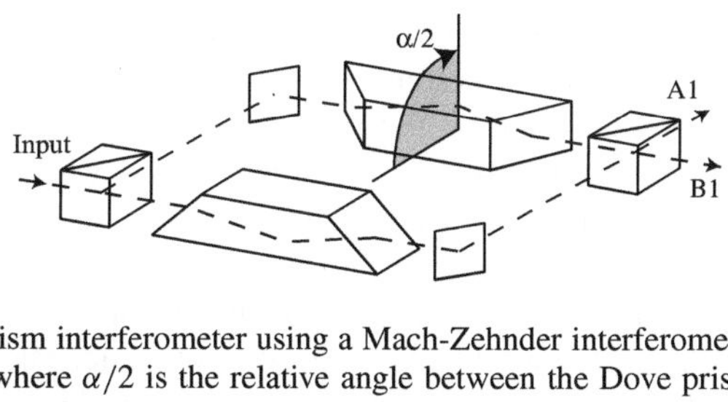

Fig. 1.15 A Dove prism interferometer using a Mach-Zehnder interferometer with a Dove prism placed in each arm, where $\alpha/2$ is the relative angle between the Dove prisms. If $\alpha/2 = \pi/2$, it sorts photons with even-ℓ into port A1 and photons with odd-ℓ into port B1 [74]

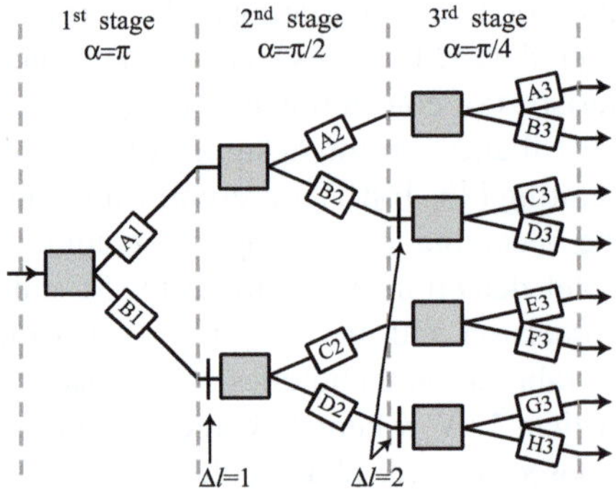

Fig. 1.16 The first three stages of a general sorting scheme, where the gray boxes represent Dove prism interferometers. The first stage introduces a phase shift of $\alpha = \pi$ to sort even ℓs into port A1 and odd ℓs into port B1. The second stage and the third stage introduce phase shifts of $\alpha = \pi/2$ and $\alpha = \pi/4$ to sort $\ell = 4n/\ell = 4n + 2$ and $\ell = 8n/\ell = 8n + 4$, respectively. It should be noted that $\triangle\ell = 1$ and $\triangle\ell = 2$ holograms are required before sorting in the second and third stages [74]

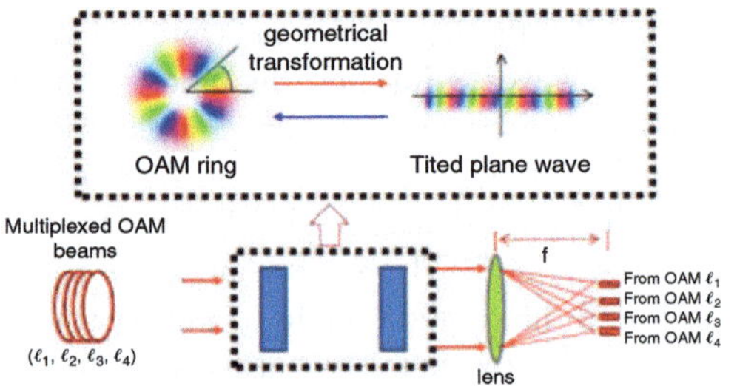

Fig. 1.17 Principle of the OAM mode sorter. By using geometric transformations, ring beams with helical wavefronts are mapped to rectangular beams with tilted wavefronts [143]

requires $N - 1$ interferometers. Therefore, such an interferometer device is too complex for practical applications with a large number of OAM modes (Fig. 1.17).

After passing through a lens, a plane wave converges into a point, whose lateral position on the focal plane depends on the lateral phase gradient of the plane wave. Note that the two plane waves need at least a phase difference of 2π to make the distance between the two focal points larger than the Rayleigh resolution limit. Using this idea, [12] proposes a mode conversion method for sorting OAM modes. The *mode sorter* consists of a converter and a corrector, both implemented on the SLM. The converter implements the conversion from a Cartesian coordinate system to a log-polar coordinate system, so that a point (x, y) on the input plane is converted into a point (u, v) on the output plane, where the conversion law is $u = -a \ln(\sqrt{x^2 + y^2}/b)$ and $v = a \arctan(y/x)$, where a and b are two independently chosen parameters. The corrector is used to compensate the phase distortion caused by the optical path length during the coordinate transformation. Thus, a ring beam with a helical phase front on the input plane is mapped to a rectangular beam with a tilted wavefront on the output plane. The rectangular beams with different lateral phase gradients are focused by the lens on different lateral positions on the focal plane, effectively separating the different OAM modes. In

[73] an OAM mode sorter composed of two custom refractive optical elements is designed to achieve higher conversion efficiency. Nevertheless, there is a limit to the mode sorter due to the overlap between the focal points of two adjacent OAM modes. In [96] it is shown that by using a series of unitary optical transformations, which generate multiple copies of the rectangular beams, it is possible to achieve a larger separation between two OAM modes, trading a greater efficacy with higher system complexity.

In addition to the above methods, there are other techniques for separating OAM modes. In [67] it is proposed a method based on the use of q-plates. The radius ratio method can be used to detect certain special optical OAM [60, 85]. Photonic integrated circuits are well compatible with single-mode fibers while achieving optical OAM multiplexing and demultiplexing [38, 123].

Four common methods for detecting optical OAM are compared in Table 1.5. As mentioned above, also in this case each method has its own advantages and disadvantages. From the perspective of conversion efficiency, the Dove prism interferometer is the best method, but it is bad in terms of flexibility and system complexity, so that it is a not suitable technology for detecting a large number of OAM modes. Inverse SPPs and mode sorters perform better in both respects. A drawback of inverse SPPs is that they can only detect a specific OAM mode, and it is necessary to combine beam splitters and multiple inverse SPPs to achieve composite mode detection. Mode sorters enable composite mode sort, but they cannot effectively separate two adjacent OAM modes. The holographic grating has a low conversion efficiency, but it can be easily loaded onto a SLM. In current optical OAM experimental systems, the holographic grating loaded on a SLM is a relatively common method. In addition, inverse phase holograms are also frequently used in practice. Some novel phase holograms have also been tried to detect optical OAM with high efficiency.

Table 1.5 Comparison of optical OAM detection methods

Features	Inverse SPPs	Holographic gratings	Dove prism interferometers	Mode sorters
Cost	Low	Low	Low	Low
Conversion efficiency	High	Low, no higher than $1/N$	High, near 100%	Relatively high
OAM mode	Single mode detection	Composite mode detection; the number is limited by the size of the grating	Composite mode sort; the more the number, the more complicated the device	Composite mode sort; adjacent OAM mode cannot be effectively separated
Flexibility	Low	High	Low	High
System complexity	Low	Low	High	Low
Market readiness	Yes	Yes	Yes	Yes (If implemented on SLMs)

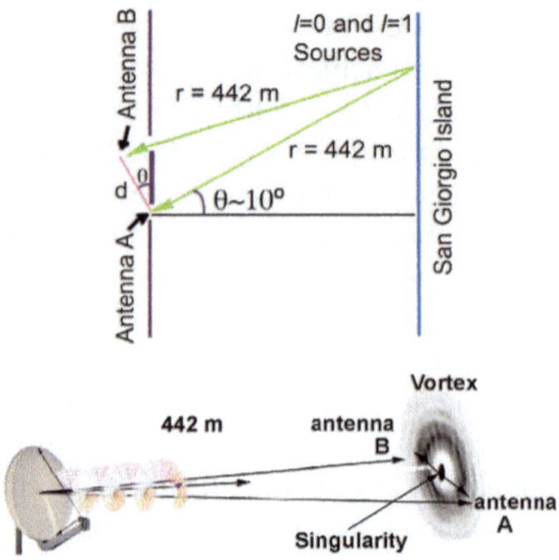

Fig. 1.18 Sketch of the famous Venice experiment of radio OAM communications, in which two receive antennas A and B are aligned with the phase singularity [129]

1.3.2 Radio OAM

As in optics, the detection of OAM at radio frequencies can be achieved by employing carefully designed *inverse SPPs* [154]. It is also possible to detect the radio OAM mode by observing the interference pattern of a vortex wave and a plane wave. Nevertheless, many detection methods in optics are no longer applicable at radio frequencies, and some new methods have been proposed.

In [129] the *interference phase detection* is used to separate two electromagnetic beams with topological charges of $\ell = 0$ and $\ell = 1$. In the experiment setup, two Yagi antennas A and B are connected by a 180° phase-shift cable to form a phase interferometer, designed to reduce background interference in a real environment. As shown in Fig. 1.18, if the device is fully aligned, the vortex electromagnetic beam with $\ell = 1$ will produce a phase difference of 180° between receive antennas A and B. This phase difference is compensated by the cable line phase delay, so that the received interference intensity is maximized. For the beam with $\ell = 0$, there is only the phase delay of the cable, and the intensity is minimal. In this way, it is possible to separate beams with different wavefront characteristics. When the interference phase detection method is used for more radio OAM modes, the receiving device becomes very complicated.

In 2010, Mohammadi et al. proposed two methods to estimate far-field OAM modes in radio beams, the *single-point estimation method* and the *phase gradient method* [99]. The single-point estimation method uses an approximation for OAM in the far-field to estimate the OAM from the measurements at a single point of the vertical and the transverse components of the electric field. This method is efficient for the detection of OAM with low-mode numbers. Compared with the single-point estimation method, the phase gradient method is more intuitive and effective. As shown in Fig. 1.19, the measurement setup consists of two test points separated by an angle β on a circumference at the receiving end. If the measured phases are ϕ_1 and ϕ_2, respectively, the OAM mode will be $\ell = (\phi_1 - \phi_2)/\beta$. In order to accurately

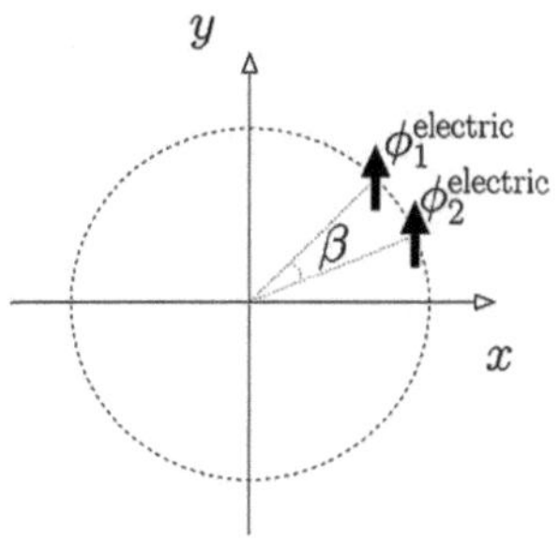

Fig. 1.19 Illustration of the phase gradient method. The OAM mode can be estimated by computing the phase difference of the electric field components at two sampling points separated by an angle β, i.e., $\ell = (\phi_1 - \phi_2)/\beta$ [99]

detect the OAM mode, the angle β between the two test points should satisfy the relationship $\beta < \pi/|\ell|$. The phase gradient method requires that the center of the receiving circle is aligned with the center axis of the vortex beam; otherwise, there will be large errors. Generally, it can only be used to detect single-mode OAM beams, but with some modifications, it can also achieve accurate measurement of two mixed modes [150].

UCAAs are used not only for radio OAM generation but also for OAM detection [145]. The condition for detection is that the two arrays are aligned, i.e, the cross section of the receiving circular array C should be perpendicular to the beam axis, and its center should coincide with the axis. UCAAs can demultiplex several radio OAM modes. The whole process of OAM demultiplexing can be seen as some sort of spectral analysis of the received vortex electromagnetic waves. When the mode number ℓ' of the circular array C is different from one of the beam modes, the result of the spectral analysis tends to 0; conversely, it tends to 1. By selecting different ℓ' values, different information can be extracted from the OAM multiplexed beam. When multiple topological charge values are selected at the same time, composite mode detection can be realized. In general, the detection process needs to receive information of the entire wavefront. Due to the divergence of OAM beams, a large receiving array is required to capture the effective power of the OAM beam in the far field, but this might be difficult to achieve. An answer to the problem of OAM detection in the far field is *partial aperture sampling receiving* (PASR) [56], which uses partial wavefront information to detect and distinguish different radio OAM modes, as shown in Fig. 1.20, and has been verified by computational simulation. The PASR, which is a combination of the partial angular receiving aperture method for OAM demultiplexing in optics [178] and a sampling receiving scheme, has a receiving arc which is $1/P$ of a circumference and employs M antennas evenly distributed on the arc as signal sampling points. The angle between adjacent antennas is $2\pi/MP$. To ensure strict orthogonality and realize non-crosstalk separation of the two OAM modes ℓ_{n1} and ℓ_{n2}, they must satisfy simultaneously the two conditions: $\text{mod}\,(|\ell_{n1} - \ell_{n2}|, P) = 0$ and $\text{mod}\,(|\ell_{n1} - \ell_{n2}|, MP) \neq 0$. Although the PASR has certain limitations, it can greatly simplify the scale of the receiving end and is robust to non-ideal OAM beams.

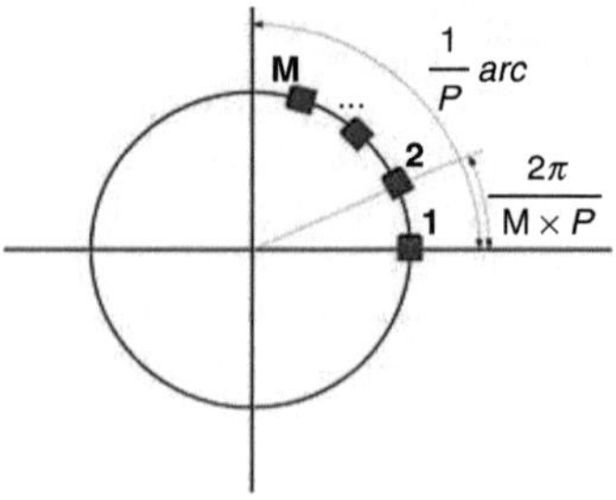

Fig. 1.20 The PASR scheme with M antennas uniformly distributed on a $1/P$ arc [56]

Table 1.6 Comparison of radio OAM detection methods

Features	Inverse SPPs	Interference phase detection	Single-point estimation method	Phase gradient method	UCAAs	PASR
Cost	Low	Low	Low	Low	High	High
Sampling range	Entire wave-front	Two points	One point	Two points	Entire wave-front	Partial wavefront
OAM mode	Single mode	Composite mode; The higher the n-umber, the more complicated the device	Single mode; It is more suitable for low mode	Single mode; There is a restriction on the use of OAM mode	Composite mode	Composite mode; There is a restrictionon the use of OAM mode
System complexity	Low	High	Low	Low	High	High
Computational complexity	Low	Low	High	Low	Low	Low

Table 1.6 compares the common detection methods for radio OAM. Unlike the case of optical OAM detection, these methods often require a heavy computational load so that also computational complexity is one of the parameters evaluated in Table 1.6. For multiple OAM modes detection, the system complexity of interference phase detection is very high. The single-point estimation method extrapolates the OAM mode in the far field from the measurements of the two components of the electric field, so its computational complexity is rather high. Therefore, these two methods are not recommended for practical implementations. In terms of complexity, the phase gradient method is probably the best method, but it can only detect single-mode OAM. Although UCAAs need to receive the entire wavefront information, it can be used for composite OAM mode detection. The PASR can be seen as an improvement of UCAAs. It has proved its advantages through computational simulation, but further experimental verification is required. In radio OAM communications, inverse SPPs are widely used to detect OAM

modes. In the particular case of line-of-sight (LOS) wireless communications, UCAAs have proved to be more suitable for demultiplexing composite OAM modes.

After accurately detecting the amplitude, phase, and other parameters of THz beams, the various applications based on THz technology could be realized. Therefore, the detection of THz waves is as important as the generation. The proposed methods for detecting THz beams can be divided into two categories: photodetection method and electronic detection method [122].

The photodetection method detects the THz signals through optical beams, and it includes the photoconductive antenna scheme, the electro-optical crystal scheme, and other schemes. The photoconductive antenna consists of two parallel electrodes plated on a semiconductor substrate [18, 69, 102]. When the pulsed optical beam illuminates on the photoconductive antenna, a large number of free carries would be generated. The THz field could be approximately regarded as a constant electric field when the lifetimes of free carriers are much shorter than the pulse period of THz. Thus, the free carriers would directionally move to generate current, and the information of the THz waves could be obtained by detecting the current.

The electro-optic crystal scheme uses the Pockels effect. When a linearly polarized light passes through the electro-optic crystal under this THz field, the light would become elliptically polarized as the external THz field has different effects for the refractive indices of ordinary light and extraordinary light. Consequently, the information of THz pulse could be reflected when detecting the difference of ordinary light and extraordinary light [151].

The photoconductive antenna scheme and electrooptical crystal scheme have a similar principle for detecting THz signals, but they also have their own characteristics. For the low-frequency THz signals, the photoconductive antenna scheme has a higher signal-to-noise ratio; for the high-frequency THz signals, the electro-optical crystal scheme has higher sensitivity.

The detection process of THz-OAM beams can be divided into two parts: the demodulation of THz-OAM beams and the detection of the demodulated THz-OAM beams. Currently, the key of the application of THz-OAM technology in practice is how to separate and detect OAM beams carrying different modes efficiently and accurately. Currently, the main methods of detecting the THz OAM beams are reverse spiral phase plate and plane wave interference method, coordinate transformation method, and electro-optical crystal-based method.

1.3.3 Acoustic OAM

As mentioned above, research on OAM detection at acoustic frequencies is still in an early stage. Due to the orthogonality between the OAM eigenstates, Shi et al. use the inner product operation to separate the multiplexed OAM modes received by a transducer array [121]. Recently, Jiang et al. used passive metamaterials to detect acoustic OAM [62]. The structure of such passive acoustic metamaterials has been described in Sect. 1.3. After that an acoustic beam is passed through a metamaterial with the topological charge of $\ell = -1$, the mode number of all multiplexed OAM modes will be reduced by 1. Therefore, m metamaterials with $\ell = -1$ can convert the component with the mode number $\ell = m$ into a plane wave, and a nonzero value can be detected at the "dark" core point of the OAM beam. This nonzero value contains the information carried by the OAM with $\ell = m$. This method can realize real-time demodulation of OAM beams and has high conversion efficiency but low flexibility and is not suitable for detection of high-mode OAM.

1.3.4 Summary and Open Challenges

Table 1.7 summarizes the most common methods employed for the detection of OAM modes. As seen from the table, for optical OAM and radio OAM, many detection methods have been proposed, while further research, investigation, and testing are needed for developing effective acoustic OAM detection.

In general, using the inverse SPP is the easiest way to detect optical OAM and radio OAM. In optical OAM communications, holographic gratings or phase holograms loaded on SLMs, which can detect multiple OAM modes simultaneously, are used in most cases. In radio OAM communications, in addition to SPPs, UCAAs have the potential to demultiplex composite OAM modes. However, OAM beam divergence increases the aperture of the receiving UCAA. Reducing the aperture and scale of the far-field receiving array is extremely challenging. The PASR and the phase gradient method are promising solutions to reduce the scale of the receiver. But how to apply them to OAM mode demultiplexing requires further research in the coming years.

1.4 Application of OAM Technology in Communications

This section focuses on the applications of OAM in communications. Since OAM modes with different values of ℓ are mutually orthogonal, vortex beams carrying different OAM modes can provide independent communication channels for efficient information transmission. As shown in Fig. 1.21, there are two main strategies to transfer information with OAM: *OAM shift keying* (OAM-SK), where the information is encoded in the value ℓ of the OAM beam, and *OAM division multiplexing* (OAM-DM), where the information is multiplexed in multimodal OAM beams. OAM-DM uses the vortex beams as the carrier of information so that data is loaded onto different OAM modes and then multiplexed and transmitted coaxially through a single aperture. At the receiver, the beams are collected by another aperture and then demultiplexed and detected for data recovery. Compared with OAM-SK, OAM-DM system has higher spectral efficiency and exhibits lower bit error rates.

Table 1.7 Summary of OAM detection methods

Methods	Introduction and features	Remarks
Holographic gratings	Holographic gratings with $\exp(-i\ell\theta)$ phase factor can convert vortex beams into plane beams, and Daman gratings can realize parallel detection of multiple OAM modes. Simple and fast; low conversion efficiency, no higher than $1/N$, where N is the total number of detected modes	Optical OAM
Dove prism interferometers	By adjusting the rotation angle of the Dove prism, the odd-even sort of OAM modes can be achieved on a single-photon level. High conversion efficiency, close to 100%; any number of OAM modes can be detected by cascading multiple interferometers while increasing the complexity of the device	Optical OAM
Mode sorters	The geometric transformation is used to map the Cartesian coordinate to the log-polar coordinate, and ring beams with helical wavefronts are mapped to rectangular beams with tilted wavefronts, which are focused by lens in different lateral positions. High conversion efficiency; adjacent OAM mode cannot be effectively separated	Optical OAM
Inverse SPPs	SPPs with the topological charge of $-\ell$. Simple structure; single mode detection; composite mode detection can be realized by combining multiple inverse SPPs and beam splitters	Optical OAM Radio OAM
Interference phase detection	The phase shift cable is used to compensate the phase difference between the receiving antennas, so that the specific OAM mode has high intensity and the other modes have low intensity. Intuitive; the receiving device is very complicated when detecting multimodal OAM	Radio OAM
Single-point estimation method	It uses an approximation for OAM in the far field to estimate the OAM from the measurements at a single point of the vertical and the transverse components of the electric field. It is efficient for the detection of OAM with low mode numbers	Radio OAM
Phase gradient method	Estimate the OAM mode by computing the phase difference of the electric field components at two test points separated by an angle β, i.e., $\ell = (\phi_1 - \phi_2)/\beta$. Low complexity; small scale; the angle β between the test points should satisfy $\beta < \pi/(\lvert\ell\rvert)$	Radio OAM
UCAAs	A receiving circular array C with $\ell = \ell'$ can extract the information carried by the beam with $\ell = \ell'$ from the multiplexed OAM beam. The detection process needs to receive information of the entire wavefront	Radio OAM
PASR	Demultiplexing is implemented by utilizing partial wavefront information and the orthogonality between OAM modes. Small scale; any two OAM modes ℓ_{n1} and ℓ_{n2} must satisfy $\bmod(\lvert\ell_{n1} - \ell_{n2}\rvert, P) = 0$ and $\bmod(\lvert\ell_{n1} - \ell_{n2}\rvert, MP) \neq 0$	Radio OAM
Metamaterials	M metamaterials with $\ell = -1$ can convert the component with the mode number $\ell = m$ into a plane wave. Real time; high conversion efficiency; low flexibility; specific working frequency; it is not suitable for detection of high-mode OAM	Acoustic OAM

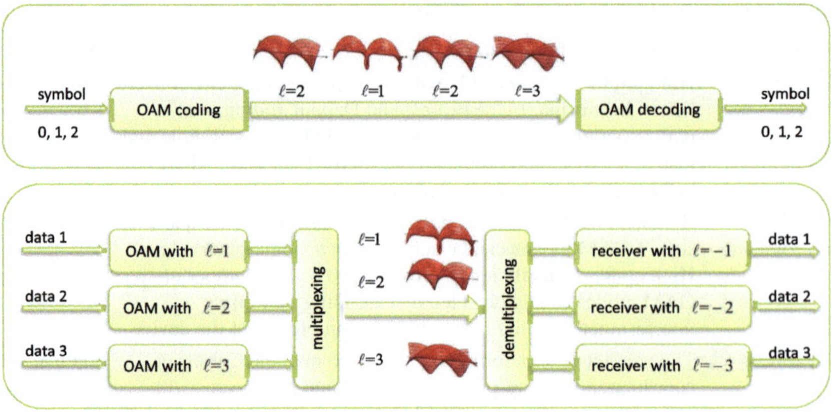

Fig. 1.21 Schematic diagram of OAM-SK and OAM-DM communications: in OAM-SK communication (top), the information symbols are encoded in the topological charges of OAM beams, while in OAM-DM communication (bottom), the information is multiplexed in multimodal OAM beams

OAM-DM as a mode-division multiplexing (MDM) is a subset of space-division multiplexing (SDM) and can be compatible with different modulation formats, such as M-ary amplitude-shift keying (M-ASK), M-ary phase-shift keying (M-PSK), and M-ary quadrature amplitude modulation (M-QAM), as well as other multiplexing techniques such as frequency-division/wavelength-division multiplexing (FDM/WDM) and polarization-division multiplexing (PDM), thereby further improving the communication system capacity from another dimension. In the OAM-DM system, N OAM waves carrying information are multiplexed, and the obtained field can be expressed as

$$U_{MUX}(r,\theta,t) = \sum_{p=1}^{N} S_p(t) A_p(r) e^{i\ell_p\theta}, \tag{1.7}$$

where $S_p(t)$ is the modulated data signal on the pth OAM mode and $A_p(r)$ is the complex electric field amplitude of the pth OAM mode. Then, at the receiving end, the multiplexed OAM wave is multiplied by an anti-helical phase factor $\exp(-i\ell_q\theta)$, which is achieved by using an inverse SPP or an antenna array. The OAM wave with $\ell = \ell_q$ is converted into a plane wave and can be easily separated from other vortex waves with $\ell = \ell_p - \ell_q$, thus achieving demultiplexing. Eventually, the data information carried by the OAM wave with $\ell = \ell_q$ can be obtained.

OAM communications still face many challenges and technical issues, such as misalignment and OAM beam divergence. OAM waves are required to be transmitted and received coaxially in OAM-DM system, so that precise alignment is required between the transmitter and receiver, i.e., the center of the receiver

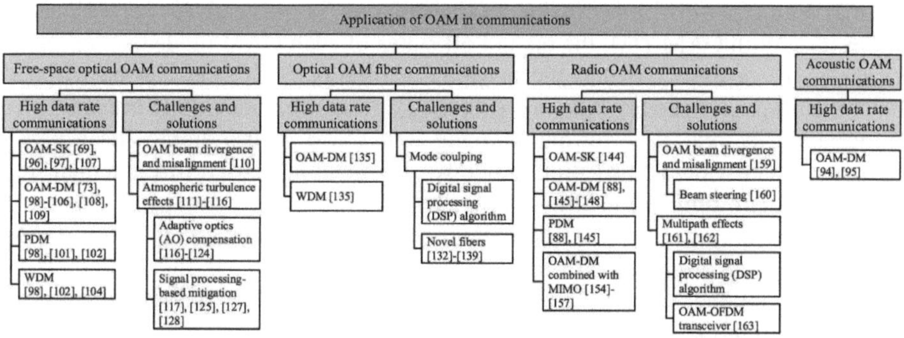

Fig. 1.22 Taxonomy diagram of applications of OAM in communications

coincides with the center of the transmitted beam, and the receiver is perpendicular to the line connecting the centers. Moreover, since optical components usually have a limited aperture size, OAM beam divergence during propagation can result in received power loss in the far field, limiting the link achievable distance. The atmospheric turbulence, mode coupling, and multipath effects should be considered in free-space optical links, fiber links, and radio communication links, respectively. For acoustic OAM communications, it is clear that underwater creatures and turbulence have a great impact on links, although there is currently no research to discuss them. A taxonomy diagram of the applications discussed in this section, together with main challenges and solutions for each application, is shown in Fig. 1.22.

1.4.1 Free-Space Optical OAM Communications

In 2004, Gibson et al. [48] demonstrated for the first time that OAM-SK modulation can be used for free-space optical communications with good results. The transmitter maps the information data on the OAM beams' topological charge, which belongs to the set $\mathcal{L} = \{-16, -12, -8, -4, +4, +8, +12, +16\}$. At the receiving end, two vertically stacked forked gratings are used to detect the transmitted OAM modes. To compensate for small perturbations in alignment, a Gaussian beam with $\ell = 0$ is used as a reference signal. In [71], 16 superimposed OAM modes are employed to transmit information over a 3 km intra-city link. At the receiver, the beam topological charge is recovered by a non-coherent detection scheme aided by an artificial neural network. This experimental scheme has recently been extended to a 143 km free-space link between the two Canary Islands of La Palma and Tenerife [72].

If an OAM-SK system with L different topological charges can transmit up to $\log_2(L)$ bits per beam, an OAM-DM system that multiplexes L modes can transmit up to L bits per beam. For example, consider the case with $\mathcal{L} = \{\ell_0, \ell_1\}$, i.e.,

$L = 2$. With OAM-SK, it is possible to transmit one bit per beam (0 if $\ell = \ell_0$ and 1 if $\ell = \ell_1$). With OAM-DM, the on and off states of each OAM mode can represent either a "1" or a "0" so that the two-bit set of $\{00, 01, 10, 11\}$ can be mapped on the multiplexed modes $\{00, 0\ell_1, \ell_0 0, \ell_0\ell_1\}$. Using this idea, [83] transmits the information in free space by multiplexing four OAM modes employing a phase hologram loaded on a SLM. Subsequently, this method is used to implement the encoding of two-dimensional images [65, 184].

In order to meet the ever-increasing demands for higher data rates, OAM multiplexing can be combined with different modulation formats and different multiplexing techniques to achieve high-speed communication in multiple dimension. A data link multiplexing the signal in three dimensions is shown in Fig. 1.23. Three OAM beams with $\mathcal{L} = \{\ell_1, \ell_2, \ell_3\}$, carrying the modulation information data 1, data 2, and data 3, are multiplexed into one multimodal beam. A second OAM beam with the same set of modes $\mathcal{L}$ carrying the data streams data 4, data 5, and data 6 is multiplexed with the first one employing orthogonal polarizations. Finally, three polarization-and-OAM-multiplexed beams carrying three different sets of information streams, data 1–6, data 7–12, and data 13–18, are transmitted at three different wavelengths λ_1, λ_2 and λ_3. In a recent laboratory experiment [137], a transmission rate of 2.56 Tbps and a spectral efficiency of 95.7 bps/Hz have been achieved by using 20 × 4 Gbps 16-QAM signals on eight OAM modes, two polarization states, and two sets of concentric rings. In [58], a transmission rate of 100.8 Tbps is achieved by transmitting 100 Gbps quadrature phase-shift keying (QPSK) signals on 12 OAM modes, 2 polarization states, and 42 wavelengths. In a similar fashion, in [138], a transmission rate of 1.036 Pbps with a spectral efficiency of 112.6 bps/Hz has been achieved by multiplexing 54.139 Gbps OFDM-8QAM signals over 368 wavelengths, 2 polarization states, and 26 OAM modes. In addition to these lab-scale high-rate data experiments, a 400 Gbps transmission rate has been

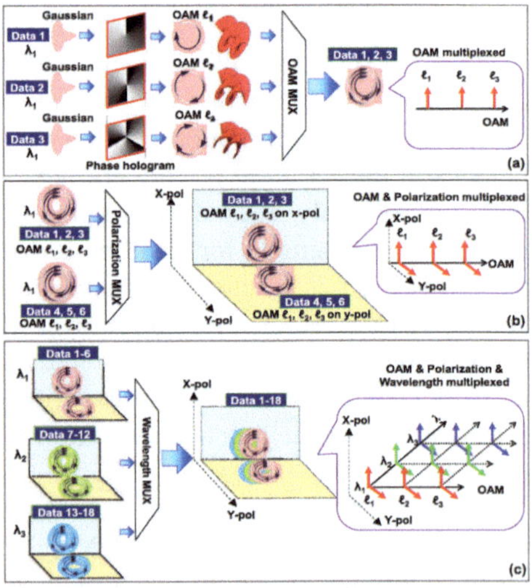

Fig. 1.23 Concept of using three-dimensional multiplexing to increase the multiplexed data channels. (**a**), (**b**), and (**c**) are performed successively to achieve OAM-DM, PDM, and WDM, respectively [58]

achieved over an outdoor link of about 120m by using 100 Gbps QPSK signals on 4 OAM modes [114].

Multimodal LG beams with different topological charges are often employed to provide a set of orthogonal OAM modes for information transmission. Most of the literature on the LG beam consider azimuthal index $|\ell| > 0$ and radial index $p = 0$. Because the radial index p can be used as a radial degree of freedom just as ℓ can provide an azimuthal degree of freedom, LG beams with $p > 0$ have received some attention [132, 147] recently. Indeed, LG beams with different p and ℓ form a complete set of orthogonal mode bases. By multiplexing LG beams with different p, the system capacity and transmission rate can be improved. In addition to LG beams, Bessel beams with non-diffractive properties, because of their self-healing properties after encountering an obstruction, also have a great application potential in free-space optical communications [1, 33, 44]. Perfect vortex beams, obtained by the Fourier transformation of Bessel beams, have the attractive feature that the beam radius is independent of the OAM mode and have recently used with success in a free-space optical communication link [183].

Due to the unique wavefront of OAM beams, there are some challenges in designing an OAM-based communication link in free space, such as *beam divergence* and *misalignment*. Based on diffraction theory, it is known that divergence occurs when a collimated OAM beam propagates in free space. Divergence increases as OAM mode number increases. Figure 1.24 shows the concept of OAM beam divergence in a free space communication link. It can be seen that divergence leads to system power loss, especially when the size of the receiving aperture is limited. A suitable transmitted beam size can be designed for a specific OAM mode number and transmitted distance to achieve the minimum received beam diameter [148]. However, in long-distance propagation, a larger receiving aperture is still expected to capture more received power. Besides beam divergence, the misalignment between transmitter and receiver, which can result in both power loss and mode crosstalk, also needs to be considered. Misalignment errors generally include lateral displacement and receiver angle errors [148], as shown in Fig. 1.25. It can be found through simulation that large lateral displacements and large receiver angle errors cause a high power leakage into other modes, resulting in severe crosstalk. Misalignment effects can be mitigated by increasing OAM mode spacing, but this will result in higher received power loss due to beam divergence. The trade-off between system power loss and crosstalk needs to be considered in link design. When the system power loss dominates (i.e., small lateral displacement and receiver

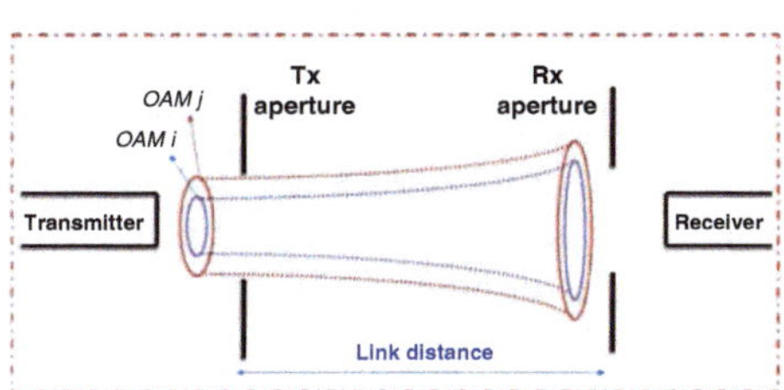

Fig. 1.24 Concept of OAM beam divergence in free-space optical communications. Tx, transmitted; Rx, receiver [148]

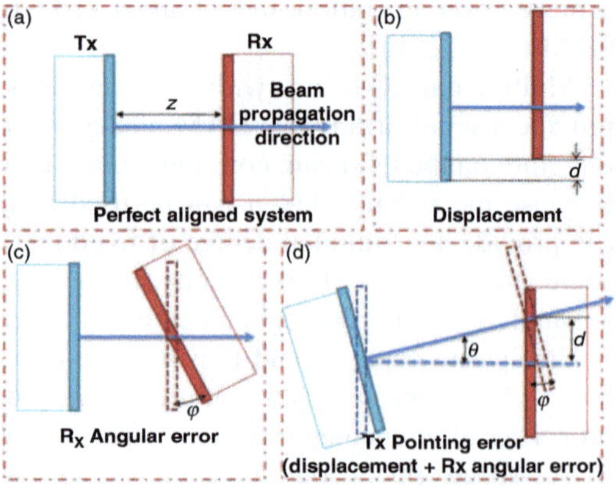

Fig. 1.25 Alignment between the transmitter and the receiver for (**a**) a perfectly aligned system, (**b**) a system with lateral displacement d, (**c**) a system with a receiver angular error φ, and (**d**) a system with a transmitter pointing error θ. Tx, transmitted; Rx, receiver [148]

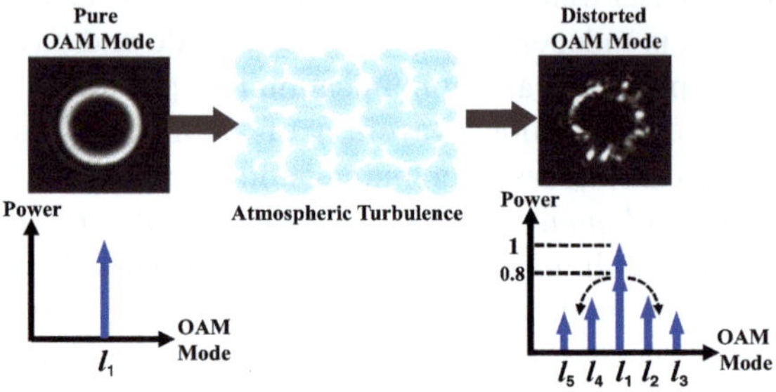

Fig. 1.26 Concept diagram of the effects of atmospheric turbulence on an OAM beam. A distorted OAM mode can be decomposed into multiple OAM modes [108]

angular error), small mode spacing is used, and when crosstalk dominates (i.e., large lateral displacement and receiver angular error), large mode spacing is used.

For a free-space optical OAM link, another critical challenge is *atmospheric turbulence*, caused by changes of the atmosphere refractive index due to temperature and pressure non-uniformities. Atmospheric turbulence can destroy the helical phase front of vortex beams. Figure 1.26 shows its effects on an OAM beam, including received energy fluctuations and crosstalk between OAM channels. The effects have been quantitatively analyzed using the Kolmogorov spectral statistical model [103, 134] and experimentally verified by simulating turbulence in laboratory [108, 118]. In [108], a thin phase screen plate mounted on a rotating stage and placed in the middle of the optical path is used as a turbulence emulator. The pseudorandom phase distribution due to the thin plate obeys the Kolmogorov spectral statistics and is characterized by a parameter r_0. The strength of the simulated turbulence can be varied by using a plate with a different r_0 or by adjusting the size of the beam incident on the plate or the number of passes through the plate. It is found that, as the turbulence strength increases, the power of the transmitted OAM mode leaks into neighboring modes and for strong turbulence tends to be equally distributed among modes, which will result in severe crosstalk at the receiver. The effects of atmospheric turbulence on different vortex beams, including LG and Bessel beams, have also been numerically calculated by analyzing their OAM spectra [39], and it has been shown that Bessel beams suffer more than LG beams in passing through atmosphere turbulence. The non-diffractive property of Bessel beams is affected by a strong turbulence [78].

Many approaches for mitigating the turbulence effects have been proposed and are divided into two main categories: *adaptive optics (AO) compensation* and *signal processing-based mitigation* [81]. AO compensation corrects the distorted OAM wavefronts in the optical domain, and signal processing-based mitigation utilizes digital signal processing (DSP) algorithms, such as MIMO equalization, to reduce the signal degradation in the electrical domain.

AO compensation systems usually work in a closed-loop configuration. A typical working iteration is to first sense the wavefront of the distorted OAM beam and then, on the base of the received feedback, generate an error correction pattern and apply it to the beam to undo the distortion. Based on whether there is a wavefront sensor (WFS) to measure the distorted OAM wavefront, AO compensation can be further divided into WFS-based and non-WFS types.

The main challenge for the WFS-based AO compensation is the difficulty to correctly measure the wavefront of the helical phase front of OAM beams. Recently, a modified AO system that employs a Gaussian probe beam has been proposed to overcome this problem [110]. As shown in Fig. 1.27, the Gaussian probe and the multiplexed OAM beams are coaxially propagated through the atmospheric turbulence so that they all suffer the same distortion. At the receiver, the distorted Gaussian probe is separated and sent to the WFS for the estimation of wavefront distortion and the retrieval of the required correction pattern. At the transmitter, the two wavefront controllers are updated with the feedback correction pattern. It should be noted that, for efficient separation, the Gaussian probe should occupy a separate orthogonal channel, which can be an orthogonal polarization [110] or a separate wavelength [113]. However, the use of an orthogonal polarization channel sacrifices the polarization degree of freedom for multiplexing. It has been shown that using a separate wavelength channel, the compensation performance degrades slowly with the increase in the probe's wavelength offset from the OAM beams [113]. Moreover, the AO system shown in Fig. 1.27 can be used not only to post-compensate but also to pre-compensate the distorted multiplexed OAM beams in the bidirectional free-space optical communication link [111]. The AO

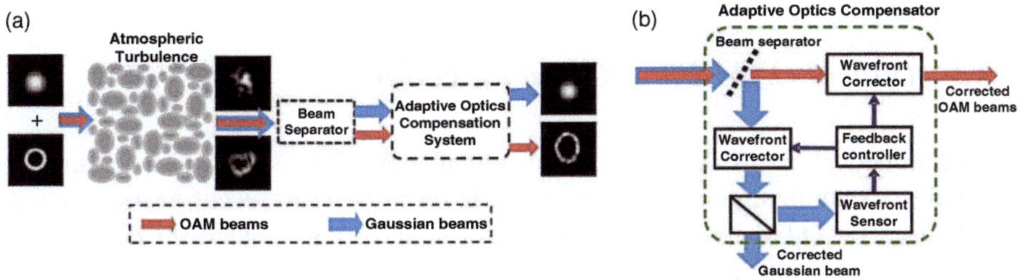

Fig. 1.27 (**a**) Schematic of an AO compensation system for OAM beams using a Gaussian probe beam for wavefront sensing and (**b**) detailed implementation of the AO system [81]

compensation has been experimentally demonstrated for multiple Bessel beams through atmospheric turbulence and obstructions. Fortunately, the self-healing property of Bessel beams can be recuperated after turbulence compensation [78].

In addition to WFS sensing, the distortion wavefronts can also be retrieved from measured intensity profiles by using phase retrieval algorithms, such as the Gerchberg-Saxton (GS) algorithm [40, 41, 107] and stochastic-parallel-gradient-descent (SPGD) algorithm [149]. The GS algorithm, a well-known iterative algorithm, can be used to analyze the original and distorted probe Gaussian intensity patterns to obtain a correction phase pattern, which can be used for pre-compensation of the distorted OAM beam through the same turbulence [40]. If the complex amplitude of the transmitted vortex beam, including the "doughnut" amplitude and the helical phase (or mixed helical phase), is known, this method can be generalized to pre-turbulence compensation without the addition of a Gaussian probe [41]. Moreover, the phase correction pattern can also be derived by using the distorted OAM beam intensity pattern with the SPGD algorithm [149].

Signal processing-based algorithms at the receiver can also help mitigate atmospheric turbulence effects and partially shift the complexity of the optical subsystem to the electrical domain, providing a complementary approach to AO compensation systems. A 4×4 adaptive MIMO equalization system has been implemented in a 4-channel OAM multiplexed free-space optical link to reduce crosstalk caused by weak turbulence [59]. The used MIMO DSP is similar to the one that has been used in few-mode and multi-mode fiber multiplexing systems for mitigating the mode coupling effects [117]. Experimental results show that all data channels can be recovered and the system power penalty is reduced after MIMO equalization. However, MIMO equalization is not universally useful. Under strong turbulence distortion, outage may occur because crosstalk between channels exceeds a certain threshold or severe power attenuation causes some channels to be barely detectable, in which case MIMO equalization will be ineffective [144]. A modified method is proposed in [115] to improve system performance under strong turbulence. The proof-of-concept experiment shows that, in an OAM-based spatial diversity free-space optical link, the OAM channel can be recovered under strong turbulence distortion by using a diversity reception strategy assisted with MIMO equalization [115].

1.4.2 Optical OAM Fiber Communications

As mentioned above, OAM free-space optical communication links are greatly affected by beam divergence and atmospheric turbulence. In recent years, research has focused on OAM-based optical fiber communications, which do not need to take these effects into account. Conventional optical fiber communications utilize SDM by using multicore fibers (MCFs) [119, 181] and few-mode fibers (FMFs) [106] to increase system capacity and spectral efficiency. Generally, multicore fibers require complex manufacturing, while multi-mode fibers face the problem of mode coupling caused by random perturbations or other nonidealities. OAM modes, just like linear polarization (LP) modes in fibers, can be used as an orthogonal basis for data transmission in FMFs. Like LP modes, also OAM modes face the challenge of *mode coupling*, which leads to channel crosstalk. One possible solution is to use DSP algorithms based on complex MIMO equalization. The other is to use a specially designed FMF, which is called a vortex fiber. The vortex fiber lifts the near degeneracy between OAM modes and the parasitic TE_{01} and TM_{01} by modifying the fiber refractive index profile, minimizing mode crosstalk [105]. Therefore, by using vortex fibers, OAM multiplexing technology has the potential to increase the throughput of optical fiber communication systems with low complexity DSP algorithms or even without any DSP algorithm.

Using a vortex fiber with a characteristic high-index ring around the fiber core, Bozinovic et al. have performed experiments of OAM mode transmission in fibers 20 m and 900 m long [15, 16]. In the experiments, a microbend grating is used before the vortex fiber to achieve the conversion from the fundamental modes to the desired HE_{21} modes. The linear combination of the odd and even modes of HE_{21} with a $\pm\pi/2$ phase shift between them results in OAM modes with $\ell = \pm 1$. In a later experiment, two OAM modes with $\ell = \pm 1$ and two LP modes, each mode carrying a 50 Gbaud QPSK signal, are simultaneously propagated to achieve a transmission rate of 400 Gbps in a 1.1-km-long vortex fiber [17]. At the output of the fiber, the measured mode crosstalk between two OAM modes is approximately −20 dB. In addition, WDM has also been added to further increase the capacity of the system. By using two OAM modes over 10 wavelengths in a vortex fiber, 20 channels, each transmitting a 20 Gbaud 16-QAM signal, are established, and a transmission rate of 1.6 Tbps is achieved. These experiments show that OAM can provide an additional degree of freedom for data multiplexing in fiber networks.

At present, research on OAM-based optical fiber communications mainly focuses on the design of fibers that supports stable transmission of multiple OAM modes. In 2012, Birnbaum et al. designed a ring fiber with 0.05 up-doping to support up to 10 OAM modes while maintaining radial single-mode conditions [166]. In 2013, Li et al. designed a multi-OAM-mode multi-ring fiber (MOMRF) that supports multi-mode OAM transmissions [76]. As shown in Fig. 1.28, the fiber consists of seven rings, each supporting 18 OAM modes. Mode crosstalk and inter-ring crosstalk are reduced by increasing the effective refractive index and the distance between rings. In addition, it is compatible with WDM and advanced multilevel

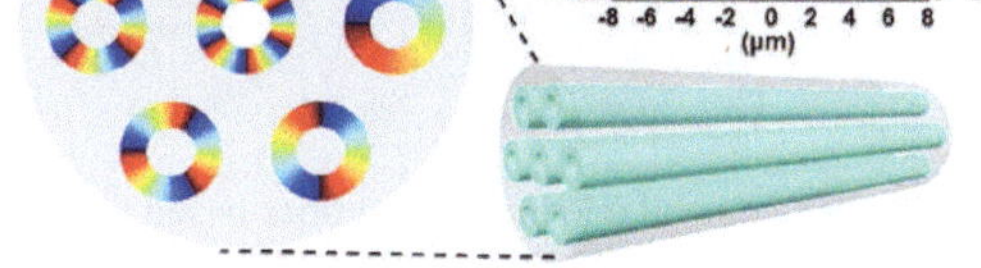

Fig. 1.28 (**a**) 3D structure and cross-section of multi-OAM-mode multi-ring fiber (MOMRF), in which the ring-to-ring distance is Λ. (**b**) Index profile of single ring (black) and mode profile of TE_{01} mode (red) in the ring [76]

amplitude/phase modulation formats, which makes it possible to realize a total transmission capacity in the range of the petabits-per-second and hundred-bits-per-second-per-hertz aggregate spectral efficiency. In addition to these ring fibers, some new types of microstructured fibers have also been proposed for multi-mode OAM transmission [170, 180].

Recently, some studies have explored the potential of the conversion between LP modes and OAM modes in fibers. This can be used as a complement to the OAM generation method, and it is easier to couple than the methods mentioned above. In [167], Zeng et al. designed a novel all-fiber OAM generator, which is cascaded by a mode-selective coupler and a few mode-polarization maintaining fiber, to convert LP_{01} mode to OAM mode. In [77], a series of LP_{11} modes with microphase difference distribution are generated by twisting a few-mode fiber-long period grating (FMF-LPG); these LP_{11} modes are then superimposed in a fiber to generate OAM modes. In [80], Li et al. proposed and demonstrated a controllable broadband fiber-based OAM converter. As shown in Fig. 1.29, the converter consists of a single-mode fiber (SMF), a two-mode fiber (TMF) with specific offsets and tilt angles, two polarization controllers (PCs), and a polarizer. The input end of TMF is stuck to a standard SMF. By adjusting the two PCs, an input fundamental mode LP_{01} can be selectively converted to high-order LP_{11} modes or OAM modes with $\ell = \pm 1$. The purity of the OAM mode is ensured by adjusting the state of the two PCs and the polarizer. The experimental results show that the extinction ratio (used to evaluate the mode purity) of the generated OAM mode and other modes is greater than 20 dB in a wide wavelength range (1480 nm to 1640 nm).

1.4.3 Radio OAM Communications

The application of OAM to radio communications is expected to be a possible solution to the problem of spectrum scarcity, and it has, accordingly, received widespread attention. The first OAM-based wireless communication experiment that successfully separated two radio signals at the same frequency was presented

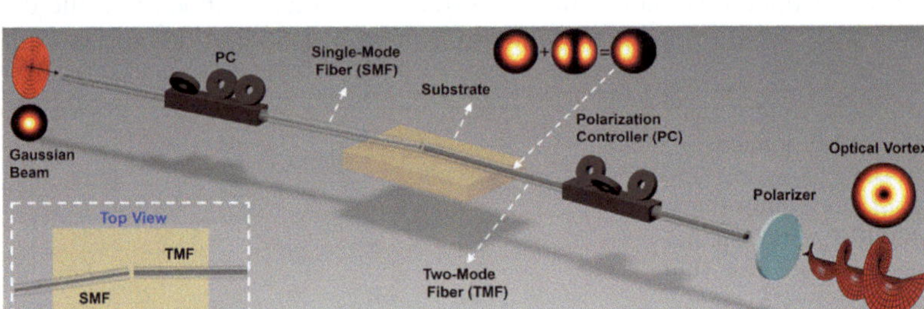

Fig. 1.29 Controllable broadband fiber-based OAM converter. The converter consists of a SMF, a TMF with specific offsets and tilt angles, two PCs, and a polarizer, and the input end of TMF is stuck to a standard SMF [80]

in [129]. A Yagi antenna and a spiral parabolic antenna were used to transmit a plane electromagnetic wave and a vortex electromagnetic wave, respectively, at a distance of 442 meters. Subsequently, a 4 Gbps uncompressed video transmission link over a 60 GHz OAM radio channel has been implemented [88]. In the RF or millimeter wave bands, provided that there is a LOS link so to guarantee the correct transmitter-receiver alignment, OAM can be used to encode information [3] and can also be multiplexed and combined with other technologies [154, 158, 177]. In [154] are reported the results of an experiment in the 28 GHz band that has achieved a 32 Gbit/s capacity and about 16 bit/s/Hz spectral efficiency by combining four OAM modes each carrying 4 Gbit/s 16-QAM modulated signals with two polarization states. In recent experiments, multiple radio OAM beams carrying multiple data streams have been generated by employing some specially designed structures, such as thin metamaterial plates based on rectangular apertures [176] and dual OAM mode antennas based on ring resonators [57].

It is well-known that MIMO is a key technology that can greatly increase the capacity of wireless communication systems by using multiple transmitting and receiving antennas. The relationship between MIMO SDM and OAM multiplexing has caused a heated debate in the academia. It has been pointed out that radio OAM systems based on UCAA are a subset of MIMO systems and do not really provide additional capacity gain [35, 127]. But in practice, there are differences between the two approaches. For example, in a keyhole channel, traditional MIMO systems have very poor performance, while UCAA-based OAM benefits of the fact that the keyhole does not change the helical phase structure of OAM [23]. In terms of capacity over a LOS link, a UCAA-based OAM system is equivalent to a traditional MIMO system from the perspective of channel spatial multiplexing [101]. However, OAM receivers have a lower complexity because unlike traditional MIMO systems, the inherent orthogonality between OAM modes can mitigate inter-channel interference [172]. A sketch of a system employing OAM-MDM in a LOS MIMO channel is shown in Fig. 1.30, where phase shifters networks (PSNs) are used to multiplex OAM modes at the transmitter and demultiplex OAM modes with a reduced computational load at the receiver. Moreover, as described in Sect. 1.3, radio OAM can be generated not only by employing antenna arrays but also by using SPPs or spiral parabolic antennas or other single transmit antennas. Combining OAM

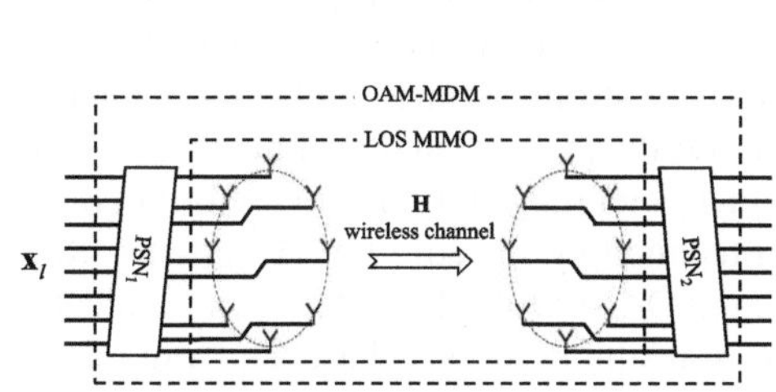

Fig. 1.30 System model for a UCAA-based OAM-MDM in a LOS MIMO channel. The OAM-MDM system uses PSN_1 and PSN_2 for OAM modes multiplexing and demultiplexing, respectively. The vectors $\mathbf{x}_l$ and $\mathbf{y}_l$ are the complex transmitted and received data vectors, respectively, and the matrix $\mathbf{H}$ is the LOS MIMO channel matrix [172]

with traditional MIMO technology can result in higher-capacity gains [45, 171] or provide a more flexible system design [109, 116]. A 2 × 2 antenna aperture architecture, where each aperture multiplexes two OAM modes, is implemented in the 28 GHz band to achieve a 16 Gbit/s transmission rate. At the receiving end, MIMO signal processing is employed to mitigate inter-channel interference [116]. Such a system architecture incorporating OAM multiplexing and MIMO technology can also be used for free-space optical communication links [112].

As it happens for optical OAM, radio OAM communication systems are affected by *beam divergence* and *misalignment* between the transmitter and the receiver. Since wavelengths at radio frequencies are much larger than at light frequencies, the effects of atmospheric turbulence on radio OAM decrease and can be usually neglected. Unfortunately, for the same reason, the problem of beam divergence becomes more important and represents a bigger challenge in wireless communication systems than in free-space optical communication systems, limiting the achievable distance of OAM radio links. Like in free-space optical communications, radio OAM communication systems also require perfect alignment between the transmitter and the receiver [169]. To prevent large performance degradation due to misalignment, one possible solution in UCAA-based OAM links is to use a beam steering approach [25]. By adding additional phases, the UCAA can transmit/receive OAM beams in the desired steering direction. In the non-parallel case, receive beam steering can compensate the phase change caused by oblique angle at the receive UCAA. The off-axis case, where the transmitter and receiver are parallel but not around the same axis, can be decomposed into two non-parallel cases by introducing a virtual UCAA that is perpendicular to and in the middle of the connection between the transmit and receive UCAA centers. Therefore, in the off-axis or the more general misalignment cases, transmit and receive beam steering need to be used simultaneously. The beam steering approach has been demonstrated through simulation [25], and the results show that after applying beam steering in the non-parallel and off-axis case, the misaligned OAM channel capacity shows almost no loss with respect to the case of perfect alignment.

It is well-known that *multipath effects* caused by beam spreading and reflections of surrounding objects must be considered in wireless communication systems. In OAM-based wireless communications, the multipath effects caused by specular reflection from a plane parallel to the propagation path have been discussed in [156] and [157]. Both simulation and experimental results show that OAM channels with larger ℓ have stronger intra- and inter-channel interference due to OAM beam divergence. DSP equalization algorithms have the potential to mitigate multipath effects. Another potential solution is to use orthogonal frequency-division multiplexing (OFDM) technology in radio OAM systems. Recently, a transceiver architecture for broadband OAM-OFDM wireless communication systems has been proposed in [24]. The transmitter uses a baseband digital two-dimensional fast Fourier transform (2-D FFT) algorithm and a UCAA to generate OAM-OFDM signals. After multipath channel propagation, the signal is collected by another UCAA and processed using an inverse FFT (IFFT) algorithm. By using the 2-

D FFT/IFFT algorithms, system implementation complexity can be reduced, and multipath effects are mitigated.

The OAM of electromagnetic wave is a new degree of freedom, which can increase communication capacity because of its infinite mutually orthogonal modes [143]. As a new communication dimension, OAM technology can not only improve communication capacity, but its mode can be also used as a symbol of information for communication. Therefore, the applications of OAM technology in communications can be divided into two categories: OAM shift keying and OAM mode multiplexing.

Traditional communication coding realizes data transmission through high-speed modulation of amplitude, frequency, phase, and polarization of electromagnetic wave. In the OAM shift keying communication systems, the mode of OAM is used as a new parameter for data transmission. When the N modes of OAM are used for communication, the carrier beam is loaded with $log_2 N$ bit information. Compared with traditional binary coding, the OAM shift keying can effectively improve the security of communication systems. In 2004, Gibson et al. used SLM to generate optical OAM beams and used eight modes of the OAM beams to encode information [49]. This is the first free space optical communication based on OAM shift keying. In addition, they also verified that the OAM encoding has the advantages of anti-eavesdropping. In 2014, Jinsong Liu et al. completed OAM shift keying communication experiments in the THz region [141]. They mounted four SPPs with different modes on a 2D rotation stage and generated OAM beams with different modes by controlling the angle of the rotation stage. In the case of switching 1000 states per second, the data transmission rate of this system can reach 2000 bit/s. Obviously, limited by the speed of the rotation stages motor, OAM shift keying cannot be used in high-speed communications temporarily.

The OAM mode multiplexing uses the orthogonal characteristic of different OAM modes to realize the simultaneous transmission of multiple channels of information. Thus, the communication system based on OAM mode multiplexing has been a research hotspot in the communication field. In 2012, Jian Wang et al. proposed an optical communication system based on OAM mode multiplexing [137]. They combined polarization multiplexing with OAM mode multiplexing to achieve the transmission of 32 channels information simultaneously, and the data transmission rate reached 2.56 Tbit/s. In the field of millimeter wave communication, OAM mode multiplexing has also played an important role in increasing the communication rate. In 2014, Yan et al. proposed a similar communication system in the microwave region, which achieved 32 Gbit/s of information transmission rate at 28 GHz [155].

1.4.4 Acoustic OAM Communications

In the field of acoustics, OAM-based communication technology is still in its infancy. In 2017, Shi et al. [121] used 8 acoustic OAM modes to transmit the

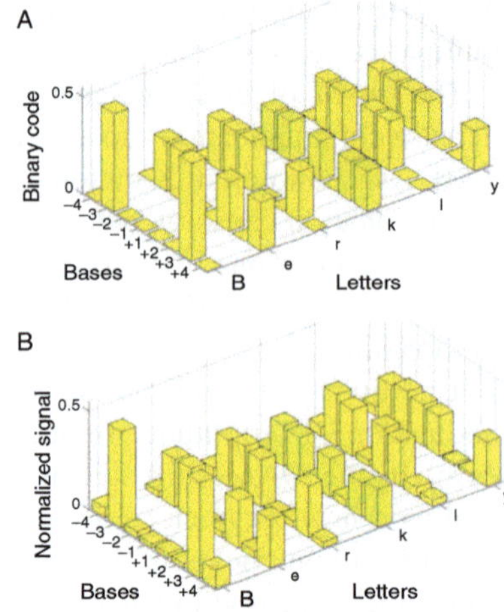

Fig. 1.31 (**a**) Encoding of the letters of the word "Berkly." Each letter is mapped on 1 byte (8 bits) of information, and each byte contains the same amount of total amplitude, and this amplitude amount is equally distributed into the vortex beams. (**b**) Decoding after transmission with eight OAM modes [121]

letters of the word "Berkly" in ASCII binary protocol, achieving a high spectral efficiency of 8.0 ± 0.4 bit/s/Hz. In details, each OAM mode carries 1-bit information of the eight bits that map each letter, as shown in Fig. 1.31a. These 8 OAM orthogonal bases are multiplexed and transmitted by a single transducer array. At the receiver, another transducer array receives the signal and then demultiplexes it by exploiting the orthogonality between OAM modes. The recovered signal is shown in Fig. 1.31b. Instead of using OAM modes to encode data, Jiang et al. directly loaded the data onto the acoustic OAM channels and established a real-time information transmission system based on passive metamaterials [62]. These studies have shown that in acoustic communications, OAM can be a viable alternative to more traditional technologies.

Since microwaves and mid- and far-infrared radiations are strongly absorbed underwater and light is easily obstructed and scattered by small particles in the oceans, sound waves are the only information carrier for underwater long-distance (more than 200 m) communications. Accordingly, OAM can be applied to underwater acoustic communications to further increase system capacity and spectrum efficiency. In addressing link design for underwater acoustic communications, underwater creatures and turbulence effects should also be considered in addition to the inherent OAM beam divergence and misalignment problems. However, very little research and experimental data are available in current scientific literature. Nevertheless, considered the potential capacity gains of OAM, theoretical analysis and experimental verification are expected to be developed in the near future.

1.4.5 Discussion and Lessons Learned

OAM has shown great potential in the communications community due to the inherent orthogonality between modes. Integrating OAM into existing communication systems is expected to further improve spectral efficiency and meet ever-increasing data rate demands. Moreover, the typical circular symmetry of OAM waves makes them very easy to be included in many communication system components. However, OAM integration presents some challenges and technical issues, such as beam divergence, misalignment, and atmospheric turbulence effects in free-space optical links, mode coupling in fiber links, and multipath effects in radio communication links. These technical issues and their solutions must be considered when building an OAM-based communication system.

Analyzing OAM implementation on the basis of the frequency bands used, we can summarize the most important characteristics of OAM systems:

- For free-space optical communication systems, SLMs loaded with phase holograms are recommended in transmitters and receivers to flexibly multiplex and demultiplex OAM beams. The effects of atmospheric turbulence on free-space optical OAM links must be considered in practical environments. Therefore, AO compensation and DSP-based mitigation techniques play a significant role in improving system performance. Optical OAM fiber communication systems can use common optical OAM generation and detection methods, such as SPPs and SLMs. Mode conversion in fibers can also be used to generate OAM. It should be noted that common OAM generators and detectors are required to be well compatible with fibers. Mode coupling is a major challenge in optical OAM fiber links. Novel fibers that support stable transmission of multiple OAM modes and reduce mode crosstalk should be developed in the future.
- In radio and acoustic communication systems, the flexibility of UCAAs and UCTAs makes them perform well in generating and detecting OAM waves. However, for generating high-frequency OAM waves, SPPs are more suitable than UCAAs and UCTAs. In radio OAM communication systems, OAM beam divergence and misalignment between the transmitter and receiver will greatly limit link achievable distance and degrade system performance. OAM beam convergence and system misalignment compensation schemes are required for long-distance transmission. Acoustic OAM communication system is still in an initial stage, and the feasibility of acoustic OAM communications, especially underwater, needs further experimental verification. The impact of the underwater environment on OAM links also needs further research.

After considering all possible challenges and technical issues, we are at a stage where experiments and research on OAM systems should be more oriented to practice testing than to develop proof of concept. At present, most OAM experiments are performed in the controlled environments of laboratories. Testing beyond laboratory distances and practical deployment of OAM systems should be the focus of future research activities. Another important issue is that, since all communication systems keep moving toward low costs, small sizes, and high data rates, all OAM system components must achieve a reduction in cost and size in the near future.

References

1. N. Ahmed, Z. Zhao, L. Li, H. Huang, M.P.J. Lavery, P. Liao, Y. Yan, Z. Wang, G. Xie, Y. Ren, A. Almaiman, A.J. Willner, S. Ashrafi, A.F. Molisch, M. Tur, A.E. Willner, Mode-division-multiplexing of multiple Bessel-Gaussian beams carrying orbital-angular-momentum for obstruction-tolerant free-space optical and millimetre-wave communication links. Sci. Rep. **6**, 22082 (2016)
2. L. Allen, M.W. Beijersbergen, R.J. Spreeuw, J.P. Woerdman, Orbital angular momentum of light and the transformation of Laguerre-Gaussian laser modes. Phys. Rev. A At. Mol. Opt. Phys. **45**(11), 8185–8189 (1992)
3. B. Allen, A. Tennant, Q. Bai, E. Chatziantoniou, Wireless data encoding and decoding using OAM modes. Electron. Lett. **50**(3), 232–233 (2014)
4. Q. Bai, A. Tennant, B. Allen, M.U. Rehman, Generation of orbital angular momentum (OAM) radio beams with phased patch array, in *Proceedings of the Loughborough Antennas Propagation Conference* (2013), pp. 410–413
5. X. Bai, X. Liang, R. Jin, J. Geng, Generation of OAM radio waves with three polarizations using circular horn antenna array, in *Proceedings of the 9th European Conference on Antennas Propagation* (2015)
6. V.Y. Bazhenov, M.V. Vasnetsov, M.S. Soskin, Laser beams with screw dislocations in their wavefronts. JETP Lett. **52**(8), 429–431 (1990)
7. V.Y. Bazhenov, M.S. Soskin, M.V. Vasnetsov, Screw dislocations in light wavefronts. J. Mod. Opt. **39**(5), 985–990 (1992)
8. M.W. Beijersbergen, L. Allen, H. van der Veen, J.P. Woerdman, Astigmatic laser mode converters and transfer of orbital angular momentum. Opt. Commun. **96**(1), 123–132 (1993)
9. M.W. Beijersbergen, R.P.C. Coerwinkel, M. Kristensen, J.P. Woerdman, Helical-wavefront laser beams produced with a spiral phaseplate. Opt. Commun. **112**(5), 321–327 (1994)
10. A. Bennis, R. Niemiec, C. Brousseau, K. Mahdjoubi, O. Emile, Flat plate for OAM generation in the millimeter band, in *Proceedings of the 7th European Conference on Antennas and Propagation* (2013), pp. 3203–3207
11. G.C. Berkhout, M.W. Beijersbergen, Method for probing the orbital angular momentum of optical vortices in electromagnetic waves from astronomical objects. Phys. Rev. Lett. **101**(10), 100801 (2008)
12. G.C. Berkhout, M.P. Lavery, J. Courtial, M.W. Beijersbergen, M.J. Padgett, Efficient sorting of orbital angular momentum states of light. Phys. Rev. Lett. **105**(15), 153601 (2010)
13. G. Biener, A. Niv, V. Kleiner, E. Hasman, Formation of helical beams by use of Pancharatnam-Berry phase optical elements. Opt. Lett. **27**(21), 1875–1877 (2002)
14. Z. Bomzon, G. Biener, V. Kleiner, E. Hasman, Space-variant Pancharatnam-Berry phase optical elements with computer-generated subwavelength gratings. Opt. Lett. **27**(13), 1141–1143 (2002)
15. N. Bozinovic, P. Kristensen, S. Ramachandran, Long-range fiber-transmission of photons with orbital angular momentum, in *Proceedings of the Conference on Lasers and Electro-Optics* (2011), p. CTuB1
16. N. Bozinovic, P. Kristensen, S. Ramachandran, S. Golowich, Control of orbital angular momentum of light with optical fibers. Opt. Lett. **37**(13), 2451–2453 (2012)
17. N. Bozinovic, Y. Yue, Y. Ren, M. Tur, P. Kristensen, H. Huang, A.E. Willner, S. Ramachandran, Terabit-scale orbital angular momentum mode division multiplexing in fibers. Science **340**(6140), 1545–1548 (2013)
18. N.M. Burford, M.O. El-Shenawee, Review of terahertz photoconductive antenna technology. Opt. Eng. **56**(1), 010 901–010 901 (2017)
19. C. Chaccour, R. Amer, B. Zhou, W. Saad, On the reliability of wireless virtual reality at terahertz (thz) frequencies, in *2019 10th IFIP International Conference on New Technologies, Mobility and Security (NTMS)* (2019), pp. 1–5

20. J.A.C.R. Chakkittakandy et al., Terahertz spectroscopy to identify the polymorphs in freeze-dried mannitol. J. Pharm. Sci. **99**(2), 932–940 (2010)
21. M.L.N. Chen, L.J. Jiang, W.E.I. Sha, Artificial perfect electric conductor-perfect magnetic conductor anisotropic metasurface for generating orbital angular momentum of microwave with nearly perfect conversion efficiency. J. Appl. Phys. **119**(6), 064506 (2016)
22. R. Chen, H. Xu, M. Moretti, J. Li, Beam steering for the misalignment in uca-based oam communication systems. IEEE Wireless Commun. Lett. **7**(4), 582–585 (2018)
23. R. Chen, H. Xu, W. Yang, X. Wang, J. Li, On the performance of OAM in keyhole channels. IEEE Wireless Commun. Lett. **8**(1), 1–4 (2018)
24. R. Chen, W. Yang, H. Xu, J. Li, A 2-D FFT-based transceiver architecture for OAM-OFDM systems with UCA antennas. IEEE Trans. Veh. Technol. **67**(6), 5481–5485 (2018)
25. R. Chen, H. Xu, M. Moretti, J. Li, Beam steering for the misalignment in UCA-based OAM communication systems. IEEE Wireless Commun. Lett. **7**(4), 582–585 (2018)
26. R. Chen, H. Zhou, M. Moretti, X. Wang, J. Li, Orbital angular momentum waves: Generation, detection, and emerging applications. IEEE Commun. Surv. Tutorials **22**(2), 840–868 (2020)
27. R. Chen, H. Zhou, W.-X. Long, M. Moretti, Spectral and energy efficiency of line-of-sight oam-mimo communication systems. China Commun. **17**(9), 119–127 (2020)
28. L. Cheng, W. Hong, Z.-C. Hao, Generation of electromagnetic waves with arbitrary orbital angular momentum modes. Sci. Rep. **4**, 4814 (2014)
29. G. Chengshan, L. Leilei, W. Huitian, Characterizing topological charge of optical vortices by using an annular aperture. Opt. Lett. **34**(23), 3686–3688 (2009)
30. M.S.M.Z. Chowdhury et al., 6G wireless communication systems: applications, requirements, technologies, challenges, and research directions. IEEE Open J. Commun. Soc. **1**, 957–975 (2020)
31. J.E. Curtis, B.A. Koss, D.G. Grier, Dynamic holographic optical tweezers. Opt. Commun. **207**(1), 169–175 (2002)
32. C. Deng, W. Chen, Z. Zhang, Y. Li, Z. Feng, Generation of OAM radio waves using circular Vivaldi antenna array. Int. J. Antennas Propag. **2013**(2), 607–610 (2013)
33. J. Du, J. Wang, High-dimensional structured light coding/decoding for free-space optical communications free of obstructions. Opt. Lett. **40**(21), 4827–4830 (2015)
34. J.L. Ealo, J.C. Prieto, F. Seco, Airborne ultrasonic vortex generation using flexible ferroelectrets. IEEE T. Ultrason. Ferr. **58**(8), 1651–1657 (2011)
35. O. Edfors, A.J. Johansson, Is orbital angular momentum (OAM) based radio communication an unexploited area? IEEE Trans. Antennas Propag. **60**(2), 1126–1131 (2012)
36. H. Esfahlani, H. Lissek, J.R. Mosig, Generation of acoustic helical wavefronts using metasurfaces. Phys. Rev. B **95**(2), 024312 (2017)
37. B. Ferguson, X.-C. Zhang, Materials for terahertz science and technology. Nat. Mat. **1**(1), 26–33 (2002)
38. N.K. Fontaine, C.R. Doerr, L.L. Buhl, Efficient multiplexing and demultiplexing of free-space orbital angular momentum using photonic integrated circuits, in *Proceedings of the Optical Fiber Communication Conference Expo.* (2012), p. OTu1I.2
39. S. Fu, C. Gao, Influences of atmospheric turbulence effects on the orbital angular momentum spectra of vortex beams. Photonics Res. **4**(5), B1–B4 (2016)
40. S. Fu, S. Zhang, T. Wang, C. Gao, Pre-turbulence compensation of orbital angular momentum beams based on a probe and the Gerchberg-Saxton algorithm. Opt. Lett. **41**(14), 3185–3188 (2016)
41. S. Fu, T. Wang, S. Zhang, Z. Zhang, Y. Zhai, C. Gao, Non-probe compensation of optical vortices carrying orbital angular momentum. Photonics Res. **5**(3), 251–255 (2017)
42. R. Gaffoglio, A. Cagliero, A.D. Vita, B. Sacco, OAM multiple transmission using uniform circular arrays: Numerical modeling and experimental verification with two digital television signals. Radio Sci. **51**(6), 645–658 (2016)
43. C. Gao, S. Zhang, S. Fu, T. Wang, Integrating 5×5 dammann gratings to detect orbital angular momentum states of beams with the range of -24 to $+24$. Appl. Opt. **55**(7), 1514–1517 (2016)

44. A. Gatto, M. Tacca, P. Martelli, P. Boffi, M. Martinelli, Free-space orbital angular momentum division multiplexing with Bessel beams. J. Opt. **13**(6), 064018 (2011)
45. X. Ge, R. Zi, X. Xiong, Q. Li, L. Wang, Millimeter wave communications with OAM-SM scheme for future mobile networks. IEEE J. Sel. Areas Commun. **35**(9), 2163–2177 (2017)
46. G. Gibson, J. Courtial, M. Vasnetsov, S. Barnett, S. Franke-Arnold, M. Padgett, Increasing the data density of free-space optical communications using orbital angular momentum, in *Proceedings of the SPIE* (2004), pp. 367–373
47. P. Genevet, N. Yu, F. Aieta, J. Lin, M.A. Kats, R. Blanchard, M.O. Scully, Z. Gaburro, F. Capasso, Ultra-thin plasmonic optical vortex plate based on phase discontinuities. Appl. Phys. Lett. **100**(1), 013101 (2012)
48. G. Gibson, J. Courtial, M. Vasnetsov, M.J. Padgett, S. Franke-Arnold, S.M. Barnett, V. Pas'Ko, Free-space information transfer using light beams carrying orbital angular momentum. Opt. Express **12**(22), 5448–5456 (2004)
49. J.G. Gibson et al., Free-space information transfer using light beams carrying orbital angular momentum. Opt. Express **12**(22), 5448–5456 (2004)
50. S. Gspan, A. Meyer, S. Bernet, M. Ritsch-Marte, Optoacoustic generation of a helicoidal ultrasonic beam. J. Acoust. Soc. Am. **115**(3), 1142–1146 (2004)
51. M.T.P. Han et al., A direct comparison between terahertz time-domain spectroscopy and far-infrared fourier transform spectroscopy. J. Appl. Phys. **89**(4), 2357–2359 (2001)
52. J.-l.S.T. He et al., Applications of terahertz spectroscopy in illicit drugs detection. Spectrosc. Spectral Anal. **33**(9), 2348–2353 (2013)
53. N.R. Heckenberg, R. Mcduff, C.P. Smith, A.G. White, Generation of optical phase singularities by computer-generated holograms. Opt. Lett. **17**(3), 221–223 (1992)
54. B.T. Hefner, P.L. Marston, Acoustical helicoidal waves and Laguerre-Gaussian beams: applications to scattering and to angular momentum transport. J. Acoust. Soc. Am. **103**(5), 2971–2971 (1998)
55. B.T. Hefner, P.L. Marston, An acoustical helicoidal wave transducer with applications for the alignment of ultrasonic and underwater systems. J. Acoust. Soc. Am. **106**(6), 3313–3316 (1999)
56. Y. Hu, S. Zheng, Z. Zhang, H. Chi, X. Jin, X. Zhang, Simulation of orbital angular momentum radio communication systems based on partial aperture sampling receiving scheme. IET Microwaves, Antennas Propag. **10**(10), 1043–1047 (2016)
57. X. Hui, S. Zheng, Y. Chen, Y. Hu, X. Jin, H. Chi, X. Zhang, Multiplexed millimeter wave communication with dual orbital angular momentum (OAM) mode antennas. Sci. Rep. **5**, 10148 (2015)
58. H. Huang, G. Xie, Y. Yan, N. Ahmed, Y. Ren, Y. Yue, D. Rogawski, M.J. Willner, B.I. Erkmen, K.M. Birnbaum, 100 Tbit/s free-space data link enabled by three-dimensional multiplexing of orbital angular momentum, polarization, and wavelength. Opt. Lett. **39**(2), 197–200 (2014)
59. H. Huang, Y. Cao, G. Xie, Y. Ren, Y. Yan, C. Bao, N. Ahmed, M.A. Neifeld, S.J. Dolinar, A.E. Willner, Crosstalk mitigation in a free-space orbital angular momentum multiplexed communication link using 4×4 MIMO equalization. Opt. Lett. **39**(15), 4360–4363 (2014)
60. P. Jia, Y. Yang, C.J. Min, H. Fang, X.-C. Yuan, Sidelobe-modulated optical vortices for free-space communication. Opt. Lett. **38**(4), 588–590 (2013)
61. X. Jiang, J. Zhao, S. Liu, B. Liang, X. Zou, J. Yang, C. Qiu, J. Cheng, Broadband and stable acoustic vortex emitter with multi-arm coiling slits. Appl. Phys. Lett. **108**(20), 203501 (2016)
62. X. Jiang, B. Liang, J. Cheng, C. Qiu, Twisted acoustics: metasurface-enabled multiplexing and demultiplexing. Adv. Mater. **30**(18), 1800257 (2018)
63. N. Jiménez, R. Picó, V. Sánchez-Morcillo, V. Romero-García, L.M. García-Raffi, K. Staliunas, Formation of high-order acoustic Bessel beams by spiral diffraction gratings. Phys. Rev. E **94**(5), 053004 (2016)
64. N. Jiménez, V. Romero-García, L.M. García-Raffi, F. Camarena, K. Staliunas, Sharp acoustic vortex focusing by Fresnel-spiral zone plates. Appl. Phys. Lett. **112**(20), 204101 (2018)
65. C. Kai, P. Huang, F. Shen, H. Zhou, Z. Guo, Orbital angular momentum shift keying based optical communication system. IEEE Photonics J. **9**(2), 1–10 (2017)

66. K. Kano, Y. Kozawa, S. Sato, Generation of a purely single transverse mode vortex beam from a He-Ne laser cavity with a spot-defect mirror. Int. J. Opt. **2012**, 1–6 (2012)
67. E. Karimi, B. Piccirillo, E. Nagali, L. Marrucci, E. Santamato, Efficient generation and sorting of orbital angular momentum eigenmodes of light by thermally tuned q-plates. Appl. Phys. Lett. **94**(23), 231124 (2009)
68. E. Karimi, S.A. Schulz, I.D. Leon, H. Qassim, J. Upham, R.W. Boyd, Generating optical orbital angular momentum at visible wavelengths using a plasmonic metasurface. Light Sci. Appl. **3**, e167 (2014)
69. N. Khiabani, Y. Huang, Y.-C. Shen, S. Boyes, Theoretical modeling of a photoconductive antenna in a terahertz pulsed system. IEEE Trans. Antennas Propag. **61**(4), 1538–1546 (2013)
70. N. Kou, S. Yu, L. Li, Generation of high-order Bessel vortex beam carrying orbital angular momentum using multilayer amplitude-phase-modulated surfaces in radiofrequency domain. Appl. Phys. Express **10**(1), 016701 (2016)
71. M. Krenn, R. Fickler, M. Fink, J. Handsteiner, M. Malik, T. Scheidl, R. Ursin, A. Zeilinger, Communication with spatially modulated light through turbulent air across vienna. New J. Phys. **16**(11), 113028 (2014)
72. M. Krenn, J. Handsteiner, M. Fink, R. Fickler, R. Ursin, M. Malik, A. Zeilinger, Twisted light transmission over 143 km. Proc. Natl. Acad. Sci. **113**(48), 13 648–13 653 (2016)
73. M.P.J. Lavery, D.J. Robertson, G.C.G. Berkhout, G.D. Love, M.J. Padgett, J. Courtial, Refractive elements for the measurement of the orbital angular momentum of a single photon. Opt. Express **20**(3), 2110–2115 (2012)
74. J. Leach, M.J. Padgett, S.M. Barnett, S. Franke-Arnold, J. Courtial, Measuring the orbital angular momentum of a single photon. Phys. Rev. Lett. **88**(25), 257901 (2002)
75. J. Leach, J. Courtial, K. Skeldon, S.M. Barnett, S. Franke-Arnold, M.J. Padgett, Interferometric methods to measure orbital and spin, or the total angular momentum of a single photon. Phys. Rev. Lett. **92**(1), 013601 (2004)
76. S. Li, J. Wang, Multi-orbital-angular-momentum multi-ring fiber for high-density space-division multiplexing. IEEE Photonics J. **5**(5), 7101007 (2013)
77. Y. Li, L. Jin, H. Wu, S. Gao, Y.-H. Feng, Z. Li, Superposing multiple LP modes with micro phase difference distributed along fiber to generate OAM mode. IEEE Photonics J. **9**(2), 1–9 (2017)
78. S. Li, J. Wang, Adaptive free-space optical communications through turbulence using self-healing Bessel beams. Sci. Rep. **7**, 43233 (2017)
79. Y. Li, G. Guo, Q. Ma, J. Tu, D. Zhang, Deep-level stereoscopic multiple traps of acoustic vortices. J. Appl. Phys. **121**(16), 164901 (2017)
80. S. Li, Z. Xu, R. Zhao, L. Shen, C. Du, J. Wang, Generation of orbital angular momentum beam using fiber-to-fiber butt coupling. IEEE Photonics J. **10**(4), 1–7 (2018)
81. S. Li, S. Chen, C. Gao, A.E. Willner, J. Wang, Atmospheric turbulence compensation in orbital angular momentum communications: advances and perspectives. Opt. Commun. **408**, 68–81 (2018)
82. J. Lin, X. Yuan, S.H. Tao, R. E. Burge, Synthesis of multiple collinear helical modes generated by a phase-only element. J. Opt. Soc. Am. A **23**(5), 1214–1218 (2006)
83. J. Lin, X. Yuan, S.H. Tao, R.E. Burge, Multiplexing free-space optical signals using superimposed collinear orbital angular momentum states. Appl. Opt. **46**(21), 4680–4685 (2007)
84. M. Lin, Y. Gao, P. Liu, J. Liu, Theoretical analyses and design of circular array to generate orbital angular momentum. IEEE Trans. Antennas Propag. **65**(7), 3510–3519 (2017)
85. J. Long, R. Liu, F. Wang, Y. Wang, P. Zhang, H. Gao, F. Li, Evaluating Laguerre-Gaussian beams with an invariant parameter. Opt. Lett. **38**(16), 3047–3049 (2013)
86. S. Maccalli, G. Pisano, S. Colafrancesco, B. Maffei, M.W. Ng, M. Gray, Q-plate for millimeter-wave orbital angular momentum manipulation. Appl. Opt. **52**(4), 635–639 (2013)
87. F.E. Mahmouli, S. Walker, Orbital angular momentum generation in a 60GHz wireless radio channel, in *Proceedings of the 20th Telecommunications Forum* (2012), pp. 315–318

88. F.E. Mahmouli, S.D. Walker, 4-Gbps uncompressed video transmission over a 60-GHz orbital angular momentum wireless channel. IEEE Wireless Commun. Lett. **2**(2), 223–226 (2013)
89. A. Mair, A. Vaziri, G. Weihs, A. Zeilinger, Entanglement of the orbital angular momentum states of photons. Nature **412**(6844), 313–316 (2001)
90. R. Marchiano, J.L. Thomas, Synthesis and analysis of linear and nonlinear acoustical vortices. Phys. Rev. E **71**(6), 066616 (2005)
91. A.R.A. Markelz et al., Pulsed terahertz spectroscopy of DNA, bovine serum albumin and collagen between 0.1 and 2.0 thz. Chem. Phys. Lett. **320**(1–2), 42–48 (2000)
92. L. Marrucci, C. Manzo, D. Paparo, Optical spin-to-orbital angular momentum conversion in inhomogeneous anisotropic media. Phys. Rev. Lett. **96**(16), 163905 (2006)
93. L. Marrucci, E. Karimi, S. Slussarenko, B. Piccirillo, E. Santamato, E. Nagali, F. Sciarrino, Spin-to-orbital conversion of the angular momentum of light and its classical and quantum applications. J. Opt. **13**(6), 064001 (2011)
94. C. Maurer, A. Jesacher, S. Fürhapter, S. Bernet, M. Ritsch-Marte, Tailoring of arbitrary optical vector beams. New J. Phys. **9**(3), 78 (2007)
95. M. Mirhosseini, O.S. Magaña Loaiza, C. Chen, B. Rodenburg, M. Malik, R.W. Boyd, Rapid generation of light beams carrying orbital angular momentum. Opt. Express **21**(25), 30 196–30 203 (2013)
96. M. Mirhosseini, M. Malik, Z. Shi, R.W. Boyd, Efficient separation of the orbital angular momentum eigenstates of light. Nat. Comm. **4**, 2781 (2013)
97. K.J. Mitchell, S. Turtaev, M.J. Padgett, T. Čižmár, D.B. Phillips, High-speed spatial control of the intensity, phase and polarisation of vector beams using a digital micro-mirror device. Opt. Express **24**(25), 29 269–29 282 (2016)
98. S.M. Mohammadi, L.K.S. Daldorff, J.E.S. Bergman, R.L. Karlsson, B. Thidé, K. Forozesh, T.D. Carozzi, B. Isham, Orbital angular momentum in radio–a system study. IEEE Trans. Antennas Propag. **58**(2), 565–572 (2010)
99. S.M. Mohammadi, L.K.S. Daldorff, K. Forozesh, B. Thidé, J.E.S. Bergman, B. Isham, R. Karlsson, T.D. Carozzi, Orbital angular momentum in radio: Measurement methods. Radio Sci. **45**(4), 1–14 (2010)
100. R.D. Muelas-Hurtado, J.L. Ealo, J.F. Pazos-Ospina, K. Volke-Sepúlveda, Generation of multiple vortex beam by means of active diffraction gratings. Appl. Phys. Lett. **112**(8), 084101 (2018)
101. M. Oldoni, F. Spinello, E. Mari, G. Parisi, C.G. Someda, F. Tamburini, F. Romanato, R.A. Ravanelli, P. Coassini, B. Thidé, Space-division demultiplexing in orbital-angular-momentum-based MIMO radio systems. IEEE Trans. Antennas Propag. **63**(10), 4582–4587 (2015)
102. S.-G. Park, Y. Choi, Y.-J. Oh, K.-H. Jeong, Terahertz photoconductive antenna with metal nanoislands. Opt. Express **20**(23), 25 530–25 535 (2012)
103. C. Paterson, Atmospheric turbulence and orbital angular momentum of single photons for optical communication. Phys. Rev. Lett. **94**(15), 153901 (2005)
104. Y.Y.J. Qin et al., The detection of agricultural products and food using terahertz spectroscopy: a review. Appl. Spectrosc. Rev. **48**(6), 439–457 (2013)
105. S. Ramachandran, P. Kristensen, M.F. Yan, Generation and propagation of radially polarized beams in optical fibers. Opt. Lett. **34**(16), 2525–2527 (2009)
106. S. Randel, R. Ryf, A. Sierra, P.J. Winzer, A.H. Gnauck, C.A. Bolle, R.J. Essiambre, D.W. Peckham, A. Mccurdy, R. Lingle, 6×56-Gb/s mode-division multiplexed transmission over 33-km few-mode fiber enabled by 6×6 MIMO equalization. Opt. Express **19**(17), 16 697–16 707 (2011)
107. Y. Ren, H. Huang, J.-Y. Yang, Y. Yan, N. Ahmed, Y. Yue, A.E. Willner, K. Birnbaum, J. Choi, B. Erkmen, S. Dolinar, Correction of phase distortion of an OAM mode using GS algorithm based phase retrieval, in *Proceedings of the Conference Lasers and Electro-Optics* (2012), p. CF3I.4
108. Y. Ren, H. Huang, G. Xie, N. Ahmed, Y. Yan, B.I. Erkmen, N. Chandrasekaran, M.P.J. Lavery, N.K. Steinhoff, M. Tur, S. Dolinar, M. Neifeld, M.J. Padgett, R.W. Boyd, J.H. Shapiro,

A.E. Willner, Atmospheric turbulence effects on the performance of a free space optical link employing orbital angular momentum multiplexing. Opt. Lett. **38**(20), 4062–4065 (2013)

109. Y. Ren, L. Li, G. Xie, Y. Yan, Y. Cao, H. Huang, N. Ahemd, M.J. Lavery, Z. Zhao, C. Zhang, M. Tur, M. Padgett, G. Caire, A.F. Molisch, A.E. Willner, Experimental demonstration of 16 Gbit/s millimeter-wave communications using MIMO processing of 2 OAM modes on each of two transmitter/receiver antenna apertures, in *Proceedings of the IEEE Global Communications Conference* (2014), pp. 3821–3826
110. Y. Ren, G. Xie, H. Huang, C. Bao, Y. Yan, N. Ahmed, M.P.J. Lavery, B.I. Erkmen, S. Dolinar, M. Tur, M.A. Neifeld, M.J. Padgett, R.W. Boyd, J.H. Shapiro, A.E. Willner, Adaptive optics compensation of multiple orbital angular momentum beams propagating through emulated atmospheric turbulence. Opt. Lett. **39**(10), 2845–2848 (2014)
111. Y. Ren, G. Xie, H. Huang, N. Ahmed, Y. Yan, L. Li, C. Bao, M.P.J. Lavery, M. Tur, M.A. Neifeld, R.W. Boyd, J.H. Shapiro, A.E. Willner, Adaptive-optics-based simultaneous pre- and post-turbulence compensation of multiple orbital-angular-momentum beams in a bidirectional free-space optical link. Optica **1**(6), 376–382, (2014)
112. Y. Ren, Z. Wang, G. Xie, L. Li, Y. Cao, C. Liu, P. Liao, Y. Yan, N. Ahmed, Z. Zhao, A. Willner, N. Ashrafi, S. Ashrafi, R.D. Linquist, R. Bock, M. Tur, A.F. Molisch, A.E. Willner, Free-space optical communications using orbital-angular-momentum multiplexing combined with MIMO-based spatial multiplexing. Opt. Lett. **40**(18), 4210–4213 (2015)
113. Y. Ren, G. Xie, H. Huang, L. Li, N. Ahmed, Y. Yan, M.P.J. Lavery, R. Bock, M. Tur, M.A. Neifeld, R.W. Boyd, J.H. Shapiro, A.E. Willner, Turbulence compensation of an orbital angular momentum and polarization-multiplexed link using a data-carrying beacon on a separate wavelength. Opt. Lett. **40**(10), 2249–2252 (2015)
114. Y. Ren, Z. Wang, P. Liao, L. Li, G. Xie, H. Huang, Z. Zhao, Y. Yan, Experimental characterization of a 400 Gbit/s orbital angular momentum multiplexed free-space optical link over 120 m. Opt. Lett. **41**(3), 622–625 (2016)
115. Y. Ren, Z. Wang, G. Xie, L. Li, A.J. Willner, Y. Cao, Z. Zhao, Y. Yan, N. Ahmed, N. Ashrafi, S. Ashrafi, R. Bock, M. Tur, A.E. Willner, Atmospheric turbulence mitigation in an OAM-based MIMO free-space optical link using spatial diversity combined with MIMO equalization. Opt. Lett. **41**(11), 2406–2409 (2016)
116. Y. Ren, L. Li, G. Xie, Y. Yan, Y. Cao, H. Huang, N. Ahmed, Z. Zhao, P. Liao, C. Zhang, G. Caire, A.F. Molisch, M. Tur, A. E. Willner, Line-of-sight millimeter-wave communications using orbital angular momentum multiplexing combined with conventional spatial multiplexing. IEEE Trans. Wireless Commun. **16**(5), 3151–3161 (2017)
117. D.J. Richardson, J.M. Fini, L.E. Nelson, Space-division multiplexing in optical fibres. Nat. Photonics **7**(5), 354–362 (2013)
118. B. Rodenburg, M.P.J. Lavery, M. Malik, M.N. O'Sullivan, M. Mirhosseini, D.J. Robertson, M. Padgett, R.W. Boyd, Influence of atmospheric turbulence on states of light carrying orbital angular momentum. Opt. Lett. **37**(17), 3735–3737 (2012)
119. J. Sakaguchi, Y. Awaji, N. Wada, A. Kanno, T. Kawanishi, T. Hayashi, T. Taru, T. Kobayashi, M. Watanabe, Space division multiplexed transmission of 109-Tb/s data signals using homogeneous seven-core fiber. J. Lightw. Technol. **30**(4), 658–665 (2012)
120. X.Y. Shen et al., Optical vortices 30 years on: OAM manipulation from topological charge to multiple singularities. Light Sci. Appl. **8**(1), 1–29 (2019)
121. C. Shi, M. Dubois, Y. Wang, X. Zhang, High-speed acoustic communication by multiplexing orbital angular momentum. Proc. Natl. Acad. Sci. **114**(28), 7250–7253 (2017)
122. F. Sizov, A. Rogalski, THz detectors. Prog. Quantum Electron. **34**(5), 278–347 (2010)
123. T. Su, R.P. Scott, S.S. Djordjevic, N.K. Fontaine, D.J. Geisler, X. Cai, S.J.B. Yoo, Demonstration of free space coherent optical communication using integrated silicon photonic orbital angular momentum devices. Opt. Express **20**(9), 9396–9402 (2012)
124. X. Sun, Y. Du, Y. Fan, M. Sun, The design of array antenna based on multi-modal OAM vortex electromagnetic wave, in *Progress in Electromagnetic Research Symposium* (2016), pp. 2786–2791

125. H.I. Sztul, R.R. Alfano, Double-slit interference with Laguerre-Gaussian beams. Opt. Lett. **31**(7), 999–1001 (2006)
126. I.B.P. Taday et al., Terahertz pulse spectroscopy of biological materials: L-glutamic acid. J. Biol. Phys. **29**(2), 109–115 (2003)
127. M. Tamagnone, C. Craeye, J. Perruisseau-Carrier, Comment on 'Encoding many channels on the same frequency through radio vorticity: first experimental test'. New J. Phys. **14**(11), 118001 (2012)
128. F. Tamburini, E. Mari, B. Thidé, C. Barbieri, F. Romanato, Experimental verification of photon angular momentum and vorticity with radio techniques. Appl. Phys. Lett. **99**(20), 204102 (2011)
129. F. Tamburini, E. Mari, A. Sponselli, B. Thidé, A. Bianchini, F. Romanato, Encoding many channels in the same frequency through radio vorticity: first experimental test. New J. Phys. **14**(3), 033001 (2012)
130. A. Tennant, B. Allen, Generation of radio frequency OAM radiation modes using circular time-switched and phased array antennas, in *Proceedings of the Loughborough Antennas Propagation Conference* (2012)
131. B. Thidé, H. Then, J. Sjöholm, K. Palmer, J. Bergman, T.D. Carozzi, Y.N. Istomin, N.H. Ibragimov, R. Khamitova, Utilization of photon orbital angular momentum in the low-frequency radio domain. Phys. Rev. Lett. **99**(8), 087701 (2007)
132. A. Trichili, C. Rosales-Guzmán, A. Dudley, B. Ndagano, A.B. Salem, M. Zghal, A. Forbes, Optical communication beyond orbital angular momentum. Sci. Rep. **6**, 27674 (2016)
133. G.A. Turnbull, D.A. Robertson, G.M. Smith, L. Allen, M.J. Padgett, The generation of free-space Laguerre-Gaussian modes at millimetre-wave frequencies by use of a spiral phaseplate. Opt. Commun. **127**(4), 183–188 (1996)
134. G.A. Tyler, R.W. Boyd, Influence of atmospheric turbulence on the propagation of quantum states of light carrying orbital angular momentum. Opt. Lett. **34**(2), 142–144 (2009)
135. J.M. Vaughan, D.V. Willetts, Temporal and interference fringe analysis of TEM_{01}* laser modes. J. Opt. Soc. Am. **73**(8), 1018–1021 (1983)
136. K. Volke-Sepúlveda, A.O. Santillán, R.R. Boullosa, Transfer of angular momentum to matter from acoustical vortices in free space. Phys. Rev. Lett. **100**(2), 024302 (2008)
137. J. Wang, J.-Y. Yang, I.M. Fazal, N. Ahmed, Y. Yan, H. Huang, Y. Ren, Y. Yue, S. Dolinar, M. Tur, A.E. Willner, Terabit free-space data transmission employing orbital angular momentum multiplexing. Nat. Photonics **6**(7), 488–496 (2012)
138. J. Wang, S. Li, M. Luo, J. Liu, L. Zhu, C. Li, D. Xie, Q. Yang, S. Yu, J. Sun, X. Zhang, W. Shieh, A.E. Willner, N-dimentional multiplexing link with 1.036-Pbit/s transmission capacity and 112.6-bit/s/Hz spectral efficiency using OFDM-8QAM signals over 368 WDM pol-muxed 26 OAM modes, in *Proceedings of the European Conference on Optical Communication* (2014), pp. 1–3
139. W. Wang, Y. Li, Z. Guo, R. Li, J. Zhang, A. Zhang, S. Qu, Ultra-thin optical vortex phase plate based on the metasurface and the angular momentum transformation. J. Opt. **17**(4), 045102 (2015)
140. T. Wang, M. Ke, W. Li, Q. Yang, C. Qiu, Z. Liu, Particle manipulation with acoustic vortex beam induced by a brass plate with spiral shape structure. Appl. Phys. Lett. **109**(12), 123506 (2016)
141. X. Wei, C. Liu, Z. Zhang, L. Zhu, J. Wang, K. Wang, Z. Yang, J. Liu, Orbit angular momentum encoding at 0.3 thz via 3d printed spiral phase plates, in *Infrared, Millimeter-Wave, and Terahertz Technologies III*, vol. 9275 (SPIE, Bellingham, 2014), pp. 260–265
142. G.P. Williams, Filling the THz gaphigh power sources and applications. Rep. Progr. Phys. **69**(2), 301 (2005)
143. A.E. Willner, H. Huang, Y. Yan, Y. Ren, N. Ahmed, G. Xie, C. Bao, L. Li, Y. Cao, Z. Zhao, J. Wang, M.P.J. Lavery, M. Tur, S. Ramachandran, A.F. Molisch, N. Ashrafi, S. Ashrafi, Optical communications using orbital angular momentum beams. Adv. Opt. Photonics **7**(1), 66–106 (2015)

144. P.J. Winzer, G.J. Foschini, MIMO capacities and outage probabilities in spatially multiplexed optical transport systems. Opt. Express **19**(17), 16 680–16 696 (2011)
145. H. Wu, Y. Yuan, Z. Zhang, J. Cang, UCA-based orbital angular momentum radio beam generation and reception under different array configurations, in *Proceedings of the 6th International Conference Wireless Communication Signal Process* (2014)
146. R. Wunenburger, J.I.V. Lozano, E. Brasselet, Acoustic orbital angular momentum transfer to matter by chiral scattering. New J. Phys. **17**(10), 103022 (2015)
147. G. Xie, Y. Ren, Y. Yan, H. Huang, N. Ahmed, L. Li, Z. Zhao, C. Bao, M. Tur, S. Ashrafi, A.E. Willner, Experimental demonstration of a 200-Gbit/s free-space optical link by multiplexing Laguerre-Gaussian beams with different radial indices. Opt. Lett. **41**(15), 3447–3450 (2016)
148. G. Xie, L. Li, Y. Ren, H. Huang, Y. Yan, N. Ahmed, Z. Zhao, M.P.J. Lavery, N. Ashrafi, S. Ashrafi, R. Bock, M. Tur, A.F. Molisch, A.E. Willner, Performance metrics and design considerations for a free-space optical orbital-angular-momentum-multiplexed communication link. Optica **2**(4), 357–365 (2015)
149. G. Xie, Y. Ren, H. Huang, M.P.J. Lavery, N. Ahmed, Y. Yan, C. Bao, L. Li, Z. Zhao, Y. Cao, M. Willner, M. Tur, S.J. Dolinar, R.W. Boyd, J.H. Shapiro, A.E. Willner, Phase correction for a distorted orbital angular momentum beam using a Zernike polynomials-based stochastic-parallel-gradient-descent algorithm. Opt. Lett. **40**(7), 1197–1200 (2015)
150. M. Xie, X. Gao, M. Zhao, W. Zhai, W. Xu, J. Qian, M. Lei, S. Huang, Mode measurement of a dual-mode radio frequency orbital angular momentum beam by circular phase gradient method. IEEE Antennas Wireless Propag. Lett. **16**, 1143–1146 (2017)
151. L. Xu, X.-C. Zhang, D. Auston, Terahertz beam generation by femtosecond optical pulses in electro-optic materials. Appl. Phys. Lett. **61**(15), 1784–1786 (1992)
152. J. Xue, L. Yong, L. Bin, C. Jianchun, Z. Likun, Convert acoustic resonances to orbital angular momentum. Phys. Rev. Lett. **117**(3), 034301 (2016)
153. Y. Yan, G. Xie, H. Huang, M.J. Lavery, N. Ahemd, C. Bao, Y. Ren, A.F. Molisch, M. Tur, M. Padgett, A.E. Willner, Demonstration of 8-mode 32-Gbit/s millimeter-wave free-space communication link using 4 orbital-angular-momentum modes on 2 polarizations, in *Proceedings of the IEEE International Conference on Communication* (2014), pp. 4850–4855
154. Y. Yan, G. Xie, M.P.J. Lavery, H. Huang, N. Ahmed, C. Bao, Y. Ren, Y. Cao, L. Li, Z. Zhao, A.F. Molisch, M. Tur, M.J. Padgett, A.E. Willner, High-capacity millimetre-wave communications with orbital angular momentum multiplexing. Nat. Commun. **5**, 4876 (2014)
155. G.Y. Yan et al., High-capacity millimetre-wave communications with orbital angular momentum multiplexing. Nat. Commun. **5**(1), 1–9 (2014)
156. Y. Yan, L. Li, G. Xie, C. Bao, P. Liao, H. Huang, Y. Ren, N. Ahmed, Z. Zhao, M. P. Lavery, N. Ashrafi, S. Ashrafi, S. Talwar, S. Sajuyigbe, M. Tur, A.F. Molisch, A.E. Willner, Experimental measurements of multipath-induced intra- and inter-channel crosstalk effects in a millimeter-wave communications link using orbital-angular-momentum multiplexing, in *Proceedings of the IEEE International Conference on Communication* (2015), pp. 1370–1375
157. Y. Yan, L. Li, G. Xie, C. Bao, P. Liao, H. Huang, Y. Ren, N. Ahmed, Z. Zhao, Z. Wang, N. Ashrafi, S. Ashrafi, S. Talwar, S. Sajuyigbe, M. Tur, A.F. Molisch, A.E. Willner, Multipath effects in millimetre-wave wireless communication using orbital angular momentum multiplexing. Sci. Rep. **6**, 33482 (2016)
158. Y. Yan, L. Li, Z. Zhao, G. Xie, Z. Wang, Y. Ren, N. Ahmed, S. Sajuyigbe, S. Talwar, M. Tur, N. Ashrafi, S. Ashrafi, A.F. Molisch, A.E. Willner, 32-Gbit/s 60-GHz millimeter-wave wireless communication using orbital angular momentum and polarization multiplexing, in *Proceedings of the IEEE International Conference on Communication* (2016)
159. L. Yang, Q. Ma, J. Tu, D. Zhang, Phase-coded approach for controllable generation of acoustical vortices. J. Appl. Phys. **113**(15), 154904 (2013)
160. A. M. Yao, M.J. Padgett, Orbital angular momentum: origins, behavior and applications. Adv. Opt. Photonics **3**(2), 161–204 (2011)
161. L. Ye, C. Qiu, J. Lu, K. Tang, H. Jia, M. Ke, S. Peng, Z. Liu, Making sound vortices by metasurfaces. AIP Adv. **6**(8), 085007 (2016)

162. N. Yu, P. Genevet, M.A. Kats, F. Aieta, J.P. Tetienne, F. Capasso, Z. Gaburro, Light propagation with phase discontinuities: generalized laws of reflection and refraction. Science **334**(6054), 333–337 (2011)
163. S. Yu, L. Li, G. Shi, C. Zhu, X. Zhou, Y. Shi, Design, fabrication, and measurement of reflective metasurface for orbital angular momentum vortex wave in radio frequency domain. Appl. Phys. Lett. **108**(12), 121903 (2016)
164. S. Yu, L. Li, G. Shi, C. Zhu, Y. Shi, Generating multiple orbital angular momentum vortex beams using a metasurface in radio frequency domain. Appl. Phys. Lett. **108**(24), 241901 (2016)
165. S. Yu, L. Li, G. Shi, Dual-polarization and dual-mode orbital angular momentum radio vortex beam generated by using reflective metasurface. Appl. Phys. Express **9**(8), 082202 (2016)
166. Y. Yue, Y. Yan, N. Ahmed, J.-Y. Yang, L. Zhang, Y. Ren, H. Huang, K.M. Birnbaum, B.I. Erkmen, S. Dolinar, M. Tur, A.E. Willner, Mode properties and propagation effects of optical orbital angular momentum (OAM) modes in a ring fiber. IEEE Photonics J. **4**(2), 535–543 (2012)
167. X. Zeng, Y. Li, Q. Mo, W. Li, Y. Tian, Z. Liu, J. Wu, Experimental investigation of LP_{11} mode to OAM conversion in few mode-polarization maintaining fiber and the usage for all fiber OAM generator. IEEE Photonics J. **8**(4), 1–7 (2016)
168. N. Zhang, X.C. Yuan, R.E. Burge, Extending the detection range of optical vortices by dammann vortex gratings. Opt. Lett. **35**(20), 3495–3497 (2010)
169. Y. Zhang, W. Feng, N. Ge, On the restriction of utilizing orbital angular momentum in radio communications, in *8th International Conference on Communications and Networking in China* (2013), pp. 271–275
170. H. Zhang, W. Zhang, L. Xi, X. Tang, X. Zhang, X. Zhang, A new type circular photonic crystal fiber for orbital angular momentum mode transmission. IEEE Photon. Technol. Lett. **28**(13), 1426–1429 (2016)
171. Z. Zhang, S. Zheng, Y. Chen, X. Jin, H. Chi, X. Zhang, The capacity gain of orbital angular momentum based multiple-input-multiple-output system. Sci. Rep. (6), 25418 (2016)
172. W. Zhang, S. Zheng, X. Hui, R. Dong, X. Jin, H. Chi, X. Zhang, Mode division multiplexing communication using microwave orbital angular momentum: an experimental study. IEEE Trans. Wireless Commun. **16**(2), 1308–1318 (2017)
173. Y. Zhang, L. Yang, H. Wang, X. Zhang, X. Jin, Transforming surface wave to propagating OAM vortex wave via flat dispersive metasurface in radio frequency. IEEE Antennas Wireless Propag. Lett. **17**(1), 172–175 (2018)
174. C. Zhang, K. Ota, J. Jia, M. Dong, Breaking the blockage for big data transmission: Gigabit road communication in autonomous vehicles. IEEE Commun. Mag. **56**(6), 152–157 (2018)
175. Z. Zhao, J. Wang, S. Li, A.E. Willner, Metamaterials-based broadband generation of orbital angular momentum carrying vector beams. Opt. Lett. **38**(6), 932–934 (2013)
176. Z. Zhao, Y. Ren, G. Xie, Y. Yan, L. Li, H. Huang, C. Bao, N. Ahmed, M.P. Lavery, C. Zhang, N. Ashrafi, S. Ashrafi, S. Talwar, S. Sajuyigbe, M. Tur, A.F. Molisch, A.E. Willner, Experimental demonstration of 16-Gbit/s millimeter-wave communications link using thin metamaterial plates to generate data-carrying orbital-angular-momentum beams, in *Proceedings of the IEEE International Conference Communication* (2015), pp. 1392–1397
177. Z. Zhao, Y. Yan, L. Li, G. Xie, Y. Ren, N. Ahmed, Z. Wang, C. Liu, A.J. Willner, P. Song, H. Hashemi, H. Yao, D. Macfarlane, R. Henderson, N. Ashrafi, S. Ashrafi, S. Talwar, S. Sajuyigbe, M. Tur, A.F. Molisch, A.E. Willner, A dual-channel 60 GHz communications link using patch antenna arrays to generate data-carrying orbital-angular-momentum beams, in *Proceedings of the IEEE International Conference Communication* (2016)
178. S. Zheng, X. Hui, J. Zhu, H. Chi, X. Jin, S. Yu, X. Zhang, Orbital angular momentum mode-demultiplexing scheme with partial angular receiving aperture. Opt. Express **23**(9), 12 251–12 257 (2015)
179. H. Zhou, L. Shi, X. Zhang, J. Dong, Dynamic interferometry measurement of orbital angular momentum of light. Opt. Lett. **39**(20), 6058–6061 (2014)

180. G. Zhou, G. Zhou, C. Chen, M. Xu, C. Xia, Z. Hou, Design and analysis of a microstructure ring fiber for orbital angular momentum transmission. IEEE Photonics J. **8**(2), 1–12 (2016)
181. B. Zhu, T.F. Taunay, M. Fishteyn, X. Liu, S. Chandrasekhar, M.F. Yan, J.M. Fini, E.M. Monberg, F.V. Dimarcello, 112-Tb/s space-division multiplexed DWDM transmission with 14-b/s/Hz aggregate spectral efficiency over a 76.8-km seven-core fiber. Opt. Express **19**(17), 16 665–16 671 (2011)
182. L. Zhu, X. Wei, J. Wang, Z. Zhang, Z. Li, H. Zhang, S. Li, K. Wang, J. Liu, Experimental demonstration of basic functionalities for 0.1-THz orbital angular momentum (OAM) communications, in *Proceedings of the Optical Fiber Communications Conference Expo.* (2014), p. M3K.7
183. F. Zhu, S. Huang, W. Shao, J. Zhang, M. Chen, W. Zhang, J. Zeng, Free-space optical communication link using perfect vortex beams carrying orbital angular momentum (OAM). Opt. Commun. **396**, 50–57 (2017)
184. L. Zou, X. Gu, L. Wang, High-dimensional free-space optical communications based on orbital angular momentum coding. Opt. Commun. **410**, 333–337 (2018)

Chapter 2
Circular Array-Based Radio Frequency Point-to-Point OAM Communications

2.1 Introduction

The precondition of achieving mode multiplexing is to generate multiple OAM modes simultaneously. Typical approaches of using planar phase plate [2, 7], spiral phase plate (SPP) [20], helicoidal parabolic antenna [14], and uniform circular array (UCA) [15] are easy to generate single-mode OAM wave due to simple antenna structures. However, radiating multi-mode OAM beams from a single aperture still has big challenges. So far, a few antenna structures have been proposed to generate multi-mode OAM waves by SPPs [27], solid cylinder dielectric resonator antenna [12], a tri-mode concentric circularly polarized patch antenna [30], metallic traveling-wave ring-slot structure [33], time-switched array structure [18], the uniform concentric circular array (UCCA) [11], and metasurface [17, 28]. Since antenna array is popular in modern wireless communication systems (e.g., 4G, 5G, and Beyond 5G) and easy to steer the beam direction of OAM [5, 25], we focus on the UCA instead of other complicated antenna structures for OAM generation and reception in the following chapters.

Another challenge for the practical application of LoS OAM wireless communications is that OAM requires perfect alignment between the transmit and receive antennas [24]. At least, the beam directions of the transmit and receive antennas have to be aligned to avoid large performance degradation [5]. To steer the beam direction of the receive antenna toward the angle of arrived beam, angle of arrival (AoA) estimation has to be performed first. Although the traditional AoA estimation of planar EM waves has been well studied [21–23], the AoA estimation of vortex EM waves is rarely investigated in the field of wireless communications. Interestingly, a few papers on OAM-based radar claimed that with vortex EM waves radar could obtain azimuthal super-resolution [4, 13]. Inspired by the OAM radar target detection, in this chapter, we propose the accurate AoA estimation methods for aligning the transmit and receive OAM beams in LoS OAM wireless communications.

R. Chen et al., *Circular Array-Based Radio Frequency OAM Communications*, Wireless Networks, https://doi.org/10.1007/978-3-031-37589-7_2

It is now generally accepted that the superiority of UCA-based OAM system to the MIMO system is the low transmitter and receiver complexity due to no need of channel state information (CSI) at either transmitter or receiver, and OAM does not offer any additional gains in channel capacity [8, 16, 32]. However, in practical applications, when the transmit and receive antennas are misaligned with even a small oblique angle, large performance degradation occurs. To deal with the problem, the transmit and receive beam steering approach is proposed in [5], which is shown to be able to circumvent the performance deterioration in a single-mode OAM communication system. Therefore, it is reasonable to speculate that good reception performance could also be obtained in a LoS multi-mode OAM communication system by the beam steering method with highly accurate AoA estimate.

To deal with the aforementioned challenges, we consider the LoS multi-carrier and multi-mode OAM/OAM-MIMO (MCMM-OAM/MCMM-OAM-MIMO) communication schemes in this chapter, including the generation, the distance, and AoA estimation and the reception of multi-mode OAM waves. The novelty and major contributions are summarized as follows:

- First, we verify that UCA can generate multi-mode OAM radio beams with the proposed RF analog synthesis method and baseband digital synthesis method as shown in Fig. 2.1. Through EM simulations, we show that phase distributions of UCA-based multi-mode OAM beams coincide with the theoretical superimposed phase distributions of multiple OAM modes.
- Second, we propose a UCA-based LoS MCMM-OAM communication scheme and formulate the signal models of training and data transmission phases. The channel matrix of this system is parametrized by only three parameters, i.e., azimuth angle, the elevation angle, and the distance. Then, we propose two effective methods to estimate these three parameters based on the estimating signal parameter via rotational invariance (ESPRIT) algorithm. Hence, compared

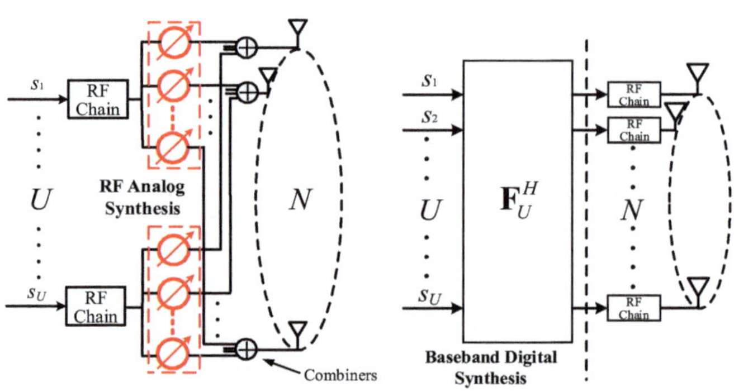

Fig. 2.1 UCA-based multi-mode OAM transmitter implemented by (**a**) RF analog synthesis with phase shifters and power combiners and (**b**) baseband digital synthesis

with the traditional MIMO-OFDM system, there is no need to estimate large channel matrices, greatly reducing the required training overhead especially at sub-6 GHz bands.

- Third, we propose an OAM reception scheme including the beam steering with the estimated AoA and the amplitude detection with the estimated distance. We prove that the proposed OAM reception scheme can eliminate the inter-mode interference induced by the misalignment error in a practical OAM channel and approaches the performance of an ideally aligned OAM channel. Furthermore, we extend the proposed methods to the UCCA-based LoS MCMM-OAM-MIMO system, for which again the channel matrices are parameterized by only three parameters.

The remainder of this chapter is organized as follows. In Sect. 2.2, we verify the RF analog and baseband digital synthesis methods for multi-mode OAM beam generation. Based on the feasibility of generating multi-mode OAM beam by baseband digital method, we model the UCA-based LoS MCMM-OAM communication system in Sect. 2.3. In Sect. 2.4, the two distance and AoA estimation methods based on the ESPRIT algorithm are proposed. With the estimated distance and AoA, an OAM reception scheme is proposed to correctly recover the transmitted data symbols in Sect. 2.5. In Sect. 2.6, the proposed methods are extended to the UCCA-based LoS MCMM-OAM-MIMO system to achieve higher SE. Simulation results are shown in Sect. 2.7.

2.2 Generation of Muti-mode OAM with UCA

The feasibility of utilizing UCA to generate OAM waves has been verified in theory [15, 19] and in practice [9, 10]. The antenna elements in a UCA are fed with the same input signal but with a successive phase shift from element to element such that after a full turn the phase has the increment of $2\pi\ell$, where ℓ is an unbounded integer termed as *topological charge or OAM mode number* [1]. In general, the phased UCA can generate only single-mode OAM beam, but with phase shifters network (PSN) and power combiners, UCA is expected to generate multi-mode OAM beam [6] as shown in Fig. 2.1a.

2.2.1 *Superimposed Radiation of Multiple OAM Modes*

The transmitter generates OAM beams with the UCA consisting of N equidistant antenna elements on a circle with radius of R_t. The phase of the n-th antenna element $\varphi_n = \ell\phi_n$, where $\phi_n = 2\pi(n-1)/N$ is the azimuthal angle of the n-th antenna element. Thus, for the point $A(\bar{r}, \bar{\varphi}, \bar{\theta})$ in spherical coordinates in the direction of the OAM beam, the electric field can be expressed as [15, 29]

$$E(\bar{\boldsymbol{r}}, \ell, k) \approx \frac{\beta_t}{4\pi} \sum_{n=1}^{N} e^{i\ell\phi_n} \frac{e^{ik|\bar{\boldsymbol{r}}-\boldsymbol{r}_n|}}{|\bar{\boldsymbol{r}}-\boldsymbol{r}_n|} \approx \frac{\beta_t}{4\pi} \frac{e^{ik\bar{r}}}{\bar{r}} \sum_{n=1}^{N} e^{-i(\boldsymbol{k}\cdot\boldsymbol{r}_n - \ell\phi_n)}$$
$$\approx \frac{\beta_t N e^{ik\bar{r}} e^{i\ell\bar{\varphi}}}{4\pi\bar{r}} i^{-\ell} J_\ell(kR_t \sin\bar{\theta}), \tag{2.1}$$

where β_t models all the constants relative to each transmit antenna element, $i = \sqrt{-1}$ is the imaginary unit, $\boldsymbol{k}$ is the wave vector, $k = |\boldsymbol{k}| = 2\pi f/c$ is the wave number corresponding to subcarrier frequency f, $\bar{\boldsymbol{r}}$ is the position vector of the point A, and $J_\ell(\cdot)$ is ℓth-order Bessel function of the first kind. The far-field approximations are $|\bar{\boldsymbol{r}} - \boldsymbol{r}_n| \approx \bar{r}$ for amplitudes and $|\bar{\boldsymbol{r}} - \boldsymbol{r}_n| \approx \bar{r} - \hat{\boldsymbol{r}} \cdot \boldsymbol{r}_n$ for phases, where $\hat{\boldsymbol{r}}$ is the unit vector of $\bar{\boldsymbol{r}}$ and $\boldsymbol{r}_n = R_t(\hat{\boldsymbol{x}} \cos\phi_n + \hat{\boldsymbol{y}} \sin\phi_n)$, $\hat{\boldsymbol{x}}$ and $\hat{\boldsymbol{y}}$ are the unit vectors of x-axis and y-axis of the coordinate system at the transmitter, respectively.

According to the pattern multiplication principle, the electric field $E(\bar{\boldsymbol{r}}, \ell, k)$ can be rewritten as

$$E(\bar{\boldsymbol{r}}, \ell, k) = E_{\text{dipole}}(\bar{r}) \Psi_\ell(\bar{\varphi}, \bar{\theta}), \tag{2.2}$$

where $\boldsymbol{E}_{\text{dipole}}(\bar{r})$ corresponds to electric dipole radiation and the array factor denoted by $\Psi_\ell(\bar{\varphi}, \bar{\theta})$ takes the form

$$\Psi_\ell(\bar{\varphi}, \bar{\theta}) = \sum_{n=1}^{N} e^{-i(\boldsymbol{k}\cdot\boldsymbol{r}_n - \varphi_n)}. \tag{2.3}$$

Considering OAM mode multiplexing or multi-mode OAM beam, the electric field at the observation point $A(\bar{r}, \bar{\varphi}, \bar{\theta})$ should be the superimposed electric fields of multiple single-mode OAM beams. Thus, for a U-mode OAM beam with each mode carrying one information symbol, the electric field vector $E^{(U)}$ can be written as

$$E^{(U)}(\bar{\boldsymbol{r}}, \ell, k) = \sum_{u=1}^{U} E(\bar{\boldsymbol{r}}, \ell_u, k) \cdot s_u = E_{\text{dipole}}(\bar{r}) \sum_{u=1}^{U} \Psi_{\ell_u}(\bar{\varphi}, \bar{\theta}) \cdot s_u$$
$$= E_{\text{dipole}}(\bar{r}) \sum_{n=1}^{N} A_n e^{-i(\boldsymbol{k}\cdot\boldsymbol{r}_n - \varphi'_n)}, \tag{2.4}$$

where s_u is the data symbol transmitted on the u-th OAM mode, A_n, and φ'_n satisfies $A_n e^{i\varphi'_n} = \sum_{u=1}^{U} e^{i\ell_u\phi_n} s_u$. Comparing (2.2) and (2.3) with (2.4), we find that radiating U-mode OAM beam can be achieved by feeding the n-th antenna with the amplitude A_n and the phase φ'_n, $n = 1, 2, \cdots, N$. Thus, the operation of multi-mode synthesis can be fulfilled in RF analog form as shown in Fig. 2.1a or in baseband digital form as shown in Fig. 2.1b. We note that other RF analog and baseband digital synthesis methods of OAM beams have been applied in [26] and [31], respectively.

2.2.2 Electromagnetic Simulations

To verify both the synthesis methods of generating the multi-mode OAM beam, we first build a UCA model in the high-frequency structure simulator (HFSS) from ANSYS as shown in Fig. 2.2. In this model, the UCA is formed by $N = 8$ linearly polarized patch antennas designed at 2.45 GHz, the radius of the UCA is 0.66λ, and $\{s_u = 1|u = 1, 2, \cdots, U\}$. To avoid excessive computational burden of finite element calculations of HFSS, the transmission distance from the UCA to the observation plane is set as 2λ. The detailed parameters are listed in Table 2.1.

Then, we compare the phase distributions of the electric fields radiated by the UCA in HFSS with the ideal phase distributions of Laguerre-Gaussian beams in Fig. 2.3. It can be seen from the figure that the phase distributions of single-mode OAM beams with mode numbers $\ell = +1, +2, +3$ generated by the designed UCA coincide with those of ideal OAM beams in principle. Thus, the designed UCA in HFSS can be used to verify the superimposed phase distributions of multi-mode OAM beams.

After that, we feed each antenna port with corresponding amplitude A_n and phase φ'_n according to (2.4) and compare the phase distributions of the electric fields radiated by the UCA in HFSS with the ideal superimposed phase distributions of two-mode OAM beams in Fig. 2.4. Comparing the subfigures at the top side and those at the bottom side, we confirm that UCA with either RF analog synthesis or baseband digital synthesis can generate multi-mode OAM beams. It is known that for the N-element UCA, at most N OAM modes can be resolved, i.e., $|\ell| < N/2$ [15]. Hence, the multi-mode OAM beam composed of more and higher-order OAM modes could also be generated by a UCA with enough number of elements theoretically.

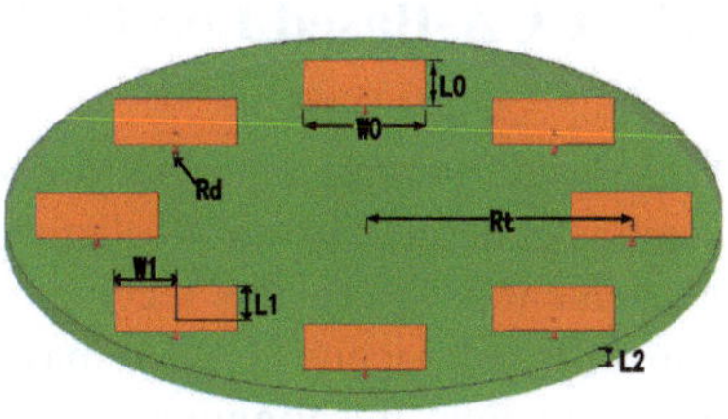

Fig. 2.2 The UCA model built in HFSS

Table 2.1 Parameters of the designed uniform circular array

Parameter	$W0$	$L0$	$W1$	$L1$	$L2$	R_t	R_d
Value(mm)	37.26	28.1	18.63	21.05	5	81	0.6

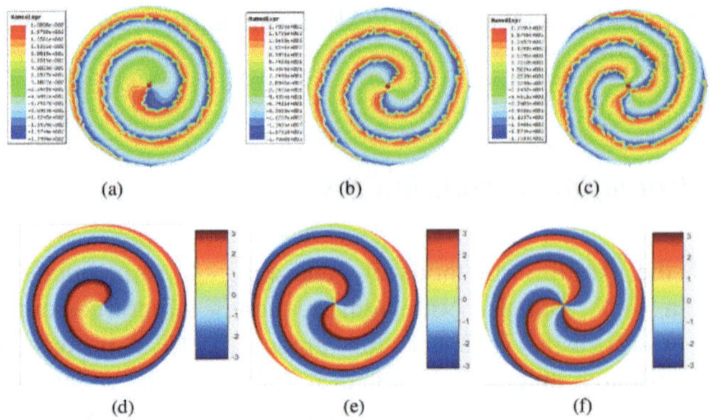

Fig. 2.3 The rotational phase of an OAM beam simulated by HFSS with mode number: (**a**) $\ell = +1$, (**b**) $\ell = +2$, (**c**) $\ell = +3$; and the ideal rotational phase of an OAM beam simulated by MATLAB with mode number: (**d**) $\ell = +1$, (**e**) $\ell = +2$, (**f**) $\ell = +3$. A change in color from blue to green, yellow, and red to blue again corresponds to a change in phase of 2π

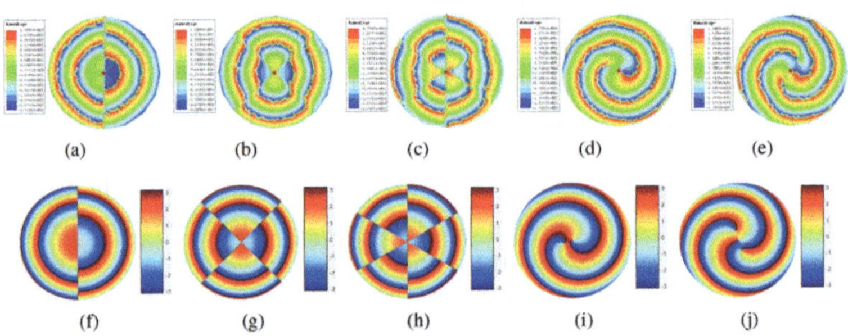

Fig. 2.4 The phase distributions of two-mode OAM beam simulated by HFSS with mode numbers: (**a**) $\ell_1 = -1, \ell_2 = +1$, (**b**) $\ell_1 = -2, \ell_2 = +2$, (**c**) $\ell_1 = -3, \ell_2 = +3$, (**d**) $\ell_1 = +1, \ell_2 = +2$, (**e**) $\ell_1 = +1, \ell_2 = +3$; and the phase distributions of two superimposed ideal OAM beams simulated by MATLAB with mode numbers: (**f**) $\ell_1 = -1, \ell_2 = +1$, (**g**) $\ell_1 = -2, \ell_2 = +2$, (**h**) $\ell_1 = -3, \ell_2 = +3$, (**i**) $\ell_1 = +1, \ell_2 = +2$, (**j**) $\ell_1 = +1, \ell_2 = +3$

2.3 UCA-Based LoS OAM Communication Systems

We consider a radio LoS OAM communication system, where a multi-mode OAM beam is generated by an N-element UCA at the transmitter and received by another N-element UCA at the receiver. In practice, the perfect alignment between the transmit and receive UCAs may be difficult to realize. Therefore, we consider more practical misalignment cases. For easier analysis, we adopt the non-parallel misalignment case [5] as shown in Fig. 2.5 but with a more general model including both elevation angle θ and azimuth angle φ.

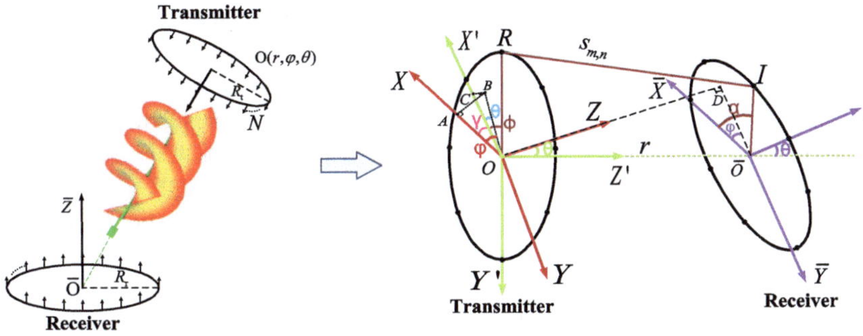

Fig. 2.5 UCA-based LoS OAM communication system

2.3.1 *Channel Model*

In free space communications, propagation through the RF channel leads to attenuation and phase rotation of the transmitted signal. This effect is modelled through multiplying by a complex constant h, whose value depends on the frequency k and the distance d between the transmit and receive antenna [8]:

$$h(k, d) = \frac{\beta}{2kd} \exp(-ikd), \tag{2.5}$$

where λ is the wavelength, $1/(2kd)$ denotes the degradation of amplitude, $\beta = \beta_r \beta_t$, β_r models all the constants relative to each receive antenna element, and the complex exponential term is the phase difference due to the propagation distance. It is worth noting that in order to mainly analyze the factors we concerned, the transmission loss caused by the polarization mismatch between the transmit and receive UCAs is ignored here.

The geometrical model of the UCA-based LoS OAM channel is illustrated in Fig. 2.5. In the geometrical model, the transmitter coordinate system $\mathrm{Z}' - \mathrm{X}'\mathrm{OY}'$ is established using the transmit UCA plane as the $\mathrm{X}'\mathrm{OY}'$ plane and the axis through the transmit UCA center O and perpendicular to the UCA plane as the Z'-axis, and the receiver coordinate system $\bar{\mathrm{Z}} - \bar{\mathrm{X}}\bar{\mathrm{O}}\bar{\mathrm{Y}}$ is established on the plane of the receive UCA by the similar approach. Since the transmit UCA is aligned with the center of the receive UCA in the non-parallel misalignment case, the coordinate of the receive UCA center can be denoted as $\bar{\mathrm{O}}(r, 0, 0)$ in $\mathrm{Z}' - \mathrm{X}'\mathrm{OY}'$. According to the geometrical model, the coordinate of the transmit UCA center is denoted as $\mathrm{O}(r, \varphi, \theta)$ in $\bar{\mathrm{Z}} - \bar{\mathrm{X}}\bar{\mathrm{O}}\bar{\mathrm{Y}}$, where r is the distance between the transmit and the receive UCA centers, φ is the azimuth angle, θ is the elevation angle, and φ *and* θ *are defined as the AoA of the transmitted OAM beams*.

To obtain the estimates of φ and θ, we have to build the coordinate system Z − XOY that is at the transmit UCA and parallel to $\bar{\mathrm{Z}} - \bar{\mathrm{X}}\bar{\mathrm{O}}\bar{\mathrm{Y}}$. According to the geometrical model, the angle between Z′-axis and $\bar{\mathrm{Z}}$-axis is θ, so is that between Z-axis and Z′-axis. Let C be a point on the X′-axis, B the projection of C on the XOY plane, and A the projection of B on the X-axis; then $\angle\mathrm{BOC} = \theta$. As the line OB is parallel to the line $\bar{\mathrm{O}}$D and X-axis is parallel to $\bar{\mathrm{X}}$-axis, $\angle\mathrm{AOB} = \varphi$, where the point D is the projection of O on the $\bar{\mathrm{X}}\bar{\mathrm{O}}\bar{\mathrm{Y}}$ plane. Denote the angle between X′-axis and X-axis as γ; then γ can be obtained as

$$\gamma = \angle\mathrm{AOC} = \arccos\left(\cos(\angle\mathrm{AOB})\cos(\angle\mathrm{BOC})\right) = \arccos\left(\cos\theta\cos\varphi\right). \quad (2.6)$$

Define the angle between the line O$\aleph$ and X′-axis as ϕ and the angle between the line $\bar{\mathrm{O}}\mathfrak{R}$ and $\bar{\mathrm{X}}$-axis as α, where $\aleph$ is the position of the n-th $(1 \le n \le N)$ element at the transmitter and $\mathfrak{R}$ is the position of the m-th $(1 \le m \le N)$ element at the receiver. According to (2.5) and the geometric relationship in Fig. 2.5, the channel coefficients from the n-th transmit antenna element to the m-th receive antenna element can be expressed as $h_{m,n} = h(k, d_{m,n})$, where the transmission distance $d_{m,n}$ is calculated as

$$\begin{aligned} d_{m,n} = \big[&R_t^2 + R_r^2 + r^2 - 2rR_r\cos\alpha\cos\varphi\sin\theta + 2rR_r\sin\alpha\sin\varphi\sin\theta \\ &- 2R_tR_r(\sin\phi\cos\alpha\sin\varphi - \cos\phi\sin\alpha\sin\varphi\cos\theta) \\ &- 2R_tR_r(\cos\phi\cos\alpha\cos\gamma + \sin\phi\sin\alpha\cos\varphi)\big]^{1/2}, \end{aligned} \quad (2.7)$$

where R_t and R_r are, respectively, the radii of the transmit and receive UCAs, $\phi = [2\pi(n-1)/N + \phi_0]$, $\alpha = [2\pi(m-1)/N + \alpha_0]$, and ϕ_0 and α_0 are, respectively, the corresponding initial angles of the first reference antenna elements in both UCAs. For easier analysis, we assume $\phi_0 = 0$ and $\alpha_0 = 0$ here.

In the end, the channel matrix of the UCA-based free space communication system can be expressed as $\mathbf{H} = [h_{m,n}]_{N\times N}$. Note that when $\theta = 0$ and $\varphi = 0$, $\mathbf{H}$ is a circulant matrix that can be decomposed by the N-dimentional Fourier matrix $\mathbf{F}_N$ as $\mathbf{H} = \mathbf{F}_N^H \boldsymbol{\Lambda} \mathbf{F}_N$, where $\boldsymbol{\Lambda}$ is a diagonal matrix with the eigenvalues of $\mathbf{H}$.

2.3.2 *Signal Model*

For higher data rate transmission, OFDM is often applied, and naturally, the OAM-OFDM communication system has been proposed in [6]. In an LoS MCMM-OAM communication system, we assume W subcarriers and U OAM modes for data transmission and $\widetilde{W}$ subcarriers and $\widetilde{U}$ OAM modes for training, $1 \le \widetilde{W} \le W$, $1 \le \widetilde{U} \le U$.

As a UCA can generate both single-mode and multi-mode OAM beams with the baseband (partial) discrete Fourier transform (DFT) matrix $\mathbf{F}_U$, the equivalent baseband signal models of transmitting multi-mode OAM data symbols and multiple single-mode OAM training symbols can be expressed as $\mathbf{F}_U^H \mathbf{s}(k_w)$ and $\mathbf{F}_{\widetilde{U}}^H \mathbf{S}'(k_w)$, respectively, where $\mathbf{s}(k_w) = [s(\ell_1, k_w), s(\ell_2, k_w), \cdots, s(\ell_U, k_w)]^T$ contains the data symbols transmitted on U OAM modes at the w-th ($1 \le w \le W$) subcarrier simultaneously and $\mathbf{S}'(k_w) = \text{diag}\{s'(\ell_1, k_w), s'(\ell_2, k_w), \cdots, s'(\ell_{\widetilde{U}}, k_w)\}$ consists of the training symbols transmitted on $\widetilde{U}$ OAM modes at the w-th ($1 \le w \le \widetilde{W}$) subcarrier sequentially.

Thus, the received baseband signal vector $\mathbf{y}(k_w)$ can be written as

$$\mathbf{y}(k_w) = \mathbf{H}(k_w)\mathbf{F}_U^H \mathbf{s}(k_w) + \mathbf{z}(k_w), \tag{2.8}$$

where $\mathbf{y}(k_w)$ is the N-dimensional received signal vector, $\mathbf{H}(k_w) = [h_{m,n}(k_w)]_{N \times N}$ is the channel matrix at the w-th subcarrier, $\mathbf{F}_U = [\mathbf{f}^H(\ell_1), \mathbf{f}^H(\ell_2), \cdots, \mathbf{f}^H(\ell_U)]^H$ is an $U \times N$ (partial) DFT matrix, $\mathbf{f}(\ell_u) = \frac{1}{\sqrt{N}}[1, e^{-i\frac{2\pi\ell_u}{N}}, \cdots, e^{-i\frac{2\pi\ell_u(N-1)}{N}}]$, $\mathbf{z}(k_w) = [z(\ell_1, k_w), z(\ell_2, k_w), \cdots, z(\ell_U, k_w)]^T$ is the complex Gaussian noise vector with zero mean and covariance matrix $\sigma_z^2 \mathbf{I}_N$, $1 \le U \le N$, $1 \le w \le W$.

Correspondingly, the received baseband training symbol matrix $\mathbf{Y}'(k_w)$ can be written as

$$\mathbf{Y}'(k_w) = \mathbf{H}(k_w)\mathbf{F}_{\widetilde{U}}^H \mathbf{S}'(k_w) + \mathbf{Z}'(k_w), \tag{2.9}$$

where $\mathbf{Y}'(k_w)$ is the $N \times \widetilde{U}$ matrix composed by the sequentially received training signals on $\widetilde{U}$ OAM modes at the w-th subcarrier and $\mathbf{Z}'(k_w)$ is the related additive noise matrix, $1 \le \widetilde{U} \le N$, $1 \le w \le \widetilde{W}$.

The UCA-based OAM receiver has the similar RF analog or baseband digital structure to the transmitter but with the opposite operations, i.e., separating different OAM modes and despiralizing each mode. Thus, the detected OAM data symbol vector $\mathbf{x}(k_w)$ can be expressed as

$$\mathbf{x}(k_w) = \mathbf{D}(k_w)\mathbf{y}(k_w) = \mathbf{D}(k_w)\left(\mathbf{H}(k_w)\mathbf{F}_U^H \mathbf{s}(k_w) + \mathbf{z}(k_w)\right), \tag{2.10}$$

where $\mathbf{x}(k_w) = [x(\ell_1, k_w), x(\ell_2, k_w), \cdots, x(\ell_U, k_w)]^T$, $x(\ell_u, k_w)$ is the detected data symbol on the u-th mode OAM at the w-th subcarrier, and $\mathbf{D}(k_w)$ is the $U \times N$ data signal detection matrix. In the case that the transmit and receive UCAs are in perfect alignment, the effective multi-mode OAM channel defined as $\mathbf{H}_{\text{OAM}}(k_w) = \mathbf{F}_U \mathbf{H}(k_w)\mathbf{F}_U^H$ becomes the diagonal matrix $\boldsymbol{\Lambda}_U(k_w)$ composed of U eigenvalues of $\mathbf{H}(k_w)$, and $\mathbf{D}(k_w) = \boldsymbol{\Lambda}_U^{-1}(k_w)\mathbf{F}_U$. Thus, $\mathbf{x}(k_w) = \mathbf{s}(k_w) + \boldsymbol{\Lambda}_U^{-1}(k_w)\mathbf{z}(k_w)$, which shows the transmitted data symbols being recovered and the mode multiplexing gain of OAM communication systems being achieved. Furthermore, how to design $\mathbf{D}(k_w)$ in a more general misalignment case will be specified in Sect. 2.5.

Corresponding to the training symbol transmitted on each OAM mode at each time slot, the received signals on all the N elements are combined at each time slot, which can be formulated as

$$\mathbf{x}'(k_w) = \mathbf{1}^T \mathbf{Y}'(k_w) = \mathbf{1}^T \left(\mathbf{H}(k_w)\mathbf{F}_{\widetilde{U}}^H \mathbf{S}'(k_w) + \mathbf{Z}'(k_w) \right), \tag{2.11}$$

where $\mathbf{x}'(k_w) = [x'(\ell_1, k_w), x'(\ell_2, k_w), \cdots, x'(\ell_{\widetilde{U}}, k_w)]$, $x'(\ell_u, k_w)$ is the combined signal corresponding to $s'(\ell_u, k_w)$, $u = 1, 2, \cdots, \widetilde{U}$, $w = 1, 2, \cdots, \widetilde{W}$.

$$\begin{aligned} x'(\ell_u, k_w) &= \sum_{m=1}^{N} \mathbf{h}_m(k_w)\mathbf{f}^H(\ell_u)s'(\ell_u, k_w) + z'(\ell_u, k_w) \\ &= \frac{\beta}{2k_w} \sum_{m=1}^{N} \sum_{n=1}^{N} \frac{e^{ik_w|\boldsymbol{d}_{m,n}|}}{|\boldsymbol{d}_{m,n}|} e^{i\ell_u\phi_n} s'(\ell_u, k_w) + z'(\ell_u, k_w) \\ &= \frac{\beta}{2k_w} \sum_{m=1}^{N} \sum_{n=1}^{N} \frac{e^{ik_w|\boldsymbol{r}-\boldsymbol{r}_n+\boldsymbol{r}_m|}}{|\boldsymbol{r}-\boldsymbol{r}_n+\boldsymbol{r}_m|} e^{i\ell_u\phi_n} s'(\ell_u, k_w) + z'(\ell_u, k_w) \\ &\approx \frac{\beta}{2k_w} \sum_{m=1}^{N} e^{i\boldsymbol{k}_w\cdot\boldsymbol{r}_m} \left[\frac{e^{ik_w r}}{r} \sum_{n=1}^{N} e^{-i(\boldsymbol{k}_w\cdot\boldsymbol{r}_n - \ell_u\phi_n)} \right] s'(\ell_u, k_w) + z'(\ell_u, k_w) \\ &\approx \sigma_{\ell_u,k_w} e^{ik_w r} e^{i\ell_u\gamma} i^{-\ell_u} J_{\ell_u}(k_w R_t \sin\theta) J_0(k_p R_r \sin\theta) + z'(\ell_u, k_w), \end{aligned} \tag{2.12}$$

2.4 Distance and AoA Estimation for OAM Beams

2.4.1 *Problem Formulation*

As in MIMO communications, in OAM, the transmitted frames may carry a combination of training and data symbols. In the case of MIMO, pilot data is used to perform channel estimation, and it is well-known that the number of training symbols per subcarrier or subchannel has to be larger than the number of transmit antennas. On the other hand, in the case of OAM, pilot data are exploited for the distance and AoA estimation, and $\widetilde{U}$ $(1 \le \widetilde{U} \le N)$ training symbols per subcarrier or subchannel are required. Accordingly, we assume the training symbols $\{s'(\ell_u, k_w)|u = 1, 2, \cdots, \widetilde{U}; w = 1, 2, \cdots, \widetilde{W}\}$ are known to the OAM receiver.

Then, based on (2.1), (2.11), and the coordinate transformation in Sect. 2.3, the combined signal $x'(\ell_u, k_w)$ in the $Z - XOY$ coordinate system can be derived in (2.12), where $\mathbf{h}_m(k_w) = [h_{m,1}(k_w), h_{m,2}(k_w), \cdots, h_{m,N}(k_w)]$, $\sigma_{\ell_u,k_w} = \frac{\beta N^2}{2k_w r} s'(\ell_u, k_w)$ is a function of r and unrelated to θ and φ, and $\boldsymbol{d}_{m,n}$ is the position vector from the n-th transmit antenna element to the m-th receive antenna element, $\boldsymbol{r}_m = R_r(\hat{\boldsymbol{x}}' \cos\alpha + \hat{\boldsymbol{y}}' \sin\alpha)$, and $\hat{\boldsymbol{x}}'$ and $\hat{\boldsymbol{y}}'$ are the unit vectors of $\bar{\text{X}}$-axis and $\bar{\text{Y}}$-axis, respectively.

Overall, the aim of the distance and AoA estimation is to obtain the distance r, the elevation angle θ, and the azimuthal angle φ from the received training signals $\{x'(\ell_u, k_w)|u = 1, 2, \ldots, \widetilde{U}; w = 1, 2, \cdots, \widetilde{W}\}$.

2.4.2 *Distance and AoA Estimation Based on 2-D ESPRIT*

According to the expression in (2.12), we propose to first estimate r and γ with 2-D ESPRIT algorithm and then extract θ and φ.

2.4.2.1 Estimation of *r* and *γ*

The ESPRIT algorithm provides an elegant means for estimating the parameters of complex sinusoidal signals embedded in white Gaussian noise. Accordingly, at the receiver, we need to extract the exponentials containing r and γ from the signal $x'(\ell_u, k_w)$ by performing the following operation:

$$\tilde{x}(\ell_u, k_w) = \frac{x'(\ell_u, k_w)}{|x'(\ell_u, k_w)|} \frac{s'(\ell_u, k_w)^*}{|s'(\ell_u, k_w)|} i^{\ell_u} \mathrm{sign}(x'(\ell_u, k_w)) = e^{ik_w r} e^{i\ell_u \gamma} + \tilde{z}(\ell_u, k_w), \tag{2.13}$$

where $\tilde{z}(\ell_u, k_w)$ is the noise, $k_w = 2\pi f_w/c$, $k_{w+1} - k_w = 1$, $\ell_{u+1} - \ell_u = 1$, $w = 1, 2, \cdots, \widetilde{W}$, and $u = 1, 2, \cdots, \widetilde{U}$. Then, all the signals received on the $\widetilde{U}$ OAM modes at the $\widetilde{W}$ subcarriers can be collected in the matrix $\tilde{\mathbf{X}}_{\widetilde{U},\widetilde{W}} = \left[\tilde{x}(\ell_u, k_w)\right]_{\widetilde{U}\times\widetilde{W}}$.

In the estimation of r, we first build the $\widetilde{U}\widetilde{W}$-dimensional vector $\mathbf{x}_C$ by stacking the columns of $\tilde{\mathbf{X}}_{\widetilde{U},\widetilde{W}}$ as

$$\mathbf{x}_C = \mathrm{cvec}\left(\tilde{\mathbf{X}}_{\widetilde{U},\widetilde{W}}\right) = [\tilde{x}(\ell_1, k_1), \tilde{x}(\ell_2, k_1), \cdots, \tilde{x}(\ell_{\widetilde{U}}, k_{\widetilde{W}})]^T,$$

so that we can express the received signal in compact form as

$$\mathbf{x}_C = \mathbf{a} + \mathbf{z}_C. \tag{2.14}$$

where $\mathbf{a} = [e^{ik_1 r} e^{i\ell_1 \gamma}, e^{ik_1 r} e^{i\ell_2 \gamma}, \cdots, e^{ik_{\widetilde{W}} r} e^{i\ell_{\widetilde{U}} \gamma}]^T$ and $\mathbf{z}_C$ is the noise vector. The covariance matrix of $\mathbf{x}_C$ can be written as

$$\mathbf{R}_{\mathbf{x}_C} = \mathbb{E}\left\{\mathbf{x}_C \mathbf{x}_C^H\right\} = \mathbf{a}\mathbf{a}^H + \mathbf{R}_{\mathbf{z}_C}, \tag{2.15}$$

where $\mathbf{R}_{\mathbf{z}_C} = \mathbb{E}\left\{\mathbf{z}_C \mathbf{z}_C^H\right\}$. Then, the eigen value decomposition (EVD) of $\mathbf{R}_{\mathbf{x}_C}$ is given by

$$\mathbf{R}_{\mathbf{x}_C} = \boldsymbol{Q}\boldsymbol{\Omega}\boldsymbol{Q}^H, \tag{2.16}$$

where $\mathbf{Q}$ is an $\widetilde{U}\widetilde{W} \times \widetilde{U}\widetilde{W}$ unitary matrix and $\boldsymbol{\Omega} = \text{diag}(\chi_1, \chi_2, \cdots, \chi_{\widetilde{U}\widetilde{W}})$. Denote $\chi_{max} = \max\{\chi_m | m = 1, 2, \cdots, \widetilde{U}\widetilde{W}\}$, and the eigenvector corresponding to χ_{max} as the signal subspace $\mathbf{q}$, which satisfies

$$\mathbf{R}_{\mathbf{x}_C}\mathbf{q} = \chi_{max}\mathbf{q}. \tag{2.17}$$

Now, the subspace spanned by $\mathbf{q}$ is the signal subspace spanned by $\mathbf{a}$ so that the following relationship holds true:

$$\mathbf{a} = \delta\mathbf{q}, \tag{2.18}$$

where δ is a nonzero scalar. If we consider the two vectors $\mathbf{a}_1$ and $\mathbf{a}_2$, obtained by taking the first and the last $\widetilde{U} \times (\widetilde{W} - 1)$ elements of $\mathbf{a}$, respectively, it is $\mathbf{a}_2 = \mathbf{a}_1\Phi$, where $\Phi = e^{ir}$. To obtain an estimate of Φ and r, we construct the two vectors $\mathbf{q}_1$ and $\mathbf{q}_2$, composed by the first and by the last $\widetilde{U} \times (\widetilde{W} - 1)$ elements of $\mathbf{q}$, respectively. Then, exploiting the fact that $\mathbf{a}_1 = \delta\mathbf{q}_1$ and $\mathbf{a}_2 = \delta\mathbf{q}_2$, one obtains

$$\mathbf{q}_2 = \mathbf{q}_1\Phi, \tag{2.19}$$

which leads to

$$e^{ir} = \Phi = \mathbf{q}_1^{\dagger}\mathbf{q}_2. \tag{2.20}$$

In the estimation of γ, we stack the rows of the matrix $\tilde{\mathbf{X}}_{\widetilde{U},\widetilde{W}}$ in the vector denoted by $\mathbf{x}_R$,

$$\mathbf{x}_R = \text{rvec}\left(\tilde{\mathbf{X}}_{\widetilde{U},\widetilde{W}}\right) = [\tilde{x}(\ell_1, k_1), \tilde{x}(\ell_1, k_2), \cdots, \tilde{x}(\ell_{\widetilde{U}}, k_{\widetilde{W}})]^T,$$

so that $\mathbf{x}_R$ can be expressed as

$$\mathbf{x}_R = \mathbf{b} + \mathbf{z}_R, \tag{2.21}$$

where $\mathbf{b} = [e^{ik_1 r}e^{i\ell_1\gamma}, e^{ik_2 r}e^{i\ell_1\gamma}, \cdots, e^{ik_{\widetilde{W}} r}e^{i\ell_{\widetilde{U}}\gamma}]^T$ and $\mathbf{z}_R$ is the noise vector. By computing the correlation matrix $\mathbf{R}_{\mathbf{x}_R} = \mathbb{E}\left\{\mathbf{x}_R\mathbf{x}_R^H\right\}$ and following the same method described for obtaining an estimate of r, we can obtain an estimate of γ.

2.4.2.2 Estimation of θ and φ

After estimating r and γ based on 2-D ESPRIT algorithm, we can obtain the elevation angle θ with r and the amplitudes $\{|x'(\ell_u, k_w)| \, |w = 1, 2, \ldots, \widetilde{W}; u = 1, 2, \cdots, \widetilde{U}\}$ through the equations

$$J_{\ell_u}(k_w R_t \sin\theta) J_0(k_w R_r \sin\theta) \approx \delta(\ell_u, k_w),\ w = 1, 2, \ldots, \widetilde{W};\ u = 1, 2, \cdots, \widetilde{U}, \tag{2.22}$$

where $\delta(\ell_u, k_w) = \frac{|x'(\ell_u, k_w)|}{|\sigma_{\ell_u, k_w}|} \mathrm{sign}(x'(\ell_u, k_w))$, the approximation neglects the effect of noise at high receive signal-to-noise ratio (SNR). Thus, all the solutions of θ can be obtained from (2.22) numerically [3].

Without loss of generality, we assume that the receiver can receive signals from the transmitter only when the transmitter is located within the main lobe of the receive UCA. Thus, $[\theta_a, \theta_b]$ is assumed to be the angle range of the receive UCA's main lobe, and the solutions obtained by (2.22) is denoted as $\{\hat{\theta}_l(\ell_u, k_w) | \hat{\theta}_l(\ell_u, k_w) \in [\theta_a, \theta_b], l = 1, 2, \ldots, L_{\ell_{\widetilde{U}}, k_{\widetilde{W}}}\}$, where $\hat{\theta}_l(\ell_u, k_w)$ is the l-th possible solution of θ corresponding to $x'(\ell_u, k_w)$. In order to determine the final estimate of θ, we first look for $\widetilde{U} \times \widetilde{W}$ solutions $\{\hat{\theta}_l(\ell_u, k_w)\}$ by the following method:

(i) Divide $[\theta_a, \theta_b]$ into $\mathcal{D}$ intervals denoted as $[\theta_{a1}, \theta_{b1}], [\theta_{a2}, \theta_{b2}], \cdots, [\theta_{a\mathcal{D}}, \theta_{b\mathcal{D}}]$, where $\theta_{a1} = \theta_a$ and $\theta_{b\mathcal{D}} = \theta_b$;
(ii) Categorize all the solutions $\{\hat{\theta}_l(\ell_u, k_w)\}$ into corresponding intervals;
(iii) Count the number of solutions in each interval, and denote it as $L_\epsilon, \epsilon = 1, 2, \cdots, \mathcal{D}$;
(iv) Let $\mathcal{L} = \max\{L_\epsilon | \epsilon = 1, 2, \cdots, \mathcal{D}\}$, if $\mathcal{L} > \widetilde{U} \times \widetilde{W}$, $\mathcal{D} = \mathcal{D} + 1$, and return to (i); else if $\mathcal{L} = \widetilde{U} \times \widetilde{W}$, $\hat{\theta}_l(\ell_u, k_w)$ within $[\theta_{a\epsilon}, \theta_{b\epsilon}]$ is used as the estimates of θ, denoted as $\hat{\theta}(\ell_u, k_w)$.

In this case, the $\widetilde{U} \times \widetilde{W}$ estimates of θ can be expressed as

$$\hat{\theta}(\ell_u, k_w) = \theta + \varepsilon(\ell_u, k_w),\ w = 1, 2, \ldots, \widetilde{W};\ u = 1, 2, \cdots, \widetilde{U},$$

where $\varepsilon(\ell_1, k_1), \varepsilon(\ell_2, k_1), \cdots, \varepsilon(\ell_{\widetilde{U}}, k_{\widetilde{W}})$ represent the estimation errors. Assume $\varepsilon(\ell_1, k_1), \varepsilon(\ell_2, k_1), \cdots, \varepsilon(\ell_{\widetilde{U}}, k_{\widetilde{W}})$ have the average variance $\mathrm{Var}(\varepsilon)$. Thus,

$$\mathrm{Var}\left(\frac{1}{\widetilde{U}\widetilde{W}} \sum_{w=1}^{\widetilde{W}} \sum_{u=1}^{\widetilde{U}} \hat{\theta}(\ell_u, k_w)\right) = \frac{\mathrm{Var}\,(\varepsilon)}{\widetilde{U}\widetilde{W}}. \tag{2.23}$$

Therefore, $\hat{\theta} = \frac{1}{\widetilde{U}\widetilde{W}} \sum_{w=1}^{\widetilde{W}} \sum_{u=1}^{\widetilde{U}} \hat{\theta}(\ell_u, k_w)$ is adopted as the estimate of θ. After obtaining $\hat{\theta}$, the estimated value of φ can be calculated from (2.6) as $\hat{\varphi} = \arccos\left(\cos\hat{\gamma} / \cos\hat{\theta}\right)$, where $\hat{\gamma}$ is the estimated value of γ. Thus, the distance and AoA of the transmitted OAM beams are obtained.

However, one major shortcoming of the above 2-D ESPRIT-based estimation algorithm is that it needs to process the amplitude of the received OAM training signals exploiting the knowledge of the parameters of transmit antennas, which may be difficult to obtain at the receiver. To break the bottleneck, we discuss an easier-to-implement distance and AoA estimation method in the following part.

2.4.3 Distance and AoA Estimation Based on M-F MT-ESPRIT

Based on (2.1) and (2.9), the training signal received by the m-th ($1 \leq m \leq N$) element of receive UCA in the Z − XOY coordinate system can be expressed as

$$\begin{aligned} x'_m(\ell_u, k_w) &\approx \frac{\beta}{2k_w} e^{i\boldsymbol{k}_w \cdot \boldsymbol{r}_m} \left[e^{ik_w r} \sum_{n=1}^{N} e^{-i(\boldsymbol{k}_w \cdot \boldsymbol{r}_n - \ell_u \phi_n)} s'(\ell_u, k_w) \right] + z'_m(\ell_u, k_w) \\ &\approx \frac{\beta N}{2k_w} e^{ik_w R_r \sin\theta \cos(\varphi - \alpha)} \left[e^{ik_w r} e^{i\ell_u \gamma} i^{-\ell_u} J_{\ell_u}(k_w R_t \sin\theta) s'(\ell_u, k_w) \right] \\ &+ z'_m(\ell_u, k_w), \end{aligned} \tag{2.24}$$

where $z'_m(\ell_u, k_w)$ is the additive noise. From (2.24), we can observe that the OAM mode number ℓ_u and the angle γ satisfy the dual relationship, so do k_w and r. Meanwhile, the elevation angle θ is associated with both ℓ_u and k_w. Therefore, we propose to first estimate r and γ with the two dual relationships and then obtain the estimates of φ and θ.

2.4.3.1 Estimation of r and γ

In the estimation of r, we first denote the u-th row of $\tilde{\mathbf{X}}_{\widetilde{U},\widetilde{W}}$ as a column vector $\tilde{\mathbf{x}}_a^u$, i.e.,

$$\tilde{\mathbf{x}}_a^u = \tilde{\mathbf{X}}_{\widetilde{U},\widetilde{W}}(u, :) = [\tilde{x}(\ell_u, k_1), \tilde{x}(\ell_u, k_2), \cdots, \tilde{x}(\ell_u, k_{\widetilde{W}})]^T.$$

Following the ESPRIT-based estimation method in Sect. 2.4.2, we can obtain an estimate of r based on the u-th row of the matrix $\tilde{\mathbf{X}}_{\widetilde{U},\widetilde{W}}$, which we denote as $\hat{r}_u$. Hence, there are $\widetilde{U}$ estimates of r in total, which can be expressed as

$$\hat{r}_u = r + \varepsilon_u, u = 1, 2, \cdots, \widetilde{U},$$

where $\{\varepsilon_u | u = 1, 2, \cdots, \widetilde{U}\}$ represent the estimation errors at $\widetilde{U}$ modes. Suppose that $\{\varepsilon_u | u = 1, 2, \cdots, \widetilde{U}\}$ have the same average variance $\mathrm{Var}(\varepsilon_r)$, thus,

$$\mathrm{Var}\left(\frac{1}{\widetilde{U}}\sum_{u=1}^{\widetilde{U}}\hat{r}_u\right)=\frac{\mathrm{Var}\left(\varepsilon_r\right)}{\widetilde{U}}. \tag{2.25}$$

Therefore, $\hat{r}=\frac{1}{\widetilde{U}}\sum_{u=1}^{\widetilde{U}}\hat{r}_u$ is adopted as the estimate of r.

In the estimation of γ, we similarly denote the w-th column of $\tilde{\mathbf{X}}_{\widetilde{U},\widetilde{W}}$ as

$$\tilde{\mathbf{x}}_b^w=\tilde{\mathbf{X}}_{\widetilde{U},\widetilde{W}}(:,w)=[\tilde{x}(\ell_1,k_w),\tilde{x}(\ell_2,k_w),\cdots,\tilde{x}(\ell_{\widetilde{U}},k_w)]^T.$$

After that, by following the same method employed for $\hat{r}$, we can obtain $\hat{\gamma}$, the estimate of γ.

2.4.3.2 Estimation of φ and θ

According to (2.24), the signal received by the reference element ($\alpha = 0$) of the receive UCA on the zero OAM mode and w-th frequency can be simplified as

$$\tilde{x}'(k_w)=-\frac{x_1'(0,k_w)}{|x_1'(0,k_w)|}\frac{s'(0,k_w)^*}{|s'(0,k_w)|}\mathrm{sign}(x_1'(0,k_w))=e^{ik_w\xi}+\tilde{z}'(k_w), \tag{2.26}$$

where $\xi=r+R_r\sin\theta\cos\varphi$ is an intermediate parameter related to φ and θ, $\tilde{z}'(k_w)$ is the corresponding noise. Then, the signals received on the $\widetilde{W}$ frequencies can be collected in the vector

$$\tilde{\mathbf{x}}_c=[\tilde{x}'(k_1),\tilde{x}'(k_2),\cdots,\tilde{x}'(k_{\widetilde{W}})]^T, \tag{2.27}$$

which can be expressed in a compact form as

$$\tilde{\mathbf{x}}_c=\mathbf{c}+\mathbf{z}_c, \tag{2.28}$$

where $\mathbf{c}=[e^{ik_1\xi},e^{ik_2\xi},\cdots,e^{ik_{\widetilde{W}}\xi}]^T$ and $\mathbf{z}_c$ is the corresponding noise vector.

The covariance matrix of $\tilde{\mathbf{x}}_c$ can be written as

$$\mathbf{R}_{\tilde{\mathbf{x}}_c}=\mathbb{E}\left\{\tilde{\mathbf{x}}_c\tilde{\mathbf{x}}_c^H\right\}=\mathbf{c}\mathbf{c}^H+\mathbf{R}_{\mathbf{z}_c}, \tag{2.29}$$

where $\mathbf{R}_{\mathbf{z}_c}=\mathbb{E}\left\{\mathbf{z}_c{\mathbf{z}_c}^{\mathrm{H}}\right\}$. Accordingly, the EVD of $\mathbf{R}_{\tilde{\mathbf{x}}_c}$ is given by

$$\mathbf{R}_{\tilde{\mathbf{x}}_c}=\mathbf{U}\boldsymbol{\Lambda}\mathbf{U}^H, \tag{2.30}$$

where $\mathbf{U}$ is an $\widetilde{W}\times\widetilde{W}$ unitary matrix and $\boldsymbol{\Lambda}=\mathrm{diag}\{\chi_1',\chi_2',\cdots,\chi_{\widetilde{W}}'\}$. Denote as $\chi_{max}'=\max\{\chi_w'|w=1,2,\cdots,\widetilde{W}\}$, and as $\mathbf{q}'$ the eigenvector corresponding to χ_{max}', so that it is

$$\mathbf{R}_{\tilde{\mathbf{x}}_c}\mathbf{q}' = \chi'_{max}\mathbf{q}'. \tag{2.31}$$

Considering that the subspace spanned by $\mathbf{q}'$ is the signal subspace spanned by $\mathbf{c}$, we can obtain

$$\mathbf{c} = \delta'\mathbf{q}', \tag{2.32}$$

where δ' is a nonzero parameter. If we consider the two vectors $\mathbf{c}_1$ and $\mathbf{c}_2$, obtained by taking the first and the last $\widetilde{W} - 1$ elements of $\mathbf{c}$, respectively, it is $\mathbf{c}_2 = \Psi\mathbf{c}_1$, where $\Psi = e^{i\xi}$. To obtain the estimate of ξ, we construct the two vectors $\mathbf{q}'_1$ and $\mathbf{q}'_2$, composed by the first and by the last $\widetilde{W} - 1$ elements of $\mathbf{q}'$, respectively. Then, exploiting the fact that $\mathbf{c}_1 = \delta'\mathbf{q}'_1$ and $\mathbf{c}_2 = \delta'\mathbf{q}'_2$, one obtains

$$\mathbf{q}'_2 = \Psi\mathbf{q}'_1, \tag{2.33}$$

which leads to

$$e^{i\hat{\xi}} = \mathbf{q}_1{}^{\dagger}\mathbf{q}_2, \tag{2.34}$$

where $\hat{\xi}$ is the estimate of ξ based on the signal vector $\tilde{\mathbf{x}}_c$.

After that, according to

$$\hat{\varphi} = \arccos\sqrt{\left(\frac{\hat{\xi} - \hat{r}}{R_r}\right)^2 + \cos\hat{\gamma}^2}, \text{ and } \quad \hat{\theta} = \arctan\frac{\hat{\xi} - \hat{r}}{R_r \cos\hat{\gamma}}, \tag{2.35}$$

the estimates of the azimuthal angle φ and the elevation angle θ can be obtained. The above method applies multiple times ESPRIT algorithm to the received training signals in OAM mode and frequency domains, thus being denoted as the mode-frequency (M-F) multi-time (MT)-ESPRIT algorithm, whose working process is shown in Fig. 2.6.

Compared with the 2-D ESPRIT-based algorithm in Sect. 2.4.2, the proposed M-F MT-ESPRIT algorithm can estimate the distance, azimuth, and elevation angles of OAM beams only by utilizing the exponentials of the received training signals, which is easier to implement in practical applications. Meanwhile, the dimension of the training signal vector processed by ESPRIT algorithm each time is greatly reduced in the M-F MT-ESPRIT algorithm. Therefore, the complexity of the M-F MT-ESPRIT algorithm is considerably lower than that of 2-D ESPRIT-based algorithm in Sect. 2.4.2. The total complexity comparison between these two algorithms is shown in Fig. 2.7.

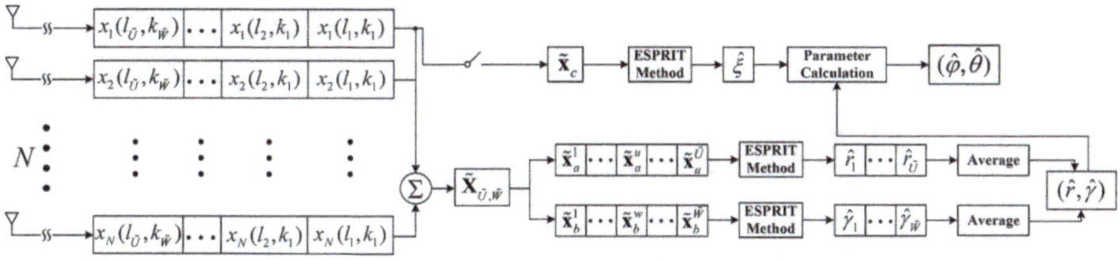

Fig. 2.6 The distance and AoA estimation of the OAM wireless communication system with M-F MT-ESPRIT method

Fig. 2.7 The complexities of the distance and AoA estimation methods vs. $\widetilde{W}$ and $\widetilde{U}$

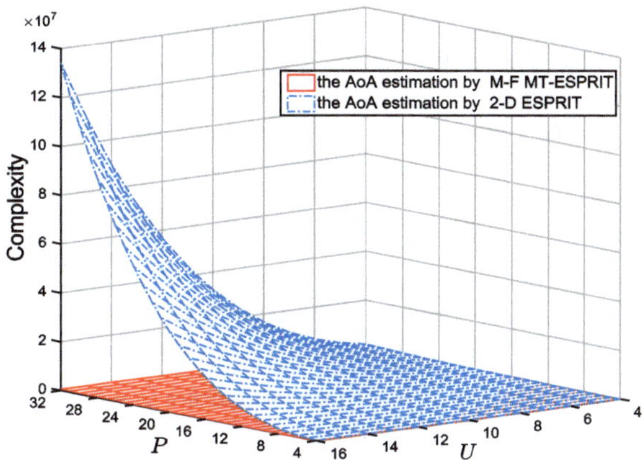

2.5 Reception of Muti-mode OAM Signals

In the following section, we first investigate the effect of φ and θ on the LoS multi-mode OAM channel and then find the effect of the proposed beam steering on inter-mode interferences. At last, we specify how to design the signal detection matrix $\mathbf{D}(k_w)$.

2.5.1 Effects of φ and θ on OAM Inter-mode Interferences

By assuming that the transmit and receive UCAs are in the far-field distance region of each other, i.e., $r \gg R_t$ and $r \gg R_r$, we can approximate $d_{m,n}$ in (2.7) as

$$d_{m,n} \overset{(a)}{\approx} \sqrt{R_t^2 + R_r^2 + r^2} - \frac{r R_r \cos\alpha \cos\varphi \sin\theta}{\sqrt{R_t^2 + R_r^2 + r^2}} + \frac{r R_r \sin\alpha \sin\varphi \sin\theta}{\sqrt{R_t^2 + R_r^2 + r^2}}$$
$$- \frac{R_t R_r (\cos\phi \cos\alpha \cos\gamma + \sin\phi \sin\alpha \cos\varphi)}{\sqrt{R_t^2 + R_r^2 + r^2}}$$

$$
\begin{aligned}
&- \frac{R_t R_r (\sin\phi\cos\alpha\sin\varphi - \cos\phi\sin\alpha\sin\varphi\cos\theta)}{\sqrt{R_t^2 + R_r^2 + r^2}} \\
&\overset{(b)}{\approx} r - \frac{R_t R_r}{r}(\cos\phi\cos\alpha\cos\gamma + \sin\phi\sin\alpha\cos\varphi) \\
&- \frac{R_t R_r}{r}(\sin\phi\cos\alpha\sin\varphi - \cos\phi\sin\alpha\sin\varphi\cos\theta) \\
&- R_r(\cos\alpha\cos\varphi\sin\theta - \sin\alpha\sin\varphi\sin\theta), \qquad (2.36)
\end{aligned}
$$

where (a) uses the method of completing a square and the condition $r \gg R_t, R_r$ same as the simple case $\sqrt{a^2 - 2b} \approx a - \frac{b}{a}, a \gg b$; (b) is directly obtained from the condition $r \gg R_t, R_r$. Then, substituting (2.36) into (2.5) and abbreviating $h(k_w, d_{m,n})$ to $h_{m,n}(k_w)$, we thus have

$$
\begin{aligned}
h_{m,n}(k_w) &\overset{(a)}{\approx} \frac{\beta}{2k_w r}\exp\Bigg(- ik_w r + ik_w R_r\cos\alpha\cos\varphi\sin\theta - ik_w R_r\sin\alpha\sin\varphi\sin\theta \\
&+ i\frac{k_w R_t R_r}{r}(\cos\phi\cos\alpha\cos\gamma + \sin\phi\sin\alpha\cos\varphi) \\
&+ i\frac{k_w R_t R_r}{r}(\sin\phi\cos\alpha\sin\varphi - \cos\phi\sin\alpha\sin\varphi\cos\theta)\Bigg), \qquad (2.37)
\end{aligned}
$$

where (a) neglects a few small terms in the denominator and thus only $2k_w r$ is left. Having (2.37), the uth-row and vth-column element $h_{\mathrm{OAM},k_w}(u, v)$ in $\mathbf{H}_{\mathrm{OAM}}(k_w)$ is calculated as

$$
\begin{aligned}
&h_{\mathrm{OAM},k_w}(u, v) \\
&= \frac{1}{N}\sum_{m=1}^{N}\sum_{n=1}^{N} h_{m,n}(k_w)\exp\left(-i\ell_u\alpha + i\ell_v\phi\right) \\
&= \eta(k_w)\sum_{m=1}^{N}\sum_{n=1}^{N}\exp\Bigg(-i\ell_u\alpha + i\ell_v\phi + ik_w R_r(\cos\alpha\cos\varphi\sin\theta - \sin\alpha\sin\varphi\sin\theta) \\
&+ i\frac{k_w R_t R_r}{r}(\cos\phi\cos\alpha\cos\gamma + \sin\phi\sin\alpha\cos\varphi + \sin\phi\cos\alpha\sin\varphi \\
&- \cos\phi\sin\alpha\sin\varphi\cos\theta)\Bigg) \\
&\overset{(a)}{\approx} \eta(k_w)\sum_{q=1}^{N}\exp\left(i\frac{2\pi q}{N}\ell_v + i\frac{k_w R_t R_r}{r}\cos\frac{2\pi q}{N}\right)
\end{aligned}
$$

$$
\begin{aligned}
&\times \sum_{m=1}^{N}\sum_{n=1}^{N} \exp\left(ik_w R_r \left(\cos\varphi \sin\theta \cos\frac{2\pi(m-1)}{N}\right.\right.\\
&- \sin\varphi \sin\theta \sin\frac{2\pi(m-1)}{N}\Bigg)\\
&+ i\frac{k_w R_t R_r}{r}\left(\sin\varphi \sin\frac{2\pi(n-1)}{N}\cos\frac{2\pi(m-1)}{N} - \sin\varphi \cos\frac{2\pi(n-1)}{N}\right.\\
&\left.\left.\times \sin\frac{2\pi(m-1)}{N}\right) - i\frac{2\pi(m-1)}{N}t\right),
\end{aligned}
\tag{2.38}
$$

where $\eta(k_w) = \frac{\beta}{2k_w r N}\exp(-ik_w r)$, $q = n - m \in \mathbb{Z}$, $t = \ell_u - \ell_v \in \mathbb{Z}$, $u, v = 1, 2, \cdots, U$, $w = 1, 2, \cdots, W$, (a) applies the approximation $\cos a \approx 1 - \frac{a^2}{2}$ for $\cos\theta$, $\cos\varphi$, and $\cos\gamma$ in the case that θ, φ, and γ are relatively small and neglects a few small terms under the condition $r \gg R_r$. Denote the second summation in (2.38) as

$$
\begin{aligned}
&\xi(\alpha, \varphi, t)\\
&= \sum_{m=1}^{N}\sum_{n=1}^{N} \exp\left(ik_w R_r \left(\cos\varphi \sin\theta \cos\frac{2\pi(m-1)}{N} - \sin\varphi \sin\theta \sin\frac{2\pi(m-1)}{N}\right)\right.\\
&+ i\frac{k_w R_t R_r}{r}\left(\sin\varphi \sin\frac{2\pi(n-1)}{N}\cos\frac{2\pi(m-1)}{N} - \sin\varphi \cos\frac{2\pi(n-1)}{N}\right.\\
&\left.\left.\times \sin\frac{2\pi(m-1)}{N}\right) - i\frac{2\pi(m-1)}{N}t\right),
\end{aligned}
\tag{2.39}
$$

Then, we can observe from (2.39) that if $\varphi = 0$, $\theta = 0$, and $t = 0$, $\xi = N$; if $\varphi = 0$, $\theta = 0$, and $t \neq 0$, then $\xi = 0$, that is to say, $\mathbf{H}_{\mathrm{OAM}}$ is diagonal in perfect alignment case. However, when $\varphi \neq 0$ or $\theta \neq 0$ and $t \neq 0$, then $\xi \neq 0$, which indicates that φ and θ result in inter-mode interferences. Moreover, even if θ and/or φ have small values, large interferences occur.

2.5.2 *Beam Steering for Inter-mode Interference Cancellation*

To alleviate the inter-mode interferences induced by the misalignment, we propose applying the beam steering to the UCA-based LoS multi-mode OAM communication systems. The beam steering approach steers the beam pattern toward the direction of the incident OAM beam and thus compensates the changed phases caused by θ and φ at the receive UCA. Through calculating the phase difference between the reference element and the m-th element of the receive UCA, the receive beam steering matrix $\mathbf{B}(k_p)$ can be designed as $\mathbf{B}(k_w) = \mathbf{1} \otimes \mathfrak{b}(k_w)$, where $\mathfrak{b}(k_w) = [e^{i\mathbb{B}_1(k_w)}, e^{i\mathbb{B}_2(k_w)}, \cdots, e^{i\mathbb{B}_N(k_w)}]$,

$$\mathbb{B}_m(k_w) = k_w R_r \left(\sin\alpha \sin\varphi \sin\theta - \cos\alpha \cos\varphi \sin\theta\right), \tag{2.40}$$

$m = 1, \cdots, N$, $w = 1, \cdots, W$. After involving these phases into the original phases in $\mathbf{F}_U$ at the receive UCA, the effective multi-mode OAM channel matrix at the w-th subcarrier becomes

$$\mathbf{H}'_{\mathrm{OAM}}(k_w) = \left(\mathbf{F}_U \odot \mathbf{B}(k_w)\right) \mathbf{H}(k_w) \mathbf{F}_U^H. \tag{2.41}$$

Theorem 1 *For a LoS MCMM-OAM communication system composed of an N-elements transmit UCA and an N-elements receive UCA, beam steering matrix $\mathbf{B}(k_w)$ can eliminate inter-mode interferences induced by θ and φ in non-parallel misalignment case.*

Proof 1 The uth-row and vth-column element $h'_{\mathrm{OAM},k_w}(u,v)$ in $\mathbf{H}'_{\mathrm{OAM}}(k_w)$ can be obtained as

$$\begin{aligned}
&h'_{\mathrm{OAM},k_w}(u,v)\\
&= \frac{1}{N}\sum_{m=1}^{N}\sum_{n=1}^{N} h_{m,n}(k_w)\exp(-i\ell_u\alpha + i\ell_v\phi + i\mathbb{B}_m(k_w))\\
&= \eta(k_w)\sum_{m=1}^{N}\sum_{n=1}^{N}\exp\Bigg(-i\ell_u\alpha + i\ell_v\phi + i\frac{k_w R_t R_r}{r}(\sin\phi\cos\alpha\sin\varphi\\
&\quad -\cos\phi\sin\alpha\sin\varphi\cos\theta) + i\frac{k_w R_t R_r}{r}(\cos\phi\cos\alpha\cos\gamma + \sin\phi\sin\alpha\cos\varphi)\Bigg)\\
&\overset{(a)}{\approx} \eta(k_w)\sum_{q=1}^{N}\exp\left(i\frac{2\pi q}{N}\ell_v + i\frac{k_w R_t R_r}{r}\cos\frac{2\pi q}{N}\right)\\
&\quad\times\sum_{m=1}^{N}\exp\left(-i\frac{2\pi(m-1)}{N}t\right),
\end{aligned} \tag{2.42}$$

where (a) applies the approximation $\cos a \approx 1 - \frac{a^2}{2}$ for $\cos\theta$, $\cos\varphi$, and $\cos\gamma$ and neglects a few small terms in the condition $r \gg R_r$. From (2.42), we find that when $t \neq 0$, even if $\varphi \neq 0$ and $\theta \neq 0$,

$$\sum_{m=1}^{N}\exp\left(-i\frac{2\pi(m-1)t}{N}\right) = \frac{1-\exp(-i2\pi t)}{1-\exp(-i\frac{2\pi}{N}t)} = 0,$$

That is to say, $\mathbf{H'}_{\text{OAM}}(k_w) = \text{diag}\{h'_{\text{OAM},k_w}(1,1), h'_{\text{OAM},k_w}(2,2), \cdots, h'_{\text{OAM},k_w}(U,U)\}$, which proves that beam steering almost completely eliminates inter-mode interferences.

2.5.3 Multi-mode OAM Signal Reception

Since after proper beam steering $\mathbf{H'}_{\text{OAM}}(k_w)$ becomes diagonal, we first consider the behavior of the diagonal elements $h'_{\text{OAM},k_w}(u,u), u = 1, 2, \cdots, U, w = 1, 2, \cdots, W$. Define $S_{k_w} = k_w R_t R_r / r$; according to (2.42), the u-th diagonal element of $\mathbf{H'}_{\text{OAM}}(k_w)$ takes the form

$$h'_{\text{OAM},k_w}(u,u) = N\eta(k_w)\sum_{q=1}^{N}\exp\left(i\frac{2\pi q}{N}\ell_u + iS_{k_w}\cos\frac{2\pi q}{N}\right). \tag{2.43}$$

Theorem 2 *For a LoS MCMM-OAM communication system composed of an N-element transmit UCA with radius R_t and an N-element receive UCA with radius R_r at a distance r ($r \gg R_t, R_r$) from each other in the non-parallel misalignment case, after beam steering, the effective channel gain of each OAM mode $h'_{\text{OAM},k_w}(u,u)$ approaches a function of OAM mode number ℓ and wave number k_w.*

Proof 2 Recall that the Taylor series expansion of the exponential function is $e^x = \sum_{g=0}^{+\infty}(x^g)/g!$. Then, we substitute it into (2.43) and have

$$\begin{aligned}
h'_{\text{OAM},k_w}(u,u) &= N\eta(k_w)\sum_{q=1}^{N}\exp\left(i\frac{2\pi q}{N}\ell_u\right)\exp\left(iS_{k_w}\cos\frac{2\pi q}{N}\right)\\
&= N\eta(k_w)\sum_{q=1}^{N}\exp\left(i\frac{2\pi q}{N}\ell_u\right)\times\sum_{g=0}^{+\infty}\frac{i^g}{g!}\left(\frac{\exp(i\frac{2\pi q}{N})+\exp\left(-i\frac{2\pi q}{N}\right)}{2}\right)^g S_{k_w}^g\\
&= N\eta(k_w)\sum_{g=0}^{+\infty}\frac{i^g}{2^g g!}\sum_{q=1}^{N}\exp\left(i\frac{2\pi q}{N}\ell_u\right)\sum_{l=0}^{g}\binom{g}{l}\exp\left(il\frac{2\pi q}{N}\right)\\
&\quad\times\exp\left(-i(g-l)\frac{2\pi q}{N}\right)S_{k_w}^g\\
&= N\eta(k_w)\sum_{g=0}^{+\infty}\frac{i^g}{2^g g!}\sum_{l=0}^{g}\binom{g}{l}\sum_{q=1}^{N}\exp\left(i(\ell_u+2l-g)\frac{2\pi q}{N}\right)S_{k_w}^g.
\end{aligned} \tag{2.44}$$

For the last summation in (2.44), it is easy to verify that

$$\sum_{q=1}^{N}\exp\left(i(\ell_u+2l-g)\frac{2\pi q}{N}\right)=\begin{cases}0, & \mathrm{Mod}_N(\ell_u+2l-g)\neq 0;\\ N, & \mathrm{Mod}_N(\ell_u+2l-g)=0.\end{cases} \tag{2.45}$$

When $0 \le g \le \min\{|\ell_u|, N-|\ell_u|\}$, we always have (recall that $0 \le l \le g$)

$$-N < \ell_u + 2l - g < N. \tag{2.46}$$

When $0 \le g < \min\{|\ell_u|, N-|\ell_u|\}$, (2.45) is always zero, i.e., the Taylor series of $h'_{\mathrm{OAM},k_w}(u,u)$ does not contain the terms $S^g_{k_w}$, and when $g = \min\{|\ell_u|, N-|\ell_u|\}$, (2.45) equals to N if and only if

$$l=\begin{cases}0, & \ell_u \ge 0;\\ g, & \ell_u < 0.\end{cases} \tag{2.47}$$

Thus, the coefficient of $S^{\min\{|\ell_u|,N-|\ell_u|\}}_{k_w}$ in (2.44) is written as

$$\begin{cases}\frac{N}{2^{|\ell_u|}}\cdot\frac{i^{|\ell_u|}}{|\ell_u|!}, & g=|\ell_u|;\\ \frac{N}{2^{(N-|\ell_u|)}}\cdot\frac{i^{(N-|\ell_u|)}}{(N-|\ell_u|)!}, & g=N-|\ell_u|.\end{cases} \tag{2.48}$$

Furthermore, when $g > \min\{|\ell_u|, N-|\ell_u|\}$, the coefficient of $S^g_{k_w}$ is much smaller due to the term $\frac{1}{2^g g!}$, and hence the term $\{S^g_{k_w}|g > \min\{|\ell_u|, N-|\ell_u|\}$ is counted into the higher-order infinitesimal of $S^g_{k_w}$ denoted as $o\{S^g_{k_w}\}$ as S_{k_w} goes to zero.

Therefore, for $r \gg R_t, R_r$, the u-th diagonal elements of $\mathbf{H}'_{\mathrm{OAM}}(k_w)$ can be approximately obtained as

$$h'_{\mathrm{OAM},k_w}(u,u) \approx \eta(k_w)\frac{N^2}{2^\tau}\frac{i^\tau}{\tau!}\cdot S^\tau_{k_w}, \tag{2.49}$$

where $\tau = \min\{|\ell_u|, N-|\ell_u|\}$. It follows that in the long-distance communication scenario, all diagonal elements $\{h'_{\mathrm{OAM},k_w}(u,u)|u = 1,2,\cdots,U\}$ can be analytically characterized, $w = 1,2,\cdots,W$.

Given that k_w, R_t, R_r, N, and ℓ_u are known to the receiver, the effective channel coefficient $h'_{\mathrm{OAM},k_w}(u,u)$ of each OAM mode is only the function of r, which leads to very simple signal detection, i.e.,

$$\begin{aligned}\mathbf{x}(k_w) &= \boldsymbol{\Gamma}^{-1}(k_w)(\mathbf{F}_U \odot \mathbf{B}(k_w))\left(\mathbf{H}(k_w)\mathbf{F}_U^H\mathbf{s}(k_w)+\mathbf{z}(k_w)\right)\\ &\approx \mathbf{s}(k_w)+\hat{\mathbf{z}}(k_w),\ w=1,2,\cdots,W,\end{aligned} \tag{2.50}$$

where $\hat{\mathbf{z}}(k_w) = \boldsymbol{\Gamma}^{-1}(k_w)(\mathbf{F}_U \odot \mathbf{B}(k_w))\mathbf{z}(k_w) = [\hat{z}(\ell_1,k_w), \hat{z}(\ell_2,k_w),\cdots,\hat{z}(\ell_U, k_w)]^T$, $\boldsymbol{\Gamma}(k_w) = \mathrm{diag}\{\zeta(\ell_1,k_w),\cdots,\zeta(\ell_U,k_w)\}$ is a diagonal matrix, and

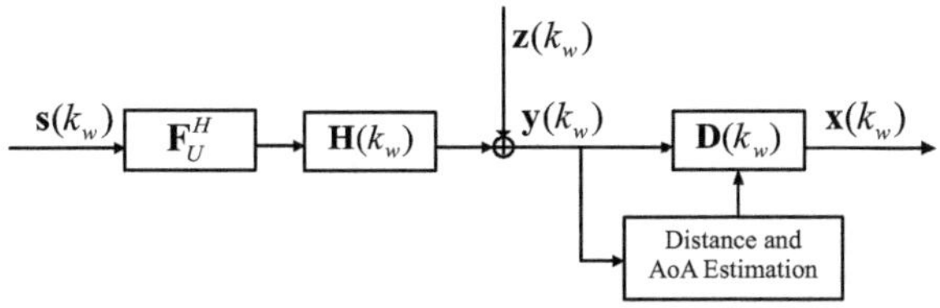

Fig. 2.8 The block diagram of a LoS MCMM-OAM communication system with the signal detection assisted by the distance and AoA estimation at the receiver

$$\zeta(\ell_u, k_w) = \eta(k_w)\frac{N^2}{2^\tau}\frac{i^\tau}{\tau!}\cdot \hat{S}_{k_w}^\tau, \quad u = 1, 2, \cdots, U, \tag{2.51}$$

$\hat{S}_{k_w} = k_w R_t R_r/\hat{r}$. Following the signal model in (2.10), the detection matrix $\mathbf{D}(k_w)$ can be designed as the cascade of the beam steering matrix $\mathbf{B}(k_w)$, the despiralization matrix $\mathbf{F}_U$, and the amplitude detection matrix $\boldsymbol{\Gamma}^{-1}(k_w)$, i.e.,

$$\mathbf{D}(k_w) = \boldsymbol{\Gamma}^{-1}(k_w)\mathbf{F}_U \odot \mathbf{B}(k_w). \tag{2.52}$$

Figure 2.8 shows the block diagram of a LoS MCMM-OAM communication system with the signal detection assisted by the distance and AoA estimation at the receiver.

Denote the estimate of r as $\hat{r}$. Then, if $\hat{r}$, $\hat{\theta}$, and $\hat{\varphi}$ are accurate enough, the effective multi-mode OAM channel matrix $\mathbf{H}'_{\mathrm{OAM}}(k_w)$ approaches diagonal according to Theorem 1. Since $\mathbf{F}_U \odot \mathbf{B}(k_w)$ is semi-unitary ($\mathbf{F}_U \odot \mathbf{B}(k_p)(\mathbf{F}_U^H \odot \mathbf{B}^H(k_w)) = \mathbf{I}$) and does not change the noise power, beam steering and amplitude detection convert the vector signal detection into multiple independent scalar signal detection greatly reducing the computational complexity. However, considering the inevitable estimation errors of r, θ, and φ in practice and the higher-order infinitesimal term in (2.49), we have to evaluate the effect of these errors on the system SE and detection performance. Following (2.50), we have

$$\begin{aligned} x(\ell_u, k_w) &= \frac{1}{\zeta(\ell_u, k_w)}\sum_{v=1}^{U} h'_{\mathrm{OAM},k_w}(u, v)s(\ell_v, k_w) + \hat{z}(\ell_u, k_w) \\ &= s(\ell_u, k_w) \\ &\quad + \frac{o\{S_{k_w}^\tau\}s(\ell_u, k_w)}{\zeta(\ell_u, k_w)} + \sum_{v\neq u} h'_{\mathrm{OAM},k_w}(u, v)\frac{s(\ell_v, k_w)}{\zeta(\ell_u, k_w)} + \frac{\hat{z}(\ell_u, k_w)}{\zeta(\ell_u, k_w)}. \end{aligned} \tag{2.53}$$

Therefore, the signal-to-interference-plus-noise ratio (SINR) of the u-th mode OAM at the w-th subcarrier can be formulated as

$$\mathrm{SINR}(\ell_u, k_w) = \frac{\zeta(\ell_u, k_w)^2\mathbb{E}\left(|s(\ell_u, k_w)|^2\right)}{\sum\limits_{v=1}^{U} |L(u, v, k_w)|^2\mathbb{E}\left(|s(\ell_v, k_w)|^2\right) + \mathbb{E}\left(|z(\ell_u, k_w)|^2\right)}, \tag{2.54}$$

where

$$L(u, v, k_w) = \begin{cases} h'_{\mathrm{OAM},k_w}(u, v), & v \neq u; \\ o\{S^{\tau}_{k_w}\}, & v = u. \end{cases} \tag{2.55}$$

Thus, the SE of the LoS MCMM-OAM communication system described in (2.50) can be written as

$$C = \left(1 - \frac{T_p \widetilde{W}}{T_c W}\right) \frac{1}{W} \sum_{w=1}^{W} \sum_{u=1}^{U} \log_2 \left(1 + \mathrm{SINR}(\ell_u, k_w)\right), \tag{2.56}$$

where T_c is the length of the coherence time over which the channel can be assumed constant and T_p is the time spent in transmitting training symbols.

2.6 UCCA-Based LoS OAM-MIMO Communications

2.6.1 LoS MCMM-OAM-MIMO Systems with UCCAs

In order to further enhance the degree of freedom and increase the multiplexing gain of the UCA-based LoS OAM communication systems, multiple concentric UCAs are exploited in the transmitter and the receiver to achieve 100 Gbps data rate [11]. The prototype of the UCCA for LoS OAM-MIMO communications is shown in Fig. 2.9.

For the aforementioned LoS MCMM-OAM system combined with MIMO (i.e, replace UCAs with UCCAs), we assume that the UCCAs at the transmitter and the receiver both consist of $\mathfrak{N}$ concentric UCAs with radii $R_{t_{\mathfrak{n}}}$ and $R_{r_{\mathfrak{m}}}$, $\mathfrak{m}, \mathfrak{n} = 1, 2, \cdots, \mathfrak{N}$. Then, the channel matrix of the UCCA-based communication system can be written as $\bar{\mathbf{H}}(k_w) = \left[\mathbf{H}_{\mathfrak{m},\mathfrak{n}}(k_w)\right]_{\mathfrak{N}\times\mathfrak{N}}$, where $\mathbf{H}_{\mathfrak{m},\mathfrak{n}}(k_w)(\mathfrak{m}, \mathfrak{n} = 1, 2, \cdots, \mathfrak{N})$ denotes the $N \times N$ channel matrix from the $\mathfrak{n}$-th UCA in the transmit UCCA to the $\mathfrak{m}$-th UCA in the receive UCCA. When $\theta = 0$ and $\varphi = 0$, $\{\mathbf{H}_{\mathfrak{m},\mathfrak{n}}(k_w) | \mathfrak{m}, \mathfrak{n} = 1, 2, \cdots, \mathfrak{N}\}$ are all circulant matrices.

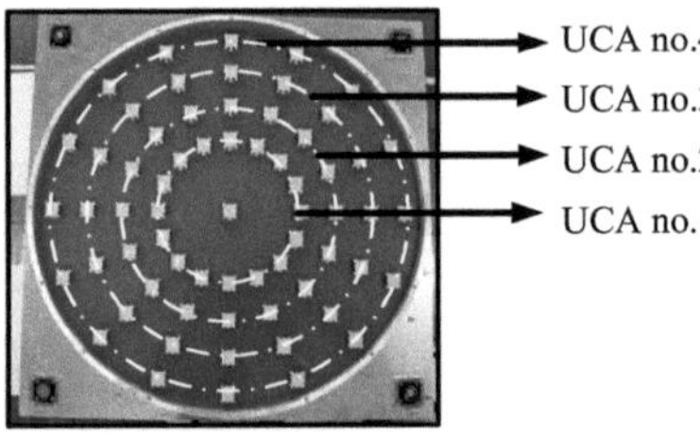

Fig. 2.9 The prototype of a uniform concentric circular array (UCCA) with 64 elements for OAM-MIMO multiplexing in [11]

2.6.2 Reception of LoS MCMM-OAM-MIMO Signals

As the distances and AoAs for every pair of the transmit and receive UCAs are exactly the same, the proposed AoA estimation and beam steering methods based on a single UCA in Sects. 2.4 and 2.5.2 can be directly utilized in the UCCA-based system. Hence, *the required number of training symbols of the LoS MCMM-OAM-MIMO system does not increase with the number of subcarriers, transmit and receive antenna elements*, which is opposite in traditional MIMO-OFDM channel estimation. The effective channel of the LoS MCMM-OAM-MIMO system can be written as

$$\mathbf{H}'_{\text{OAM-MIMO}}(k_w) = \left[\mathbf{H}'_{\text{OAM}_{\mathfrak{m},\mathfrak{n}}}(k_w)\right]_{\mathfrak{N}\times\mathfrak{N}} = \mathbf{I}_{\mathfrak{N}} \otimes \mathbf{F}_U \odot \bar{\mathbf{B}}(k_w)\bar{\mathbf{H}}(k_w)\mathbf{I}_{\mathfrak{N}} \otimes \mathbf{F}_U^H, \tag{2.57}$$

where $\bar{\mathbf{B}}(k_w) = \text{diag}\{\mathbf{B}_1(k_w), \mathbf{B}_2(k_w), \cdots, \mathbf{B}_{\mathfrak{N}}(k_w)\}$, $\mathbf{B}_{\mathfrak{m}}(k_w)$ is the receive beam steering matrix corresponding the $\mathfrak{m}$-th UCA of the receive UCCA, $\{\mathbf{H}'_{\text{OAM}_{\mathfrak{m},\mathfrak{n}}}(k_w)|\mathfrak{m}, \mathfrak{n} = 1, 2, \cdots, \mathfrak{N}\}$ denote the $U \times U$ effective MCMM-OAM channel matrix from the $\mathfrak{n}$-th UCA in the transmit UCCA to the $\mathfrak{m}$-th UCA in the receive UCCA, and $\mathbf{H}'_{\text{OAM}_{\mathfrak{m},\mathfrak{n}}}(k_w) = (\mathbf{F}_U \odot \mathbf{B}_{\mathfrak{m}}(k_w))\,\mathbf{H}_{\mathfrak{m},\mathfrak{n}}(k_w)\mathbf{F}_U^H$ is diagonal. It follows that after beam steering, only inter-mode interferences are eliminated, while co-mode interferences among different UCAs remain. Similar to (2.49), the elements of $\mathbf{H}'_{\text{OAM}_{\mathfrak{m},\mathfrak{n}}}(k_w)$ are functions of r given that k_w, $R_{t_{\mathfrak{n}}}$, $R_{r_{\mathfrak{m}}}$, N, and ℓ_u are known.

Then, the signal detection for the LoS MCMM-OAM-MIMO system could be expressed as

$$\begin{aligned}\bar{\mathbf{x}}(k_w) &= \bar{\mathbf{D}}(k_w)\bar{\mathbf{y}}(k_w)\\ &= \bar{\Gamma}^{-1}(k_w)\mathbf{I}_{\mathfrak{N}} \otimes \mathbf{F}_U \odot \bar{\mathbf{B}}(k_w)\left(\bar{\mathbf{H}}(k_w)\mathbf{I}_{\mathfrak{N}} \otimes \mathbf{F}_U^H\bar{\mathbf{s}}(k_w) + \bar{\mathbf{z}}(k_w)\right)\\ &= \bar{\mathbf{s}}(k_w) + \mathcal{I}(k_w) + \hat{\bar{\mathbf{z}}}(k_w),\end{aligned} \tag{2.58}$$

where $\bar{\mathbf{s}}(k_w) = [\mathbf{s}_1^T(k_w), \mathbf{s}_2^T(k_w), \cdots, \mathbf{s}_{\mathfrak{N}}^T(k_w)]^T$, $\bar{\mathbf{y}}(k_p) = [\mathbf{y}_1^T(k_w), \mathbf{y}_2^T(k_w), \cdots, \mathbf{y}_{\mathfrak{N}}^T(k_w)]^T$, and $\bar{\mathbf{x}}(k_w) = [\mathbf{x}_1^T(k_w), \mathbf{x}_2^T(k_w), \cdots, \mathbf{x}_{\mathfrak{N}}^T(k_w)]^T$ are the transmitted, received, and detected data symbol vectors of the UCCA-based systems, respectively, $\mathbf{s}_{\mathfrak{n}}(k_w) = [s_{\mathfrak{n}}(\ell_1, k_w), s_{\mathfrak{n}}(\ell_2, k_w), \cdots, s_{\mathfrak{n}}(\ell_U, k_w)]^T$, $\mathbf{y}_{\mathfrak{m}}(k_w) = [y_{\mathfrak{m}}(\ell_1, k_w), y_{\mathfrak{m}}(\ell_2, k_w), \cdots, y_{\mathfrak{m}}(\ell_U, k_w)]^T$ and $\mathbf{x}_{\mathfrak{m}}(k_w) = [x_{\mathfrak{m}}(\ell_1, k_w), x_{\mathfrak{m}}(\ell_2, k_w), \cdots, x_{\mathfrak{m}}(\ell_U, k_w)]^T$ are the corresponding transmitted, received and detected data symbol vectors on the $\mathfrak{n}$-th transmit and the $\mathfrak{m}$-th receive UCA, $\bar{\mathbf{z}}(k_w) = [\mathbf{z}_1^T(k_w), \mathbf{z}_2^T(k_w), \cdots, \mathbf{z}_{\mathfrak{N}}^T(k_w)]^T$ is the additive noise vector on the receive UCCA, $\mathbf{z}_{\mathfrak{n}}(k_w) = [z_{\mathfrak{n}}(\ell_1, k_w), z_{\mathfrak{n}}(\ell_2, k_w), \cdots, z_{\mathfrak{n}}(\ell_U, k_w)]^T$, $\hat{\bar{\mathbf{z}}}(k_w) = \bar{\mathbf{D}}(k_w)\bar{\mathbf{z}}(k_w)$, $\bar{\mathbf{D}}(k_w) = \bar{\Gamma}^{-1}(k_w)\mathbf{I}_{\mathfrak{N}} \otimes \mathbf{F}_U \odot \bar{\mathbf{B}}(k_w)$ is the detection matrix for $\bar{\mathbf{y}}(k_w)$, $\bar{\boldsymbol{\Gamma}}^{-1}(k_w)$ is the matrix designed for eliminating the remaining co-mode interferences among UCAs, and $\bar{\boldsymbol{\Gamma}}(k_w) = \left[\boldsymbol{\Gamma}_{\mathfrak{m},\mathfrak{n}}(k_w)\right]_{\mathfrak{N}\times\mathfrak{N}}$, $\boldsymbol{\Gamma}_{\mathfrak{m},\mathfrak{n}}(k_w) = \text{diag}\{\zeta_{\mathfrak{m},\mathfrak{n}}(\ell_1, k_w), \cdots, \zeta_{\mathfrak{m},\mathfrak{n}}(\ell_U, k_w)\}$ and

$$\zeta_{\mathfrak{m},\mathfrak{n}}(\ell_u, k_w) = \eta(k_w)\frac{N^2}{2^\tau}\frac{i^\tau}{\tau!} \cdot \hat{S}^{\tau}_{\mathfrak{m},\mathfrak{n}}(k_w),$$

$\hat{S}_{\mathfrak{m},\mathfrak{n}}(k_w) = k_w R_{t_\mathfrak{n}} R_{r_\mathfrak{m}}/\hat{r}$, $\mathfrak{m}, \mathfrak{n} = 1, 2, \cdots, \mathfrak{N}$, and $\mathcal{I}(k_w) = \left(\bar{\boldsymbol{\Gamma}}^{-1}(k_w)\mathbf{H}'_{\text{OAM-MIMO}}(k_w) - \mathbf{I}_{\mathfrak{N}U}\right)\bar{\mathbf{s}}(k_w)$ is the remaining interferences induced by the error in detection matrix. The block diagram shown in Fig. 2.8 can also be applied to the MCMM-OAM-MIMO communication system as long as $\mathbf{s}(k_w)$, $\mathbf{F}_U$, $\mathbf{H}(k_w)$, $\mathbf{z}(k_w)$, $\mathbf{y}(k_w)$, $\mathbf{D}(k_w)$, and $\mathbf{x}(k_w)$ are, respectively, replaced with $\bar{\mathbf{s}}(k_w)$, $\bar{\mathbf{F}}_U$, $\bar{\mathbf{H}}(k_w)$, $\bar{\mathbf{z}}(k_w)$, $\bar{\mathbf{y}}(k_w)$, $\bar{\mathbf{D}}(k_w)$, and $\bar{\mathbf{x}}(k_w)$, where $\bar{\mathbf{F}}_U = \mathbf{I}_{\mathfrak{N}} \otimes \mathbf{F}_U$.

In the presence of interferences and noise, if we define $\boldsymbol{R}^{\mathcal{I}}(k_w)$ and $\boldsymbol{R}^{\hat{\bar{z}}}(k_w)$ as the $\mathfrak{N}U \times \mathfrak{N}U$ covariance matrices of $\mathcal{I}(k_w)$ and $\hat{\bar{\mathbf{z}}}(k_w)$, i.e.,

$$\begin{aligned}\boldsymbol{R}^{\mathcal{I}}(k_w) &= \left(\bar{\boldsymbol{\Gamma}}^{-1}(k_w)\mathbf{H}'_{\text{OAM-MIMO}}(k_w) - \mathbf{I}\right)\mathbb{E}\left(\bar{\mathbf{s}}(k_w)\bar{\mathbf{s}}^H(k_w)\right) \\ &\quad \times \left(\bar{\boldsymbol{\Gamma}}^{-1}(k_w)\mathbf{H}'_{\text{OAM-MIMO}}(k_w) - \mathbf{I}\right)^H, \\ \boldsymbol{R}^{\hat{\bar{z}}}(k_w) &= \bar{\mathbf{D}}(k_w)\mathbb{E}\left(\bar{\mathbf{z}}(k_w)\bar{\mathbf{z}}^H(k_p)\right)\bar{\mathbf{D}}^H(k_w),\end{aligned}$$

the SINR of the u-th mode OAM generated by the $\mathfrak{n}$-th UCA at the w-th subcarrier can be formulated as

$$\overline{\text{SINR}}_{\mathfrak{n}}(\ell_u, k_w) = \frac{\mathbb{E}\left(|s_{\mathfrak{n}}(\ell_u, k_w)|^2\right)}{\left[\boldsymbol{R}^{\mathcal{I}}(k_w)\right]_{\kappa,\kappa} + \left[\boldsymbol{R}^{\hat{\bar{z}}}(k_w)\right]_{\kappa,\kappa}}, \tag{2.59}$$

where $\kappa = (\mathfrak{n} - 1)N + u$, $\mathfrak{n} = 1, 2, \cdots, \mathfrak{N}$.

Therefore, the SE of the LoS MCMM-OAM-MIMO system can be written as

$$\begin{aligned}\overline{C} &= \left(1 - \frac{T_p\widetilde{W}}{T_c W}\right)\frac{1}{W}\sum_{w=1}^{W}\sum_{u=1}^{U}\log_2\left(1 + \overline{\text{SINR}}_{\mathfrak{n}}(\ell_u, k_w)\right) \\ &\quad + \frac{1}{W}\sum_{w=1}^{W}\sum_{u=1}^{U}\sum_{\mathfrak{n}=2}^{\mathfrak{N}}\log_2\left(1 + \overline{\text{SINR}}_{\mathfrak{n}}(\ell_u, k_w)\right).\end{aligned} \tag{2.60}$$

2.6.3 Performance and Complexity Discussions

In this part, we mainly analyze the training overhead and computational complexity of the proposed LoS MCMM-OAM-MIMO system multiplexing W subcarriers and U modalities, which is similar in structure to the traditional MIMO-OFDM system using W subcarriers and $\mathfrak{N}N \times \mathfrak{N}N$ antenna elements. In traditional MIMO-OFDM communication systems especially at sub 6 GHz bands, to correctly recover the transmitted data symbols, channel estimation has to determine the $\mathfrak{N}N \times \mathfrak{N}N$ channel coefficients for each subcarrier, i.e., $WN^2\mathfrak{N}^2$ unknown variables in total. Thus, the required number of training symbols increase proportional to W, N, and $\mathfrak{N}$ in large-scale antenna systems. However, by taking advantage of the geometrical relationship between the transmit and receive UCCAs, given that the radii and the numbers of UCAs' elements of the transmit UCCA are known to the receiver, the LoS MCMM-OAM-MIMO communication system only need determining three unknown parameters, i.e., the azimuth angle φ, the elevation angle θ, and the distance r to recover the transmitted data symbols. Since $\widetilde{W}$ and $\widetilde{U}$ have only effect on the accuracy of the distance and AoA estimation, when $\hat{\varphi}$, $\hat{\theta}$, and $\hat{r}$ are accurate enough, the number of training symbols required by the LoS MCMM-OAM-MIMO system does not increase with W, N, and $\mathfrak{N}$ even in the large-scale antenna systems. Therefore, although the LoS MCMM-OAM-MIMO system has the same channel capacity as the LoS MIMO-OFDM system, it requires much less training overhead and thus improving the system SE.

For LoS MCMM-OAM-MIMO communication systems, the complexity of the proposed distance and AoA estimation methods is determined by the complexity of the EVD in ESPRIT algorithm corresponding to the values of $\widetilde{W}$ and $\widetilde{U}$; the complexity of the proposed beam steering method is determined by the complexity of calculating $\mathbf{H}'_{\text{OAM-MIMO}}(k_w)$ corresponding to the values of W, U N, and $\mathfrak{N}$; and the complexity of the proposed amplitude detection method is determined by the complexity of calculating $\bar{\mathbf{D}}(k_w)$ corresponding to the values of W, U, and $\mathfrak{N}$. The specific computational complexity of the proposed AoA estimations and signal reception for LoS MCMM-OAM-MIMO communication systems is compared with that of traditional MIMO-OFDM communication systems in Table 2.2, and the total computational complexity comparison between the LoS MCMM-OAM-MIMO communication system and MIMO-OFDM communication system is shown in Fig. 2.10. It can be seen from the figure that the complexity of LoS MCMM-OAM-MIMO system is lower than traditional MIMO-OFDM systems when $\mathfrak{N}$ and N are large. For example, when $W = 64$, $\widetilde{W} = 32$, $U = \widetilde{U} = 8$, $N = 16$, and $\mathfrak{N} = 8$, the LoS MCMM-OAM-MIMO system has lower computational complexity than the traditional MIMO-OFDM system.

Table 2.2 The complexity comparison between MCMM-OAM and MIMO-OFDM

Scheme	Complexity	
MCMM-OAM-MIMO	2-D ESPRIT estimation	$O(\widetilde{W}^3\widetilde{U}^3)$
	M-F MT-ESPRIT estimation	$O(\widetilde{U}\widetilde{W}^3) + O(\widetilde{W}\widetilde{U}^3)$
	Beam steering	$O(WUN^2\mathfrak{N}^3)$
	Amplitude detection	$O(WU^3\mathfrak{N}^3)$
MIMO-OFDM	Channel estimation	$O(WN^3\mathfrak{N}^3)$
	Signal detection	$O(WN^3\mathfrak{N}^3)$

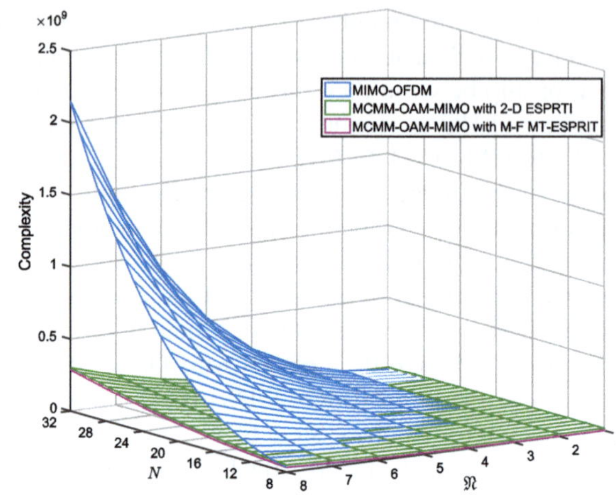

Fig. 2.10 The complexities of the MCMM-OAM-MIMO system and MIMO-OFDM system vs. N and $\mathfrak{N}$ with $W = 64$, $\widetilde{W} = 32$, $U = \widetilde{U} = 8$

2.7 Summary

In this section, we show the performances of the proposed methods by numerical simulations. We first compare the proposed two distance and AoA estimation methods at different SNRs and then compare the BER performances of the proposed OAM signal reception scheme with those of the perfect aligned OAM channel. At last, the SEs of the UCA-based LoS MCMM-OAM system, the UCCA-based LoS MCMM-OAM-MIMO system, and the traditional MIMO system are compared under different numbers of OAM training symbols. Unless otherwise stated, the SNRs in all the figures are defined as the ratio of the received signal power versus the noise power.

In Fig. 2.11, we choose $\widetilde{W} = 8$ subcarriers from 2.244 GHz to 2.578 GHz corresponding to the wave numbers $k_1, k_2, \cdots, k_8 = 47, 48, \ldots, 54$, $N = 9$, $\widetilde{U} = 8$ with $\ell_1, \ell_2, \cdots, \ell_8 = -4, -3, \cdots, +3$, $R_t = R_r = 15\lambda_1$, $\lambda_1 = 2\pi/k_1$, $(r, \varphi, \alpha) = (40\,\text{m}, 7°, 7°)$, $[\alpha_a, \alpha_b] = [2°, 8°]$ and the initial number of intervals $\mathcal{D} = 2$. These parameters are selected to ensure the receive UCA being within the main lobe of the transmit UCA at all subcarrier frequencies achieving large enough receive SNRs. Then, by using the proposed two distance and AoA estimation methods, the estimated locations of the MCMM-OAM transmitter are shown in

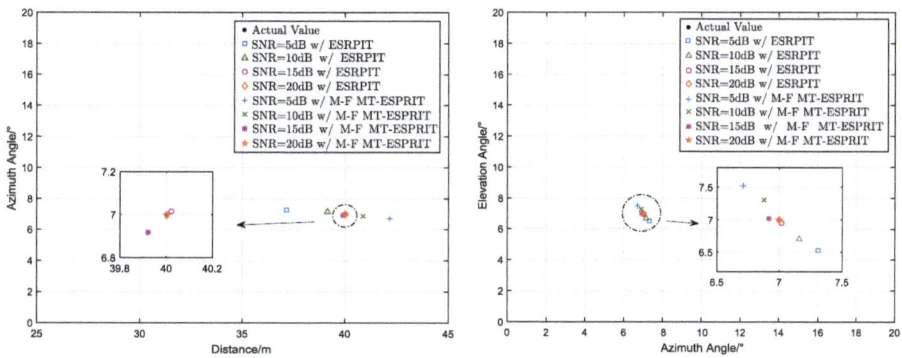

Fig. 2.11 The distance and AoA estimation results at difference SNRs

Fig. 2.11. It is obvious to see from the figure that the estimated distances and AoAs approach to the actual values with the increase of the SNR, e.g., when SNR reaches 20 dB, $(\hat{r}, \hat{\varphi}, \hat{\alpha}) = (40.000\text{m}, 6.997^\circ, 7.003^\circ)$ with 2-D ESPRIT-based algorithm and $(\hat{r}, \hat{\varphi}, \hat{\theta}) = (39.999\,\text{m}, 7.005^\circ, 6.994^\circ)$ with M-F MT-ESPRIT algorithm are very close to the actual location.

Figure 2.12 illustrates the normalized mean-squared errors (NMSEs) of $(\hat{r}, \hat{\varphi}, \hat{\theta})$ obtained by those two ESPRIT-based estimation method. The NMSE is defined as $\mathbb{E}\{(\hat{x} - x)^2/x^2\}$, where $\hat{x}$ denotes the estimate of x. As SNR increases, the NMSEs of $\hat{r}$, $\hat{\varphi}$, and $\hat{\theta}$ decline. It follows that the proposed methods are able to accurately estimate the distance and AoA of multi-mode OAM beams. Moreover, due to the reduction of the dimension of the training signal vector processed by ESPRIT algorithm each time, the NMSEs of $\hat{r}$, $\hat{\varphi}$, and $\hat{\theta}$ obtained by the M-F MT-ESPRIT method are higher than those obtained by the 2-D ESPRIT method.

Considering that the M-F MT-ESPRIT is more practical and has lower complexity, the following simulations utilize the M-F MT-ESPRIT algorithm to obtain $(\hat{r}, \hat{\varphi}, \hat{\alpha})$. Figure 2.13 compares the BERs of the proposed OAM signal detection scheme with those of the amplitude detection in perfect aligned OAM channel under $N = 9$ and 16QAM modulation. When $U = 9$, the BERs of the OAM signal detection including beam steering (BS) and amplitude detection (AD) are very close to those of the AD in perfect aligned OAM channel. It is worth noting that the BER of the proposed OAM signal detection does not increase significantly when using fewer OAM modes (i.e., $\widetilde{U} = 4$ with $\ell_1, \ell_2, \ell_3, \ell_4 = -2, -1, 0, +1$) in the distance and AoA estimation. Comparing $U = 8$ with $U = 9$, both the BERs of the OAM signal detection with $\widetilde{U} = 8$ and $\widetilde{U} = 4$ become much better because of abandoning the high-order OAM modes with smaller gains.

In Fig. 2.14, T_c is assumed to be 256 OFDM symbols, $N = 9$. Then, the SEs of the UCA-based LoS MCMM-OAM system under different values of U and $\widetilde{U}$ are shown in Fig. 2.14. It can be seen from the figure that in contrast to the reception with only AD and without BS, the SEs of the UCA-based LoS MCMM-OAM system with practical BS and AD are greatly improved in the misaligned OAM

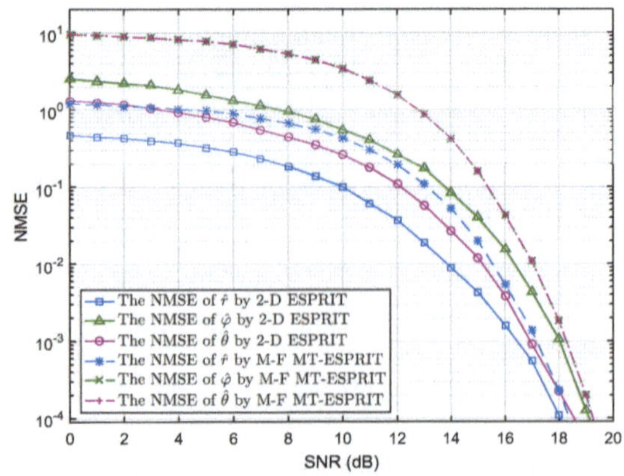

Fig. 2.12 The NMSEs of the estimated $\hat{r}$, $\hat{\varphi}$, and $\hat{\theta}$ vs. SNR at $\widetilde{W} = 8$ and $\widetilde{U} = 8$

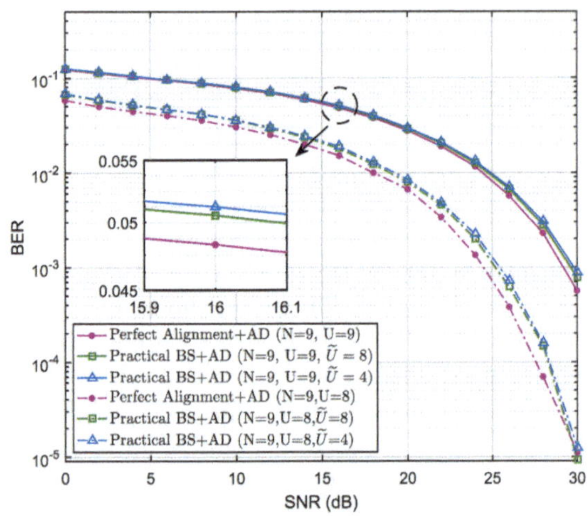

Fig. 2.13 The BER comparison of OAM detection

channel, which get close to the SEs of the OAM system in perfect aligned channel. Furthermore, the SE increases slightly when four OAM modes rather than eight OAM modes being used in the distance and AoA estimation, because the reduced number of training symbols contributes more to the SE than the detriments induced by the inaccuracy in the distance and AoA estimates. However, too few training symbols may lead to the opposite result. In addition, the SE of the UCA-based LoS MCMM-OAM system with $U = 8$ is near to that with $U = 9$, which indicates the high-order OAM modes have minor contributions to the SE of the UCA-based OAM systems in long-distance communications.

In Fig. 2.15, T_c is also assumed to be 256 OFDM symbols, $N = 16$, $\mathfrak{N} = 4$, $\widetilde{W} = 8$, $W = 64$. The SEs of the UCCA-based LoS MCMM-OAM-MIMO system and the traditional MIMO-OFDM system are compared in Fig. 2.15. We assume that the traditional channel estimation of the MIMO-OFDM system requires $N\mathfrak{N} = 64$ training symbols per subcarrier or subchannel. From the figure one can get the same conclusion as in Fig. 2.14, which verifies that the proposed distance and AoA estimation method and the OAM signal reception scheme can be applied to the

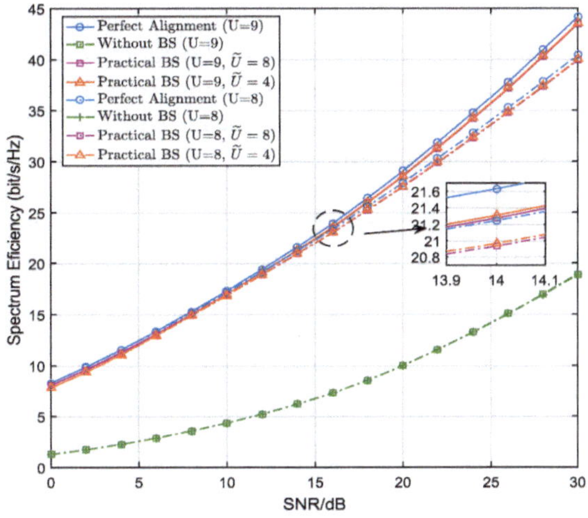

Fig. 2.14 The spectrum efficiencies of the UCA-based LoS MCMM-OAM system

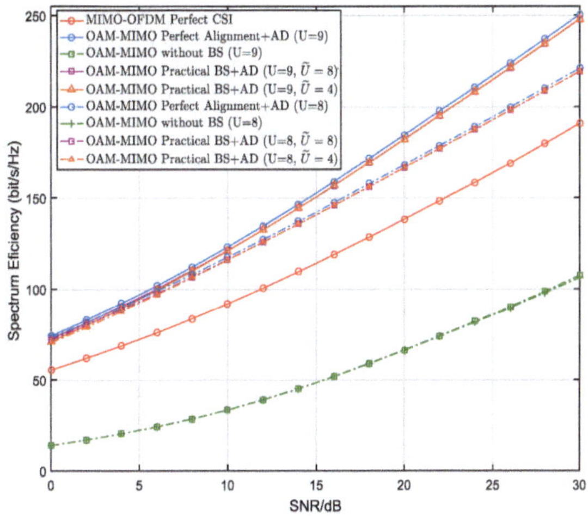

Fig. 2.15 The spectrum efficiencies of the UCCA-based LoS MCMM-OAM-MIMO system and the traditional MIMO-OFDM system

UCCA-based LoS MCMM-OAM-MIMO system. More importantly, it is obvious that the SEs of the UCCA-based LoS MCMM-OAM-MIMO system are about 20% higher than the SE of the MIMO-OFDM system with traditional channel estimation due to the greatly reduced number of training symbols.

In this chapter, we proposed an UCA-based LoS MCMM-OAM communication scheme including the generation, the distance, and AoA estimation and the reception of the multi-mode OAM beams. In terms of OAM generation, we verified that the UCA can generate multi-mode OAM radio beam with both the RF analog synthesis method and the baseband digital synthesis method. Then, we proposed two distance and AoA estimation methods based on the ESPRIT algorithm, which are shown being able to accurately estimate the location of the LoS MCMM-OAM transmitter with a flexible number of training symbols. Thereafter, we proved that the proposed beam steering method can eliminate inter-mode interferences induced by the misalignment error between the transmit and receive UCAs and the resulting diagonal elements of the effective OAM channel approaches a function of distance given other parameters are fixed. Therefore, the designed multi-mode OAM reception composed of the beam steering and the diagonal amplitude detection is shown to achieve the performance of an ideally aligned OAM channel and low complexity. At last, the proposed methods are extended to the UCCA-based LoS MCMM-OAM-MIMO system, which exhibits higher SE than the large-scale MIMO-OFDM system with traditional channel estimation.

References

1. L. Allen, M.W. Beijersbergen, R.J. Spreeuw, J.P. Woerdman, Orbital angular momentum of light and the transformation of Laguerre-Gaussian laser modes. Phys. Rev. A: At. Mol. Opt. Phys. **45**(11), 8185–8189 (1992)
2. A. Bennis, R. Niemiec, C. Brousseau, K. Mahdjoubi, O. Emile, Flat plate for OAM generation in the millimeter band, in *Proceedings of the 7th European Conference on Antennas and Propagation* (2013), pp. 3203–3207
3. J. Bremer, An algorithm for the rapid numerical evaluation of Bessel functions of real orders and arguments. Adv. Comput. Math. **45**(1), 173–211 (2019)
4. R. Chen, W.-X. Long, Y. Gao, J. Li, Orbital angular momentum-based two-dimensional super-resolution targets imaging, in *Proceedings of the IEEE Global Conference on Signal and Information Processing (GlobalSIP)* (2018), pp. 1–4
5. R. Chen, H. Xu, M. Moretti, J. Li, Beam steering for the misalignment in UCA-based OAM communication systems. IEEE Wireless Commun. Lett. 7(4), 582–585 (2018)
6. R. Chen, W. Yang, H. Xu, J. Li, A 2-D FFT-based transceiver architecture for OAM-OFDM systems with UCA antennas. IEEE Trans. Veh. Technol. 67(6), 5481–5485 (2018)
7. L. Cheng, W. Hong, Z.-C. Hao, Generation of electromagnetic waves with arbitrary orbital angular momentum modes. Sci. Rep. **4**, 4814 (2014)
8. O. Edfors, A.J. Johansson, Is orbital angular momentum (OAM) based radio communication an unexploited area? IEEE Trans. Antennas Propag. **60**(2), 1126–1131 (2012)
9. R. Gaffoglio, A. Cagliero, A.D. Vita, B. Sacco, OAM multiple transmission using uniform circular arrays: numerical modeling and experimental verification with two digital television signals. Radio Sci. **51**(6), 645–658 (2016)
10. Y. Gong, R. Wang, Y. Deng, B. Zhang, N. Wang, N. Li, P. Wang, Generation and transmission of OAM-carrying vortex beams using circular antenna array. IEEE Trans. Antennas Propag. **65**(6), 2940–2949 (2017)
11. D. Lee, H. Sasaki, H. Fukumoto, Y. Yagi, T. Kaho, H. Shiba, T. Shimizu, An experimental demonstration of 28 GHz band wireless OAM-MIMO (orbital angular momentum multi-input and multi-output) multiplexing, in *Proceedings of the IEEE 87th Vehicular Technology Conference (VTC Spring)* (2018)
12. J. Liang, S. Zhang, Orbital angular momentum (OAM) generation by cylinder dielectric resonator antenna for future wireless communication. IEEE Access **4**, 9570–9574 (2016)
13. K. Liu, Y. Cheng, Z. Yang, H. Wang, Y. Qin, X. Li, Orbital-angular-momentum-based electromagnetic vortex imaging. IEEE Antennas Wirel. Propag. Lett. **14**, 711–714 (2015)
14. E. Mari, F. Spinello, M. Oldoni, R.A. Ravanelli, F. Romanato, G. Parisi, Near-field experimental verification of separation of oam channels. IEEE Antennas Wireless Propag. Lett. **14**, 556–558 (2015)
15. S.M. Mohammadi, L.K.S. Daldorff, et al., Orbital angular momentum in radio-a system study. IEEE Trans. Antennas Propag. **58**(2), 565–572 (2010)

16. M. Oldoni, F. Spinello, E. Mari, G. Parisi, C.G. Someda, F. Tamburini, F. Romanato, R.A. Ravanelli, P. Coassini, B. Thidé, Space-division demultiplexing in orbital-angular-momentum-based mimo radio systems. IEEE Trans. Antennas Propag. **63**(10), 4582–4587 (2015)
17. Y. Shen, J. Yang, H. Meng, W. Dou, S. Hu, Generating millimeter-wave Bessel beam with orbital angular momentum using reflective-type metasurface inherently integrated with source. Appl. phys. Lett. **112**(14), 141901 (2018)
18. A. Tennant, B. Allen, Generation of radio frequency OAM radiation modes using circular time-switched and phased array antennas, in *Proceedings of the Loughborough Antennas & Propagation Conference (LAPC)* (2012)
19. B. Thidé, H. Then, J. Sjöholm, K. Palmer, J. Bergman, T.D. Carozzi, Y.N. Istomin, N.H. Ibragimov, R. Khamitova, Utilization of photon orbital angular momentum in the low-frequency radio domain. Phys. Rev. Lett. **99**(8), 087701 (2007)
20. G.A. Turnbull, D.A. Robertson, G.M. Smith, L. Allen, M.J. Padgett, The generation of free-space Laguerre-Gaussian modes at millimetre-wave frequencies by use of a spiral phaseplate. Opt. Commun. **127**(4), 183–188 (1996)
21. L. Wan, X. Kong, F. Xia, Joint range-doppler-angle estimation for intelligent tracking of moving aerial targets. IEEE Internet Things J. **5**(3), 1625–1636 (2018)
22. X.P. Wang, L.Y. Wang, X.M. Li, G. Bi, Nuclear norm minimization framework for DOA estimation in MIMO radar. Signal Process. **135**, 147–152 (2017)
23. F. Wen, Z. Zhang, K. Wang, G. Sheng, G. Zhang, Angle estimation and mutual coupling self-calibration for ULA-based bistatic MIMO radar. Signal Process. **144**, 61–67 (2018)
24. G. Xie, L. Li, Y. Ren, et al., Performance metrics and design considerations for a free-space optical orbital-angular-momentum-multiplexed communication link. Optica **2**(4), 357–365 (2015)
25. G. Xie, Y. Yan, Z. Zhao, et al., Tunable generation and angular steering of a millimeter-wave orbital-angular-momentum beam using differential time delays in a circular antenna array, in *Proceedings of the IEEE International Conference on Communications (ICC)* (2016)
26. C. Xu, S. Zheng, W. Zhang, Y. Chen, H. Chi, X. Jin, X. Zhang, Free-space radio communication employing OAM multiplexing based on Rotman lens. IEEE Microw. Wireless Compon. Lett. **26**(9), 738–740 (2016)
27. Y. Yan, G. Xie, M.P.J. Lavery, H. Huang, N Ahmed, C. Bao, Y. Ren, Y. Cao, L. Li, Z. Zhao, A.F. Molisch, M. Tur, M.J. Padgett, A.E. Willner, High-capacity millimetre-wave communications with orbital angular momentum multiplexing. Nature Commun. **5**, 4876 (2014)
28. S. Yu, L. Li, G. Shi, C. Zhu, Y. Shi, Generating multiple orbital angular momentum vortex beams using a metasurface in radio frequency domain. Appl. Phys. Lett. **108**(24), 241901 (2016)
29. T. Yuan, Y. Cheng, H. Wang, Y. Qin, Beam steering for electromagnetic vortex imaging using uniform circular arrays. IEEE Antennas Wirel. Propag. Lett. **16**, 704–707 (2017)
30. Z. Zhang, S. Xiao, Generation of multiple orbital angular momentum (OAM) modes with a circularly polarized multimode patch antenna, in *Proceedings of the IEEE MTT-S International Wireless Symposium (IWS)* (2016), pp. 1–4
31. W. Zhang, S. Zheng, Y. Chen, X. Jin, H. Chi, X. Zhang, Orbital angular momentum-based communications with partial arc sampling receiving. IEEE Commun. Lett. **20**(7), 1381–1384 (2016)
32. W. Zhang, S. Zheng, X. Hui, R. Dong, X. Jin, H. Chi, X. Zhang, Mode division multiplexing communication using microwave orbital angular momentum: An experimental study. IEEE Trans. Wirel. Commun. **16**(2), 1308–1318 (2017)
33. W. Zhang, S. Zheng, X. Hui, Y. Chen, X. Jin, H. Chi, X. Zhang, Four-OAM-mode antenna with traveling-wave ring-slot structure. IEEE Antennas Wireless Propag. Lett. **16**, 194–197 (2017)

Chapter 3
Hybrid Mechanical and Electronic Beam Steering for Point-to-Point OAM Communications

3.1 Introduction

Since OAM communication requires precise alignment of the transmit and the receive antennas, the electronic beam steering approach has been proposed for the uniform circular array (UCA)-based OAM communication system to circumvent large-performance degradation induced by small antenna misalignment in practical environments [2, 4]. However, in large-angle misalignment case, the OAM channel capacity can not be effectively compensated only by the electronic beam steering, which has not been investigated before as far as we know. Moreover, it is observed that when multiple OAM modes are transmitted simultaneously, due to the interferometry, the electric field intensities of the multi-mode OAM beam are not uniform on the ring of its main lobe. Hence, the receive signal-to-noise ratios (SNRs) are not equal at the elements of the receive UCA, which results in the OAM channel capacities being quite different at different rotation angles of the receive UCA around its axis.

In this chapter, to deal with the aforementioned problems, we first propose a hybrid mechanical and electronic beam steerable transceiver structure, which consists of a UCA fixed with mechanical rotation devices in pitch, roll, and yaw directions, respectively, as shown in Fig. 3.1. Specifically, the coarse alignment is first achieved by mechanical beam steering, and then electronic beam steering is exploited to compensate the small-scale misalignment error caused by perturbations, such as random wind, rain, or installation errors [6]. The principle of electronic beam steering for UCA-based OAM communication system is compensating the changed phases caused by oblique angle at the transmit or receive UCA through adjusting the delays of antenna elements by phase shifters; then the normal phases used only for OAM piralization/despiralization are tuned to the new phases used for both OAM piralization/despiralization and beam steering toward the direction of departure/arrival of OAM beams [2]. Furthermore, due to the interferometry, the receive UCA needs rotating to an optimal angle for maximizing the OAM channel

R. Chen et al., *Circular Array-Based Radio Frequency OAM Communications*, Wireless Networks, https://doi.org/10.1007/978-3-031-37589-7_3

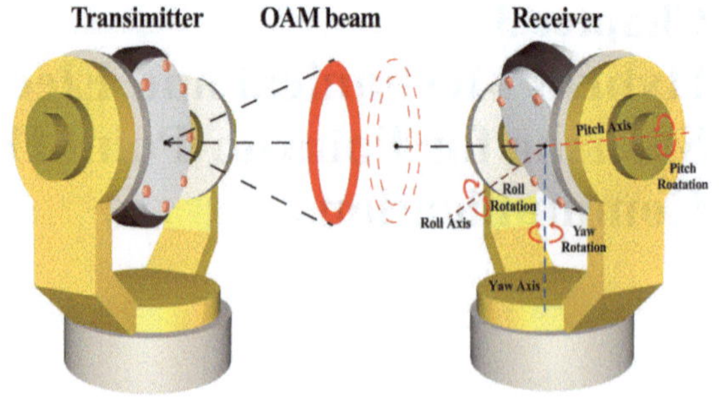

Fig. 3.1 The UCA-based LoS MCMM-OAM communication system with hybrid mechanical and electronic beam steerable transceiver

capacity. The mathematical analysis and numerical simulations show that the proposed hybrid mechanical and electronic beam steering scheme outperforms the traditional scheme of only using electronic beam steering, especially in the practical case of large misalignment error, and obtaining the maximum OAM channel capacity. The novelty and major contributions of this chapter are summarized as follows:

- First, based on the line-of-sight (LoS) OAM channel model, we show mathematically why only using electronic beam steering is not enough for compensating the performance degradation of OAM channel capacity in the large-angle misalignment case.
- Second, we propose the hybrid mechanical and electronic beam steering scheme to effectively improve the OAM channel capacity in practical environments, in which mechanical rotating devices controlled by pulse width modulation (PWM) signals as the execution unit are utilized to eliminate the large misalignment angle, while electronic beam steering is in charge of the remaining small misalignment angle caused by perturbations.
- Third, considering that the rotation angle of the receive UCA around its axis has a large effect on the receive signal-to-noise ratio (SNR) due to the interferometry, we propose a rotatable UCA structure for the OAM receiver to maximize the channel capacity, for which the simulated annealing algorithm is adopted to obtain the optimal rotation angle at first, then the servo system performs mechanical rotation, and last, the electronic beam steering is adjusted accordingly.

The remainder of this chapter is organized as follows. In Sect. 3.2, we model the UCA-based LoS OAM communication system with hybrid mechanical and electronic beam steerable transceiver. In Sect. 3.3, the effect of electronic beam steering in large-angle misalignment case is analyzed. In Sect. 3.4, to deal with the performance degradation of electronic beam steering in large-angle misalignment case and maximize the OAM channel capacity, we propose the hybrid mechanical and electronic beam steering scheme. Simulation results are shown in Sect. 3.5, and conclusions are summarized in Sect. 3.6.

3.2 UCA-Based OAM Communication Systems

In this chapter, we focus on a UCA-based line-of-sight (LoS) OAM communication system as considered in [2–7, 13, 14]. Since the transmit UCA and the receive UCA need to be aligned, beam steering approach is proposed to achieve beam alignment between the UCAs at both sides [2, 4]. As the functions of the transmit beam steering and the receive beam steering are the same, in order to evaluate the effects of mechanical beam steering and electronic beam steering, we have to separate the effects of transmit and receive beam steering and only consider one side. Therefore, we assume the transmit UCA having been aligned to the center of the receive UCA through mechanical rotation, while the receive UCA will align to the transmit UCA through yaw and pitch mechanical rotations firstly and then electronic beam steering more accurately in real time, as shown in Fig. 3.1. Thus, this geometric model falls into the non-parallel misalignment case [2, 4], as shown in Fig. 3.2.

In the considered OAM communication system, a multi-mode OAM beam is generated by an N-element UCA at the transmitter and received by another N-element UCA at the receiver. For higher data rate transmission, orthogonal frequency-division multiplexing (OFDM)-based LoS multi-carrier and multi-mode OAM (MCMM-OAM) communication scheme proposed in [4, 7] is adopted, and we assume W subcarriers and $U(U \leq N)$ OAM modes for data transmission, $\overline{W}$ subcarriers and $\overline{U}$ OAM modes for coarse angle of arrival (AoA) estimation, and $\widetilde{W}$ subcarriers and $\widetilde{U}$ OAM modes for refined AoA estimation.

3.2.1 Channel Model

The geometrical model of the UCA-based LoS OAM channel is illustrated in Fig. 3.2. In this model, the transmitter coordinate system $\mathrm{Z}-\mathrm{XOY}$ is established using the transmit UCA plane as the XOY plane and the axis through the transmit UCA center O and perpendicular to the transmit UCA plane as the Z-axis, and the receiver coordinate system $\overline{\mathrm{Z}}-\overline{\mathrm{XOY}}$ is established on the plane of the receive UCA by the similar approach, where $\overline{\mathrm{X}}$-axis and $\overline{\mathrm{Z}}$-axis correspond to the pitch axis

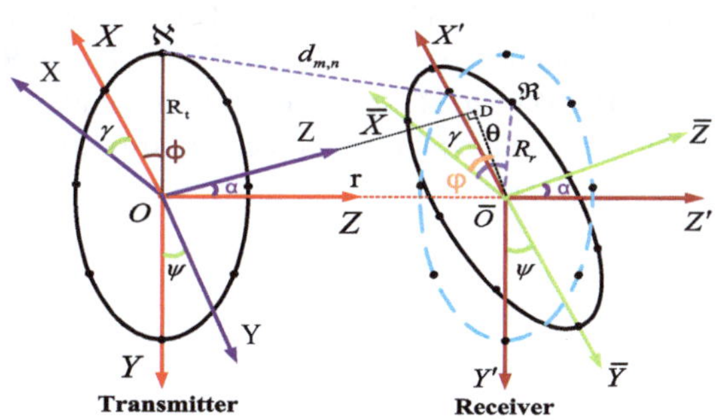

Fig. 3.2 Geometrical relationship between the transmit and receive UCAs in the non-parallel misalignment case

and the roll axis in Fig. 3.1, respectively. Since the transmit UCA is aligned with the center of the receive UCA in the non-parallel misalignment case, the spherical coordinate of the receive UCA center can be denoted as $\overline{\mathrm{O}}(r, 0, 0)$ in $\mathrm{Z} - \mathrm{XOY}$ in the non-parallel misalignment case. Suppose point D is the projection of point O on the $\overline{\mathrm{XOY}}$ plane and the spherical coordinate of the transmit UCA center is denoted as $\mathrm{O}(r, \varphi, \alpha)$ in $\overline{\mathrm{Z}} - \overline{\mathrm{XOY}}$, where r is the distance between the transmit and the receive UCA centers, φ is the azimuth angle, α is the elevation angle, and α and φ are defined as the AoA of the OAM beam [4, 7].

The coordinate system $\widetilde{\mathrm{Z}} - \widetilde{\mathrm{X}}\mathrm{O}\widetilde{\mathrm{Y}}$ is built so that its origin is at point O and $\widetilde{\mathrm{X}}\mathrm{O}\widetilde{\mathrm{Y}}$ plane is parallel to $\overline{\mathrm{XOY}}$. Meanwhile, we build the coordinate system $\mathrm{Z}' - \mathrm{X}'\overline{\mathrm{O}}\mathrm{Y}'$ so that its origin is at point $\overline{\mathrm{O}}$ and $\mathrm{X}'\overline{\mathrm{O}}\mathrm{Y}'$ plane is parallel to XOY, where Y'-axis corresponds to the yaw axis in Fig. 3.1. According to the geometrical model, the angle between $\widetilde{\mathrm{Z}}$-axis and Z-axis is α, so is that between Z'-axis and $\overline{\mathrm{Z}}$-axis. Denote the angle between the X'-axis and the $\overline{\mathrm{X}}$-axis as γ ($\gamma \in (-\pi/2, \pi/2)$), and so is that between X-axis and $\widetilde{\mathrm{X}}$-axis, where γ *is the yaw angle*. Denote the angle between the Y'-axis and the $\overline{\mathrm{Y}}$-axis as ψ ($\psi \in (-\pi/2, \pi/2)$), and so is that between Y-axis and $\widetilde{\mathrm{Y}}$-axis, where ψ *is the pitch angle* and ψ could be obtained as $\psi = \arccos\left(\frac{\cos\alpha}{\cos\gamma}\right)$ and γ and α can be obtained by the OAM AoA estimation method [4, 7]. Then, φ could be calculated as (3.1), whose proof is omitted due to limited space. Denote the angle between the line $\mathrm{O}\aleph$ and X-axis as ϕ and the angle between the line $\overline{\mathrm{O}}\mathfrak{R}$ and $\overline{\mathrm{X}}$-axis as θ, where $\aleph$ is the position of the nth ($1 \leq n \leq N$) element at the transmitter, $\mathfrak{R}$ is the position of the mth ($1 \leq m \leq N$) element at the receiver, $\phi = [2\pi(n-1)/N + \phi_0]$ and $\theta = [2\pi(m-1)/N + \theta_0]$, and ϕ_0 and θ_0 are, respectively, the corresponding initial angles of the first reference antenna elements in both UCAs.

According to Fig. 3.2, the Cartesian coordinate of the nth antenna element on the transmit UCA is $(R_t \cos\phi, R_t \sin\phi, 0)$ in $\mathrm{Z} - \mathrm{XOY}$ coordinate system, and the Cartesian coordinate of the mth antenna element on the receive UCA can be written as $(R_r \cos\theta, R_r \sin\theta, 0)$ in $\overline{\mathrm{Z}} - \overline{\mathrm{XOY}}$ coordinate system, where R_t and R_r are, respectively, the radii of the transmit and receive UCAs. In order to calculate the distance between the nth antenna element on the transmit UCA and the mth antenna element on the receive UCA, we transform the coordinate of the mth antenna element on the receive UCA in $\overline{\mathrm{Z}} - \overline{\mathrm{XOY}}$ coordinate system to the coordinate in $\mathrm{Z}' - \mathrm{X}'\overline{\mathrm{O}}\mathrm{Y}'$ coordinate system. Hence, the Cartesian coordinate of the mth antenna element on the receive UCA in $\mathrm{Z}' - \mathrm{X}'\overline{\mathrm{O}}\mathrm{Y}'$ coordinate system, denoted by (a_m, b_m, c_m), can be obtained as $[a_m, b_m, c_m]^T = \mathbf{R}_Y(\gamma)\mathbf{R}_P(\psi)[R_r \cos\theta, R_r \sin\theta, 0]^T$ [9], where $\mathbf{R}_P(\psi)$ and $\mathbf{R}_Y(\gamma)$ representing the axis rotation matrixes corresponding to the pitch direction and yaw direction, respectively, take the forms

$$\begin{cases} \varphi = \frac{\pi}{2} + \arccos\left(\frac{2\cos\gamma\sin\psi}{\sqrt{3-2\cos(2\gamma)\cos^2\psi-\cos(2\psi)}}\right), 0 \le \gamma < \frac{\pi}{2}, \\ \varphi = \frac{\pi}{2} - \arccos\left(\frac{2\cos\gamma\sin\psi}{\sqrt{3-2\cos(2\gamma)\cos^2\psi-\cos(2\psi)}}\right), -\frac{\pi}{2} < \gamma < 0, \end{cases} \tag{3.1}$$

$$\mathbf{R}_P(\psi) = \begin{bmatrix} 1 & 0 & 0 \\ 0 & \cos\psi & -\sin\psi \\ 0 & \sin\psi & \cos\psi \end{bmatrix},$$

$$\mathbf{R}_Y(\gamma) = \begin{bmatrix} \cos\gamma & 0 & \sin\gamma \\ 0 & 1 & 0 \\ -\sin\gamma & 0 & \cos\gamma \end{bmatrix}. \tag{3.2}$$

Thus, the Cartesian coordinate of the mth antenna element on the receive UCA in $\mathrm{Z}' - \mathrm{X}'\overline{\mathrm{O}}\mathrm{Y}'$ coordinate system can be written as [11]

$$\begin{cases} a_m = R_r\cos\theta\cos\gamma + R_r\sin\theta\sin\psi\sin\gamma, \\ b_m = R_r\sin\theta\cos\psi, \\ c_m = R_r\sin\theta\sin\psi\cos\gamma - R_r\cos\theta\sin\gamma. \end{cases} \tag{3.3}$$

As the coordinate system $\mathrm{Z}'-\mathrm{X}'\overline{\mathrm{O}}\mathrm{Y}'$ is parallel to Z−XOY, the Cartesian coordinate of the mth antenna element of the receive UCA in Z − XOY coordinate system, denoted by $(\tilde{a}_m, \tilde{b}_m, \tilde{c}_m)$, can be obtained as $[\tilde{a}_m, \tilde{b}_m, \tilde{c}_m]^T = [a_m, b_m, c_m]^T + [0, 0, r]^T$. With the above coordinates, we can obtain the transmission distance $d_{m,n}$ from the nth transmit antenna element to the mth receive antenna element. By assuming that the transmit and receive UCAs are in far-field distance region of each other, thus $r \gg R_t$ and $r \gg R_r$, we can approximate $d_{m,n}$ as

$$\begin{aligned} d_{m,n} &\overset{(a)}{\approx} \sqrt{R_r^2 + R_t^2 + r^2} - \frac{R_t R_r \sin\theta\cos\phi\sin\psi\sin\gamma}{\sqrt{R_r^2 + R_t^2 + r^2}} \\ &\quad - \frac{R_r R_t\left(\cos\theta\cos\phi\cos\gamma + \sin\theta\sin\phi\cos\psi\right)}{\sqrt{R_r^2 + R_t^2 + r^2}} \\ &\quad + \frac{r R_r\left(\sin\theta\sin\psi\cos\gamma - \cos\theta\sin\gamma\right)}{\sqrt{R_r^2 + R_t^2 + r^2}} \\ &\overset{(b)}{\approx} r - \frac{R_r R_t}{r}\sin\theta\cos\phi\sin\psi\sin\gamma \\ &\quad - \frac{R_r R_t}{r}\left(\cos\theta\cos\phi\cos\gamma + \sin\theta\sin\phi\cos\psi\right) \end{aligned}$$

$$+ R_r \left(\sin\theta \sin\psi \cos\gamma - \cos\theta \sin\gamma\right), \tag{3.4}$$

where (a) uses the method of completing a square and the condition $r \gg R_t, R_r$ as same as the simple case $\sqrt{a^2 - 2b} \approx a - \frac{b}{a}, a \gg b$, (b) is directly obtained from the condition of $r \gg R_t, R_r$. In LoS communications, propagation through the RF channel leads to attenuation and phase rotation of the transmitted signal. This effect is modelled through multiplying by a complex constant $h_{m,n}(w)$, whose value depends on the frequency and the distance $d_{m,n}$ between the transmit and receive antennas, as

$$\begin{aligned} h_{m,n}(w) &= \frac{\beta}{2k_w d_{m,n}} \exp\left(-ik_w d_{m,n}\right) \\ &\overset{(a)}{\approx} \frac{\beta}{2k_w r} \exp\big(i S_{k_w} \sin\theta \cos\phi \sin\psi \sin\gamma \\ &\quad + i S_{k_w} \cos\theta \cos\phi \cos\gamma + i S_{k_w} \sin\theta \sin\phi \cos\psi \\ &\quad - ik_w r - ik_w R_r (\sin\theta \sin\psi \cos\gamma - \cos\theta \sin\gamma)\big), \end{aligned} \tag{3.5}$$

where $S_{k_w} = k_w \frac{R_r R_t}{r}$, (a) neglects a few minor terms in the denominator of the amplitude term and thus only $2k_w r$ is left, $k_w = 2\pi/\lambda_w$ is the wave number at the wth subcarrier, and λ_w is the wavelength at the wth subcarrier, and $1/(2k_w d_{m,n})$ denotes the degradation of amplitude, $\beta = \beta_r \beta_t$, β_r models all the constants relative to each receive antenna element, and the complex exponential term is the phase difference due to the propagation distance. In order to mainly analyze the factors we are concerned about, the transmission loss caused by the polarization mismatch between the transmit and receive UCAs is ignored.

Then, the channel matrix of the UCA-based LoS communication system can be expressed as $\mathbf{H}(w) = [h_{m,n}(w)]_{N\times N}$. Note that when $\psi = 0$ and $\gamma = 0$, $\mathbf{H}(w)$ is a circulant matrix that can be decomposed by the N-dimentional Fourier matrix $\boldsymbol{F}_N$ as $\mathbf{H}(w) = \boldsymbol{F}_N^H \boldsymbol{\Lambda}(w) \boldsymbol{F}_N$, where $\boldsymbol{\Lambda}(w)$ is a diagonal matrix with the eigenvalues of $\mathbf{H}(w)$ on its diagonal. In [2–4, 14], the OAM channel is defined as $\mathbf{H}_{\text{OAM}}(w) = \boldsymbol{F}_N \mathbf{H}(w) \boldsymbol{F}_N^H$, which will become $\boldsymbol{\Lambda}(w)$ in perfect alignment case.

3.2.2 *Signal Model*

As a UCA can generate multi-mode OAM beams with the baseband (partial) discrete Fourier transform (DFT) matrix $\mathbf{F}_U$, the equivalent baseband signal models of transmitting multi-mode OAM data symbols could be expressed as $\mathbf{F}_U^H \mathbf{s}(w)$, where $\mathbf{F}_U^H$ is used for spiralization, $\mathbf{s}(w) = [s(w,1), s(w,2), \cdots, s(w,U)]^T$ contains the modulation symbols transmitted on U OAM modes at the wth ($1 \le w \le W$) subcarrier simultaneously, and ℓ_u is the uth OAM mode number. Thus, the received OAM signal $\mathbf{y}(w)$ can be written as

$$
\begin{aligned}
\mathbf{y}(w) &= \mathbf{F}_U\left(\mathbf{H}(w)\mathbf{F}_U^H\mathbf{s}(w) + \mathbf{z}(w)\right)\\
&= \mathbf{H}_{\mathrm{OAM}}(w)\mathbf{s}(w) + \bar{\mathbf{z}}(w),
\end{aligned}
\tag{3.6}
$$

where $\mathbf{F}_U = [\mathbf{f}^H(\ell_1), \mathbf{f}^H(\ell_2), \cdots, \mathbf{f}^H(\ell_U)]^H$ is a $U \times N$ right circularly shifted (partial) Fourier matrix used for despiralization, $\mathbf{f}(\ell_u) = \frac{1}{\sqrt{N}}\left[1, e^{-i\frac{2\pi\ell_u}{N}}, \cdots, e^{-i\frac{2\pi\ell_u(N-1)}{N}}\right]$, $\mathbf{z}(w) = [z(w,1), z(w,2), \cdots, z(w,U)]^T$ is the complex Gaussian noise vector with zero mean and covariance matrix $\sigma_z^2\mathbf{I}_N$, $\bar{\mathbf{z}}(w) = \mathbf{F}_U\mathbf{z}(w)$, $\mathbf{y}(w) = [y(w,1), y(w,2), \cdots, y(w,U)]^T$, $y(w,u)$ is the received information signal on the uth mode OAM and the wth subcarrier of receiver and can be written as

$$
\begin{aligned}
y(w,u) &= h_{\mathrm{OAM},w}(u,u)s(w,u) + \bar{z}(w,u)\\
&\quad + \sum_{u\neq v} h_{\mathrm{OAM},w}(u,v)s(w,v),
\end{aligned}
\tag{3.7}
$$

$h_{\mathrm{OAM},w}(u,v) = \mathbf{f}(\ell_u)\mathbf{H}(w)\mathbf{f}^H(\ell_v)$ is the uth-row and vth-column element of $\mathbf{H}_{\mathrm{OAM}}(w)$ and can be expressed as

$$
\begin{aligned}
h_{\mathrm{OAM},w}(u,v) = \eta(w)\sum_{m=1}^{N}\sum_{n=1}^{N}\exp\big(&-i\ell_u\theta + i\ell_v\phi\\
&+ iS_{k_w}\sin\theta\cos\phi\sin\psi\sin\gamma\\
&+ iS_{k_w}(\cos\theta\cos\phi\cos\gamma + \sin\theta\sin\phi\cos\psi)\\
&- ik_wR_r(\sin\theta\sin\psi\cos\gamma - \cos\theta\sin\gamma)\big),
\end{aligned}
\tag{3.8}
$$

$\eta(w) = \frac{\beta}{2k_w rN}\exp(-ik_w r)$, $\bar{z}(w,u) = \mathbf{f}(\ell_u)\mathbf{z}(w)$, $u, v = 1, 2, \cdots, U$ and $w = 1, 2, \cdots, W$. In (3.7), $h_{\mathrm{OAM},w}(u,u)s(w,u)$ carries useful signals. Once $\gamma \neq 0$ or $\psi \neq 0$, $\sum_{u\neq v} h_{\mathrm{OAM},w}(u,v)s(w,v) \neq 0$, which becomes interferences.

For this model, in order to alleviate the interference that is resulted by the large misalignment error in the practical case, we propose the hybrid mechanical and electronic beam steering scheme, as shown in Fig. 3.1. It can be observed from (3.5) that mechanical rotations of the UCA plane will result in the changes of $\{d_{m,n}, m = 1, \cdots, M; n = 1, \cdots, N\}$ and the amplitudes and phases of $\{h_{m,n}, m = 1, \cdots, M; n = 1, \cdots, N\}$. It is worth noting that the amplitude and phase of $h_{m,n}$ do not linearly change with the rotation angles, which are dependent on ψ, γ, and θ. Therefore, the mechanical beam steering is a nonlinear transformation for the channel matrix $\mathbf{H}(w)$. In order to represent the operation of the mechanical beam steering, we assume a nonlinear function $\mathcal{F}(\cdot)$, and then the effect of mechanical beam steering on the channel matrix can be expressed as

$$\mathbf{H}_{\mathcal{F}}(w) = \mathcal{F}\Big(\mathbf{H}(w), \hat{\psi}, \hat{\gamma}, \theta^*\Big), \tag{3.9}$$

where $\hat{\gamma}$ is the estimate of γ by the OAM AoA estimation technique; $\hat{\psi}$ is the estimate of ψ obtained with $\hat{\gamma}$ and the estimate of α, i.e., $\hat{\alpha}$; and θ^* is the optimal value of the rotation angle θ^* in roll direction for maximizing the OAM channel capacity, so $\hat{\gamma}$, $\hat{\psi}$, and θ^* are mechanical rotation angles in the pitch, yaw, and roll directions, respectively, and $\mathbf{H}_{\mathcal{F}}(w)$ is the channel matrix after mechanical rotations of the receive UCA. In contrast to the mechanical beam steering, the electronic beam steering is linear transformation [2, 4]. With the hybrid nonlinear mechanical beam steering and linear electronic beam steering, the recovered information signal vector $\mathbf{y}_H(w)$ becomes

$$\mathbf{y}_H(w) = (\mathbf{F}_U \odot \mathbf{B}_E(w)) \Big(\mathbf{H}_{\mathcal{F}}(w)\mathbf{F}_U^H \mathbf{s}(w) + \mathbf{z}(w)\Big), \tag{3.10}$$

where $\mathbf{B}_E(w)$ is the electronic beam steering matrix at the receiver. The procedure of the mechanical and electronic beam steering for the LoS OAM communication receiver will be detailed in Sect. 3.4.

3.3 The Effect of Electronic Beam Steering and Solution in Large-Angle Misalignment Case

The motivation of the proposed hybrid beam steering is from practical implementation of mmWave relays. Let us consider the case in Fig. 3.3a; whether it is OAM array or mmWave massive MIMO array, the performance of the communication is poor. However, with the proposed the hybrid beam steering scheme, which first enables the coarse alignment between the transmit array plane and the receive array plane through mechanical rotation as in Fig. 3.3b, the good performance can be expected through refined electronic beam steering.

Fig. 3.3 Two examples of transmit and receive array directions in (**a**) large angle misalignment case, (**b**) small angle misalignment case

3.3.1 The Effect of Electronic Beam Steering

In this section, we first investigate the effect of electronic beam steering, especially in large-angle misalignment case. With only electronic beam steering, the recovered information signal vector $\mathbf{y}_E(w)$ takes the form [4]

$$\begin{aligned}\mathbf{y}_E(w) &= (\mathbf{F}_U \odot \mathbf{B}_{EO}(w))\left(\mathbf{H}(w)\mathbf{F}_U^H\mathbf{s}(w) + \mathbf{z}(w)\right)\\ &= \widetilde{\mathbf{H}}_{\text{OAM}}(w)\mathbf{s}(w) + \tilde{\mathbf{z}}(w),\end{aligned} \tag{3.11}$$

where $\widetilde{\mathbf{H}}_{\text{OAM}}(w) = (\mathbf{F}_U \odot \mathbf{B}_{EO}(w))\,\mathbf{H}(w)\mathbf{F}_U^H$ is defined as the effective OAM channel matrix after electronic beam steering, the electronic beam steering matrix $\mathbf{B}_{EO}(w)$ of the receive UCA is designed as $\mathbf{B}_{EO}(w) = \mathbf{1} \otimes \mathfrak{b}_{EO}(w)$, $\mathfrak{b}_{EO}(w) = [e^{iW_1(w)}, e^{iW_2(w)}, \cdots, e^{iW_N(w)}]$ and $W_m(w) = k_w R_r(\sin\theta\sin\psi\cos\gamma - \cos\theta\sin\gamma)$, $m = 1, \cdots, N$, $w = 1, \cdots, W$, $\tilde{\mathbf{z}}(w) = (\mathbf{F}_U \odot \mathbf{B}_{EO}(w))\,\mathbf{z}(w) = [\tilde{z}(w,1), \tilde{z}(w,2), \cdots, \tilde{z}(w,U)]^T$, $\mathbf{y}_E(w) = [y_E(w,1), y_E(w,2), \cdots, y_E(w,U)]^T$, $y_E(w,u)$ is the information symbol on the uth mode OAM and the wth subcarrier after electronic beam steering and can be written as

$$\begin{aligned}y_E(w,u) &= \tilde{h}_{\text{OAM},w}(u,u)s(w,u) + \tilde{z}(w,u)\\ &\quad + \sum_{u\neq v}\tilde{h}_{\text{OAM},w}(u,v)s(w,v),\end{aligned} \tag{3.12}$$

$\tilde{h}_{\text{OAM},w}(u,v) = \left(\mathfrak{b}_{EO}(w) \odot \mathbf{f}(\ell_u)\right)\mathbf{H}(w)\mathbf{f}^H(\ell_v)$ is the uth-row and vth-column element of $\widetilde{\mathbf{H}}_{\text{OAM}}(w)$ and can be expressed as

$$\begin{aligned}\tilde{h}_{\text{OAM},w}(u,v) = \eta(w)\sum_{m=1}^{N}\sum_{n=1}^{N}\exp\Big(&-i\ell_u\theta + i\ell_v\phi\\ &+ iS_{k_w}\sin\theta\cos\phi\sin\psi\sin\gamma\\ &+ iS_{k_w}(\cos\theta\cos\phi\cos\gamma + \sin\theta\sin\phi\cos\psi)\Big).\end{aligned} \tag{3.13}$$

Hence, after electronic beam steering, the signal-to-interference-plus-noise ratio (SINR) on the uth mode and the wth subcarrier can be formulated as

$$\text{SINR}_{EO}(w,u) = \frac{\left|\tilde{h}_{\text{OAM},w}(u,u)\right|^2\mathbb{E}\left(|s(w,u)|^2\right)}{\sum\limits_{u\neq v}\left|\tilde{h}_{\text{OAM},w}(u,v)\right|^2\mathbb{E}\left(|s(w,v)|^2\right) + \sigma_z^2}, \tag{3.14}$$

where $\mathbb{E}\left(|s(w, v)|^2\right)$ is the average power of modulation symbols on the vth mode and wth subcarrier. We assume that the average power of modulation symbols on all modes and subcarriers are the same, and then the SINR approaches to the signal-to-interference ratio (SIR) at high SNR as

$$\begin{aligned}\lim_{\rho\to\infty} \text{SINR}_{EO}(w, u) &= \text{SIR}_{EO}(w, u)\\ &= \frac{\left|\tilde{h}_{\text{OAM},w}(u, u)\right|^2}{\sum\limits_{u\neq v}\left|\tilde{h}_{\text{OAM},w}(u, v)\right|^2},\end{aligned} \tag{3.15}$$

where $\rho = \mathbb{E}\left(|s(w, u)|^2\right)/\sigma_z^2$. Then, we evaluate how the SINRs change with the increase of $|\gamma|$ and $|\psi|$ of the electronic beam steering. If we substitute (3.13) into (3.14), the obtained two-dimensional function of γ and ψ is too complex to analyze. Therefore, we turn to check how $\text{SIR}_{EO}(w, u)$ changes with the increase of $|\gamma|$ when $\psi = 0$ and how $\text{SIR}_{EO}(w, u)$ changes with the increase of $|\psi|$ when $\gamma = 0$ and obtain the following results.

Proposition 1 *When $\psi = 0$, the SIR after electronic beam steering on the uth mode and wth subcarrier decreases with the increase of $|\gamma|$ as S_{k_w} goes to zero.*

Proof The proof is given in Appendix.

Proposition 2 *When $\gamma = 0$, the SIR after electronic beam steering on the uth mode and wth subcarrier decreases with the increase of $|\psi|$ as S_{k_w} goes to zero.*

Proof The proof is completely similar to Proposition 1.

Propositions 1 and 2 show that $\text{SIR}_{EO}(w, u)$ is monotonically decreasing for $|\gamma|$ and $|\psi|$ in two one-dimensional directions, respectively, on the condition that S_{k_w} approaches to zero. Moreover, we show through simulation results in Fig. 3.4a and b that $\text{SIR}_{EO}(w, u)$ is also monotonically decreasing for $|\gamma|$ and $|\psi|$ in the case S_{k_w} is larger. Furthermore, Fig. 3.4c shows that when S_{k_w} is fixed to be a value other than zero, $\text{SIR}_{EO}(w, u)$ is still monotonically decreasing for $|\gamma|$ and $|\psi|$. It follows that with the increase of oblique angles, the SINR after applying only electronic beam steering will get worse.

In order to further figure out the total effect of all the subcarriers, the MCMM-OAM system with only electronic beam steering can be evaluated according to the channel capacity

$$C_{EO} = \frac{1}{W}\sum_{w=1}^{W}\sum_{u=1}^{U}\log_2\left(1 + \text{SINR}_{EO}(w, u)\right). \tag{3.16}$$

In Fig. 3.5, we illustrate what is the condition for "small-angle misalignment" so that electronic steering is sufficient and what is the condition for "large-angle

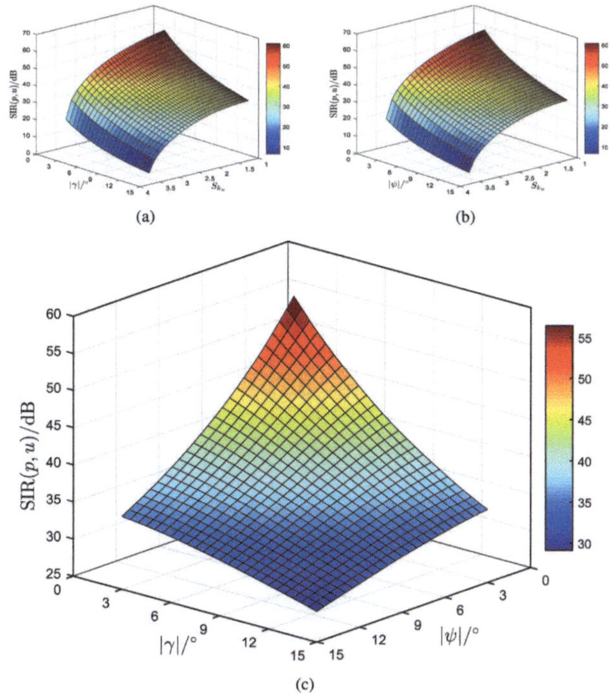

Fig. 3.4 The effects of (**a**) yaw angle γ, (**b**) pitch angle ψ, (**c**) yaw angle γ and pitch angle ψ with $S_{k_w} = 1$ on SIR. $U = 9$, $\ell_v \in [-4, 4]$, $\ell_u = 1$, $N = 10$, $R_r = R_t = 20\lambda_1$

misalignment" so that it is insufficient. It can be seen apparently from the figure that under this set of parameters, the threshold between "small angle" and "large angle" is about 20 degrees. When $|\gamma|$ and $|\psi|$ are smaller than the threshold, the electronic beam steering can well compensate for the performance loss caused by misalignment, while when $|\gamma|$ and $|\psi|$ are larger than this threshold, the electronic beam steering is not good enough especially when $|\gamma|$ and $|\psi|$ approach $\pi/2$.

3.3.2 Solution in Large-Angle Misalignment Case

The reason for the performance degradation of electronic beam steering is that with the increase of misalignment angles, the inter-mode interferences increase. In order to minimize the value of the interference term $\sum_{u \neq v} |\tilde{h}_{\mathrm{OAM},w}(u, v)|^2$, we observe from (3.13) that the smaller the $|\gamma|$ and $|\psi|$, the smaller the value of the interference term. As long as the γ and ψ are equal to zero, $\tilde{h}_{\mathrm{OAM},w}(u, v) = 0$, and interferences disappear. It follows that when the receive UCA is in the large-angle misalignment with the transmit UCA, it is the best choice to firstly steer the yaw and pitch directions of the receive UCA physically in order that $|\gamma|$ and $|\psi|$ are as small as possible. Furthermore, besides minimizing the interference, maximizing the channel capacity needs also maximizing the received signal power, that is, maximizing the

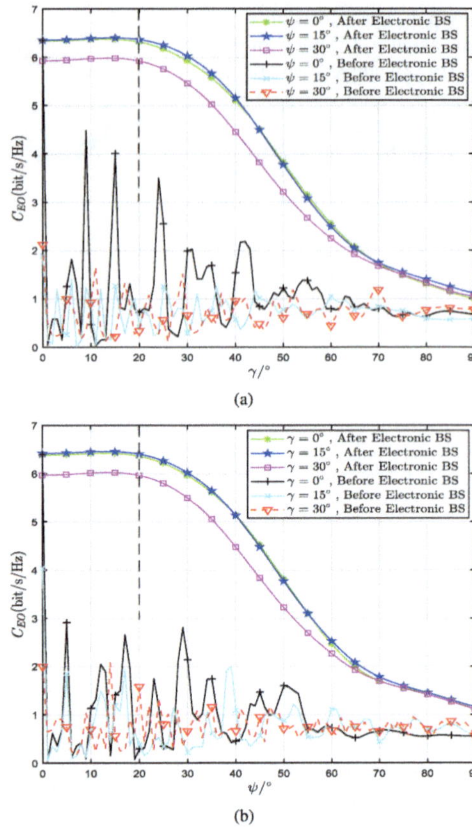

Fig. 3.5 The effects of (**a**) yaw angle γ, (**b**) pitch angle ψ on C_{EO}, $W = 6$ subcarriers from 3.9982 GHz to 4.2387 GHz, $U = 9$ with $\ell_v \in [-4, 4]$, $R_r = R_t = 20\lambda_1$, $r = 450\lambda_1$. BS: Beam steering

term $|\tilde{h}_{\mathrm{OAM},w}(u,u)|^2$. Note that when γ and ψ are fixed, for the receive UCA, only θ_0 has effect on $\tilde{h}_{\mathrm{OAM},w}(u,v)$, which means the roll rotation angle θ needs to be optimized. The above analysis could be formulated as follows:

$$
\begin{aligned}
\max C_{EO} &\Leftrightarrow \max \mathrm{SINR}_{EO} \\
&\Leftrightarrow \begin{cases} \min \sum\limits_{u \neq v} \left| \tilde{h}_{\mathrm{OAM},w}(u,v) \right|^2 \\ \max \left| \tilde{h}_{\mathrm{OAM},w}(u,u) \right|^2 \end{cases} \\
&\Leftrightarrow \begin{cases} \min \left(|\gamma| + |\psi| \right) \ \text{in yaw and pitch directions} \\ \text{Search for } \ \theta^{\star} \ \text{in roll direction.} \end{cases}
\end{aligned}
\tag{3.17}
$$

This is the motivation of proposing hybrid mechanical and electronic beam steering. Mechanical beam steering enables the misalignment angles fall into the range of "small angle" and leave them for electronic beam steering, which happens when the receiver starts running or moves to another place.

3.4 Hybrid Mechanical and Electronic Beam Steering of OAM Communication Systems

According to (3.17), the procedure of the hybrid mechanical and electronic beam steering scheme can be designed as the four successive steps shown in Fig. 3.6. Then, the hybrid beam steering can be formulated as,

$$\begin{aligned}\mathbf{y}_H(w) &= \left(\mathbf{F}_U \odot \mathbf{B}_{E_2}(w) \odot \mathbf{B}_{E_1}(w)\right)\left(\mathcal{F}_2\left(\mathcal{F}_1\left(\mathbf{H}(w), \hat{\psi}, \hat{\gamma}\right), \theta^*\right)\mathbf{F}_U^H\mathbf{s}(w) + \mathbf{z}(w)\right)\\ &= \mathbf{H}_H^{\mathrm{OAM}}(w)\mathbf{s}(w) + \hat{\mathbf{z}}(w),\end{aligned} \tag{3.18}$$

where $\mathbf{y}_H(w)$ is the recovered information signal vector with hybrid beam steering, $\mathbf{H}_H^{\mathrm{OAM}}(w) = (\mathbf{F}_U \odot \mathbf{B}_{E_2}(w) \odot \mathbf{B}_{E_1}(w))(\mathcal{F}_2(\mathcal{F}_1(\mathbf{H}(w), \hat{\psi}, \hat{\gamma}), \theta^*)\mathbf{F}_U^H)$ is defined as the effective OAM channel matrix after the hybrid steering, $\hat{\mathbf{z}}(w) = \left(\mathbf{F}_U \odot \mathbf{B}_{E_2}(w) \odot \mathbf{B}_{E_1}(w)\right)\mathbf{z}(w) = [\hat{z}(w,1), \hat{z}(w,2), \cdots, \hat{z}(w,U)]^T$, $\mathbf{B}_{E_1}(w)$ is the first electronic beam steering matrix in pitch and yaw directions, $\mathbf{B}_{E_2}(w)$ is the second beam steering matrix that represents adjusting the electronic beam in roll direction, $\mathcal{F}_1$ represents the operation of the first mechanical beam steering in pitch and yaw directions, and $\mathcal{F}_2$ represents the operation of the second mechanical beam steering in roll direction.

The implementation of mechanical beam steering relies on the servo system of mechanical rotating devices, which is controlled by PWM signals [10]. The schematic diagram of the servo system corresponding to the hybrid OAM transceiver in Fig. 3.1 is shown in Fig. 3.7, where the control unit generates three tunable PWM signals for pitch-axis, yaw-axis, and roll-axis, respectively. In the execution unit, intelligent power modules (IPMs) are used to drive steering gears in the three directions respectively, and potentiometer is used to convert the motion of steering gear to voltage and feedback to control unit. We denote the accuracy of the potentiometer as ν. The relationship between the duty cycle D of the PWM signal and the expected rotation angle ϑ can be obtained as $\vartheta = \frac{\pi}{p_e - p_s}(DK - p_0)$, where K is the signal period of PWM signal and p_s, p_0, and p_e are the pulse widths of PWM signal corresponding to the minimum angle, middle angle, and the maximum

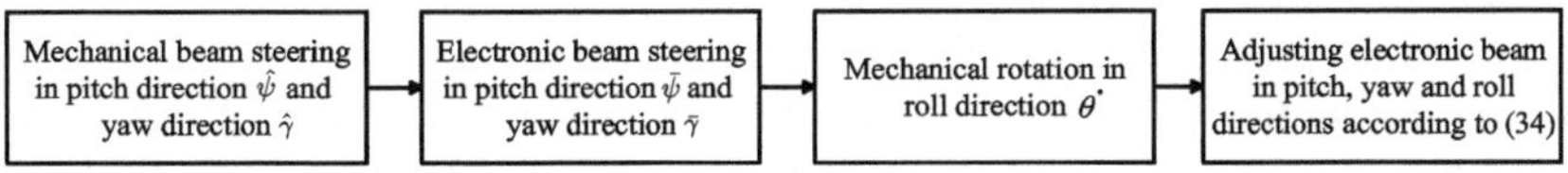

Fig. 3.6 The procedure of hybrid mechanical and electronic beam steering for the LoS MCMM-OAM receiver

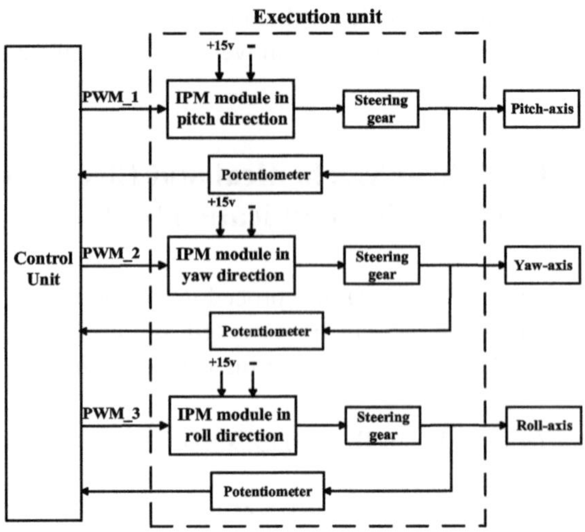

Fig. 3.7 The schematic diagram of the servo system of the mechanical rotating devices

angle of the mechanical rotating device [8]. Here, ϑ represents $\hat{\psi}$, $\hat{\gamma}$, and θ^* in pitch-axis, yaw-axis, and roll-axis, respectively.

3.4.1 Mechanical Beam Steering in Pitch and Yaw Directions

Firstly, the coarse alignment is achieved by mechanical beam steering. From Fig. 3.1 we can see that only with rotations in pitch and yaw directions can the receive UCA align to the transmit UCA. Then, according to Fig. 3.2, we can see that the mechanical beam steering in yaw and pitch directions are performed around Y′-axis and $\overline{\text{X}}$-axis, respectively. It is assumed that $\hat{\gamma}$ and $\hat{\psi}$ are the coarse estimates of γ and ψ in this step and $\bar{\gamma} = \gamma - \hat{\gamma}$ and $\bar{\psi} = \psi - \hat{\psi}$ are the remaining small misalignment errors in yaw and pitch directions, respectively. In the first-stage yaw direction, the mechanical beam steering makes the receive UCA rotate around the Y′-axis of the $\text{Z}' - \text{X}'\overline{\text{O}}\text{Y}'$ coordinate system so that the angle between the $\overline{\text{X}}$-axis and the X′-axis changes from γ to $\bar{\gamma}$. Then, in the second-stage pitch direction, the receive UCA rotates around the rotated $\overline{\text{X}}$-axis so that the angle between the $\overline{\text{Y}}$-axis and Y′-axis changes from ψ to $\bar{\psi}$. With the pitch and yaw rotations, the Cartesian coordinate of the mth antenna element on the receive UCA in $\text{Z}' - \text{X}'\overline{\text{O}}\text{Y}'$, denoted by $(\bar{a}_m, \bar{b}_m, \bar{c}_m)$, can be obtained as $[\bar{a}_m, \bar{b}_m, \bar{c}_m]^T = \mathbf{R}_Y(\bar{\gamma})\mathbf{R}_P(-\hat{\psi})\mathbf{R}_Y(-\bar{\gamma})\mathbf{R}_Y(-\hat{\gamma})[a_m, b_m, c_m]^T$, where $\mathbf{R}_P(-\hat{\psi})$ and $\mathbf{R}_Y(-\hat{\gamma})$ represent the coordinate rotation matrixes in pitch and yaw directions, and considering the second-stage pitch rotation around the rotated $\overline{\text{X}}$-axis not any axis of the $\text{Z}' - \text{X}'\overline{\text{O}}\text{Y}'$ coordinate system, the axis rotation matrix $\mathbf{R}_Y(-\bar{\gamma})$ is required after the first-stage yaw rotation for the coordinate transformation from X′-axis to $\overline{\text{X}}$-axis; in the end, $\mathbf{R}_Y(\bar{\gamma})$ performs the opposite coordinate transformation from $\overline{\mathrm{X}}$-axis to X′-axis. Thus, after the pitch and yaw rotations, the transmission distance from the nth transmit antenna element to the mth receive antenna element becomes

$$
\begin{aligned}
\bar{d}_{m,n} &= r - \frac{R_r R_t}{r} \sin\theta \cos\phi \sin\bar{\psi} \sin\bar{\gamma} \\
&\quad - \frac{R_r R_t}{r} \left(\cos\theta \cos\phi \cos\bar{\gamma} + \sin\theta \sin\phi \cos\bar{\psi} \right) \\
&\quad + R_r \left(\sin\theta \sin\bar{\psi} \cos\bar{\gamma} - \cos\theta \sin\bar{\gamma} \right).
\end{aligned} \tag{3.19}
$$

Hence, the channel matrix after the mechanical beam steering in pitch and yaw directions can be expressed as $\mathbf{H}_{\mathcal{F}_1}(w) = [\bar{h}_{m,n}(w)]_{N\times N}$, where the mth-row nth-column element $\bar{h}_{m,n}(w)$ can be written as

$$
\begin{aligned}
\bar{h}_{m,n}(w) &= \frac{\beta}{2k_w \bar{d}_{m,n}} \exp\left(-ik_w \bar{d}_{m,n}\right) \\
&\overset{(a)}{\approx} \frac{\beta}{2k_w r} \exp\Bigg(i S_{k_w} \sin\theta \cos\phi \sin\bar{\psi} \sin\bar{\gamma} + i S_{k_w} \cos\theta \cos\phi \cos\bar{\gamma} \\
&\quad + i S_{k_w} \sin\theta \sin\phi \cos\bar{\psi} - ik_w r - ik_w R_r \left(\sin\theta \sin\bar{\psi} \cos\bar{\gamma} - \cos\theta \sin\bar{\gamma} \right) \Bigg),
\end{aligned} \tag{3.20}
$$

(a) neglects a few small terms in the denominator and only $2k_w r$ is left. If only with the mechanical beam steering, the received OAM signal $\mathbf{y}_{\mathcal{F}_1}(w)$ takes the form

$$
\begin{aligned}
\mathbf{y}_{\mathcal{F}_1}(w) &= \mathbf{F}_U \left(\mathbf{H}_{\mathcal{F}_1}(w) \mathbf{F}_U^H \mathbf{s}(w) + \mathbf{z}(w) \right) \\
&= \mathbf{H}_{\mathcal{F}_1}^{\mathrm{OAM}}(w) \mathbf{s}(w) + \bar{\mathbf{z}}(w),
\end{aligned} \tag{3.21}
$$

where $\mathbf{H}_{\mathcal{F}_1}^{\mathrm{OAM}}(w) = \mathbf{F}_U \mathbf{H}_{\mathcal{F}_1}(w) \mathbf{F}_U^H$ is defined as the effective multi-mode OAM channel matrix after the first mechanical beam steering. $h_{\mathcal{F}_1,w}^{\mathrm{OAM}}(u,v) = \mathbf{f}(\ell_u) \mathbf{H}_{\mathcal{F}_1}(w) \mathbf{f}^H(\ell_v)$ is the uth-row and vth-column element of $\mathbf{H}_{\mathcal{F}_1}^{\mathrm{OAM}}(w)$ and can be calculated as

$$
\begin{aligned}
h_{\mathcal{F}_1,w}^{\mathrm{OAM}}(u,v) &= \eta(w) \sum_{m=1}^{N} \sum_{n=1}^{N} \exp\Bigg(- i\ell_u \theta + i\ell_v \phi + i S_{k_w} \sin\theta \cos\phi \sin\bar{\psi} \sin\bar{\gamma} \\
&\quad + i S_{k_w} (\cos\theta \cos\phi \cos\bar{\gamma} + \sin\theta \sin\phi \cos\bar{\psi}) \\
&\quad - ik_w R_r \left(\sin\theta \sin\bar{\psi} \cos\bar{\gamma} - \cos\theta \sin\bar{\gamma} \right) \Bigg)
\end{aligned}
$$

$$\stackrel{(a)}{\approx} \eta(w) \sum_{\delta=1}^{N} \exp\left(i \frac{2\pi\delta}{N} \ell_v + i S_{k_w} \cos \frac{2\pi\delta}{N}\right)$$
$$\times \sum_{m=1}^{N} \sum_{n=1}^{N} \exp\Bigg(- i \frac{2\pi(m-1)}{N} t + i k_w R_r \cos\theta \sin\bar{\gamma}$$
$$- i k_w R_r \sin\theta \sin\bar{\psi} \cos\bar{\gamma} \Bigg), \tag{3.22}$$

where $\delta = n - m \in \mathbb{Z}$, (a) applies the approximation $\cos a \approx 1 - \frac{a^2}{2}$ for $\cos\bar{\psi}$ and $\cos\bar{\gamma}$ in the case that $\bar{\psi}$ and $\bar{\gamma}$ are relatively small and neglects a few small terms under the condition $r \gg R_r$. As $\mathbf{H}^{\mathrm{OAM}}_{\mathcal{F}_1}(w)$ is a non-diagonal matrix, we can conclude that there is still inter-mode interferences after the first mechanical beam steering.

3.4.2 *Electronic Beam Steering in Pitch and Yaw Directions*

In order to eliminate the inter-mode interferences induced by the remaining small perturbations $\bar{\psi}$ and $\bar{\gamma}$ after previous mechanical beam steering, periodic refined AoA estimations are necessary. It is assumed that $\bar{\psi}$ and $\bar{\gamma}$ varies slower than the duration of AoA estimation and $\bar{\psi}$ and $\bar{\gamma}$ can be perfectly estimated. With the estimates of $\bar{\psi}$ and $\bar{\gamma}$, electronic beam steering can enable the receive beam to align with the direction of the arrived OAM beam. The electronic beam steering matrix can be designed as $\mathbf{B}_{E_1}(w) = \mathbf{1} \otimes \mathfrak{b}_{E_1}(w)$, where $\mathfrak{b}_{E_1}(w) = [e^{i\overline{W}_1(w)}, e^{i\overline{W}_2(w)}, \cdots, e^{i\overline{W}_N(w)}]$ and $\overline{W}_m(w) = k_w R_r \left(\sin\theta \sin\bar{\psi} \cos\bar{\gamma} - \cos\theta \sin\bar{\gamma}\right)$, $m = 1, \cdots, N$, $w = 1, \cdots, W$. After involving these phases into the original phases in $\mathbf{F}_U$ at the receive UCA, the effective multi-mode OAM channel matrix at the wth subcarrier becomes

$$\mathbf{H}^{\mathrm{OAM}}_{E_1\mathcal{F}_1}(w) = (\mathbf{F}_U \odot \mathbf{B}_{E_1}(w))\mathbf{H}_{\mathcal{F}_1}(w)\mathbf{F}_U^H. \tag{3.23}$$

Denote $h^{\mathrm{OAM}}_{E_1\mathcal{F}_1,w}(u, v) = \left(\mathfrak{b}_{E_1}(w) \odot \mathbf{f}(\ell_u)\right)\mathbf{H}_{\mathcal{F}_1}(w)\mathbf{f}^H(\ell_v)$ as the uth-row and vth-column element of $\mathbf{H}^{\mathrm{OAM}}_{E_1\mathcal{F}_1}(w)$, which can be expressed as

$$h^{\mathrm{OAM}}_{E_1\mathcal{F}_1,w}(u, v) \approx \eta(w) \sum_{m=1}^{N} \exp(-i\theta t) \times \sum_{\delta=1}^{N} \exp\left(i \frac{2\pi\delta}{N} \ell_v + i S_{k_w} \cos \frac{2\pi\delta}{N}\right), \tag{3.24}$$

$t = \ell_u - \ell_v \in \mathbb{Z}$. Note that in (3.24), we have

$$\begin{cases} \sum_{m=1}^{N} \exp(-i\theta t) = 0, & t \neq 0, \\ \sum_{m=1}^{N} \exp(-i\theta t) = N, & t = 0. \end{cases} \tag{3.25}$$

That is to say, the non-diagonal elements of $\mathbf{H}_{E_1\mathcal{F}_1}^{\mathrm{OAM}}(w)$ equal to zero approximately. It follows that after the mechanical and electronic beam steering in pitch and yaw directions, the inter-mode interferences are almost completely eliminated in the case of large-angle misalignment. At the same time, according to the conclusion in Sect. 3.3, steering the OAM beam with only a small angle in electronic way does not result in large-performance degradation. Therefore, the proposed hybrid mechanical and electronic beam steering outperforms only using electronic beam steering in a wide variety of practical scenarios.

3.4.3 *Mechanical Rotation in Roll Direction for OAM Channel Capacity Maximization*

When a multi-mode OAM beam that can be deemed as the superposition of multiple OAM modes is transmitted, we can observe the phenomenon of interferometry as shown in the simulation results of HFSS in Fig. 3.8, that is, the electric field intensities of one multi-mode OAM beam are not uniform on the ring of its main lobe. Therefore, the receive antenna elements need to be set in the maximum SNR positions on the ring of the main lobe, which means that another rotation angle in roll direction within the plane of the receive UCA should be optimized. In order to obtain the optimal rotation angle in roll direction, we first formulate the optimization problem and then apply SA algorithm to solve the problem.

3.4.3.1 Optimization Model

First, we denote the effective OAM channel matrix with twice mechanical and once electronic beam steering as $\mathbf{H}_{\mathcal{F}_2 E_1 \mathcal{F}_1}^{\mathrm{OAM}}(w)$. Since the roll rotation will result in the misalignment between the transmit and receive beams and deteriorate the performance of the OAM communication systems, we assume the second-step roll rotation can be well compensated by the fourth-step electronic beam steering. Only in this way can the effective OAM channel capacity be maximized. For this purpose, we denote the effective OAM channel matrix with twice mechanical and twice

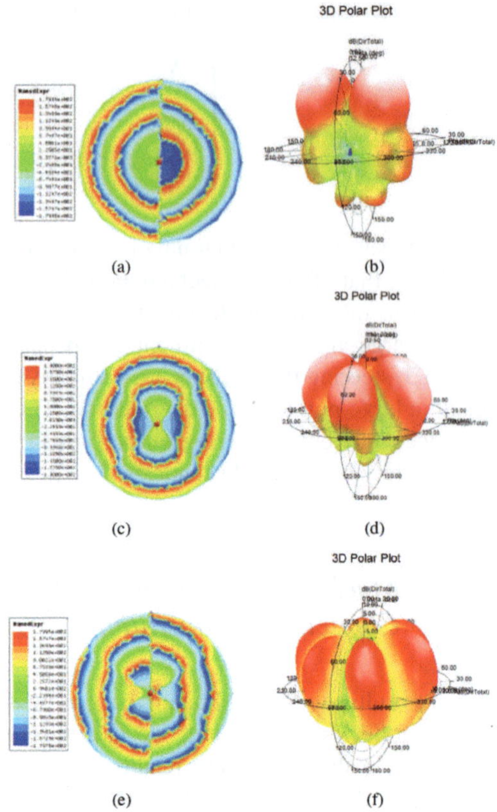

Fig. 3.8 The phase distributions (left) and radiation patterns (right) of two-mode OAM beam simulated by HFSS with mode numbers: **(a)** **(b)** $\ell_1 = +1, \ell_2 = -1$, **(c)** **(d)** $\ell_1 = +2, \ell_2 = -2$, **(e)** **(f)** $\ell_1 = +3, \ell_2 = -3$

electronic beam steering as $\mathbf{H}^{\mathrm{OAM}}_{E_2\mathcal{F}_2E_1\mathcal{F}_1}(w)$. Then, based on (3.24), the uth diagonal element of $\mathbf{H}^{\mathrm{OAM}}_{E_2\mathcal{F}_2E_1\mathcal{F}_1}(w)$ can be obtained as

$$
\begin{aligned}
&h^{\mathrm{OAM}}_{E_2\mathcal{F}_2E_1\mathcal{F}_1,w}(u,u) \\
&= N\eta(w)\sum_{\delta=1}^{N}\exp\left(i\left(\frac{2\pi\delta}{N}-\theta'\right)\ell_u + iS_{k_w}\cos\left(\frac{2\pi\delta}{N}-\theta'\right)\right), \qquad (3.26)
\end{aligned}
$$

where θ' is the rotation angle in roll direction. Therefore, the SINR at the uth mode and wth subcarrier can be formulated as $\mathrm{SINR}_{E_2\mathcal{F}_2E_1\mathcal{F}_1}(w,u) = \rho\left|h^{\mathrm{OAM}}_{E_2\mathcal{F}_2E_1\mathcal{F}_1,w}(u,u)\right|^2$. The channel capacity of the OAM communication system with hybrid beam steering can be expressed as

$$
C = \frac{1}{W}\sum_{w=1}^{W}\sum_{u=1}^{U}\log_2\left(1+\mathrm{SINR}_{E_2\mathcal{F}_2E_1\mathcal{F}_1}(w,u)\right). \qquad (3.27)
$$

Algorithm 1 Solving the optimization problem (3.28)

Input: $\theta' \in [-\frac{\pi}{N}, \frac{\pi}{N}]$, $T_{\min} \in \mathbb{R}$, $T_{\text{init}} \in \mathbb{R}$, $\zeta \in \mathbb{R}$
Output: $\theta^* \in \mathbb{R}$

procedure SA $(\theta', T_{\min}, T_{\text{init}}, \zeta)$
  $C \leftarrow$ use (3.27)
  while $T_{\text{init}} > T_{\min}$
    $\theta^* \leftarrow \theta', C_{\max} \leftarrow C, L \leftarrow 0$
    for $j = 1 : J$ **do**
      $e \leftarrow$ a small random value
      $\theta_{\text{new}} \leftarrow \begin{cases} \theta' + e, & \text{if } -\frac{\pi}{N} < \theta' + e < \frac{\pi}{N} \\ \theta' - e, & \text{otherwise} \end{cases}$
      $\theta' \leftarrow \theta_{\text{new}}, C_{\text{new}} \leftarrow$ use (3.27)
      if $C_{\text{new}} > C$ or meet the Metropolis criterion
        $L \leftarrow 1, \theta' \leftarrow \theta_{\text{new}}, C \leftarrow C_{\text{new}}$
        if $C > C_{\max}$
          $\theta^* \leftarrow \theta', C_{\max} \leftarrow C$
        end
      end
    end
    if $L \neq 0$
      $\theta' \leftarrow \theta^*, C \leftarrow C_{\max}$
    end
      $T_{\text{init}} \leftarrow T_{\text{init}}\zeta$
  end
end procedure

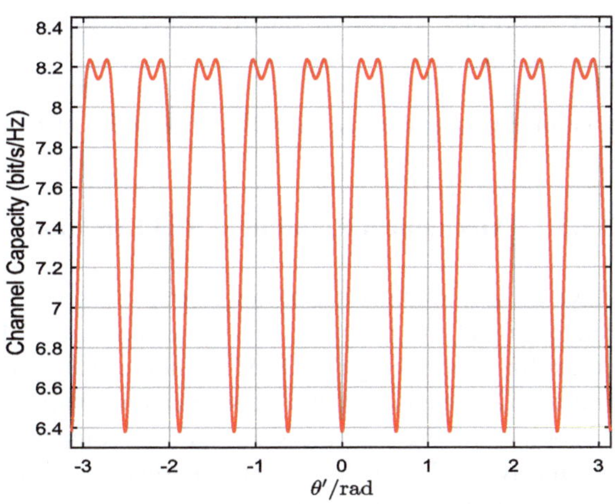

Fig. 3.9 The impact of θ' on the capacity of the effective channel $\mathbf{H}^{\text{OAM}}_{E_2\mathcal{F}_2E_1\mathcal{F}_1}$, $W = 6$ subcarriers from 3.9982 GHz to 4.2387 GHz, $U = 9$ with $\ell_u = -4, -3, \cdots, 4$, $N = 10$, $R_r = R_t = 20\lambda_1$, and $r = 450\lambda_1$

In order to see the impact of θ' on OAM channel capacity intuitively, we plot the relationship between C and θ' in Fig. 3.9. It can be seen from the figure that the OAM channel capacity varies periodically with θ' in the range of $[-\pi, \pi]$. That is because the antenna elements on the receive UCA are uniformly arranged around the center of the circle. In Fig. 3.9, $N = 10$ leads to ten cycles. Therefore, we can limit θ' to only any circle of the range to obtain the optimal rotation angle θ^* from the following optimization problem.

$$\begin{aligned} &\max \quad C \\ &s.t. \ -\tfrac{\pi}{N} \le \theta' \le \tfrac{\pi}{N} \end{aligned} \tag{3.28}$$

Since the target function C is not guaranteed to be differentiable, gradient-based procedures cannot be applied. A large number of global optimization techniques were developed to provide alternative solution methods for challenging problems, e.g., simulated annealing (SA) [12]. SA algorithm is a derivative-free promising algorithm, whose final solution quality does not strongly depend on the choice of the initial solution [1]. So, SA is adopted here to obtain the optimal rotation angle θ^* within the range $[-\frac{\pi}{N}, \frac{\pi}{N}]$. The solving procedure of the problem (3.28) is given in Algorithm 1, where T_{init} is the initial temperature, T_{min} is the lowest temperature, ζ is the cooling coefficient, J is the number of inner-layer iterations, L is used to identify whether a new value is accepted at T_{init}, C_{max} is the optimal channel capacity, C_{new} represents the new solution obtained in the algorithm, and θ_{new} represents the roll angle corresponding to C_{new}. In the end, the optimal rotation angle θ^* that maximizes the OAM channel capacity can be obtained.

Remark 1 We can observe from (3.26) and (3.28) that θ^* is only decided by $\{\eta(w), \ell_u, u = 1, \cdots, U, w = 1, \cdots, W\}$. That is to say, the mechanical beam steering $\mathcal{F}_2$ in roll direction does not depend on the electronic beam steering E_1 in pitch and yaw directions. Therefore, the procedure of hybrid mechanical and electronic beam steering can change its order from $\mathcal{F}_1, E_1, \mathcal{F}_2, E_2$ to $\mathcal{F}_1, \mathcal{F}_2, E_1, E_2$. Thus, we have $\mathbf{H}_{\mathcal{F}}(w) = \mathcal{F}_2(\mathcal{F}_1(\mathbf{H}(w), \hat{\psi}, \hat{\gamma}), \theta^*)$, which unifies (3.10) and (3.18). Therefore, the relationship between $\mathbf{B}_{E_1}(w)$, $\mathbf{B}_{E_2}(w)$, and $\mathbf{B}_E(w)$ is $\mathbf{B}_E(w) = \mathbf{B}_{E_2}(w) \odot \mathbf{B}_{E_1}(w)$.

3.4.3.2 Mechanical Rotation in Roll Direction

Having the optimal rotation angle θ^*, the servo system make the receive UCA rotate around the roll-axis. Since the mechanical beam steering in roll direction is just for maximizing the OAM channel capacity rather than aligning the transmit and receive beams, we don't consider the rotation error in θ^* for easier analysis. According to Fig. 3.2, the Cartesian coordinate of the mth antenna element on the receive UCA in $\overline{\mathrm{Z}} - \overline{\mathrm{XOY}}$ coordinate system after the mechanical rotation in roll direction, denoted by $(\breve{a}_m, \breve{b}_m, \breve{c}_m)$, can be expressed as $[\breve{a}_m, \breve{b}_m, \breve{c}_m]^T = \mathbf{R}_R(\theta^*)[R_r \cos\theta, R_r \sin\theta, 0]^T$, where $\mathbf{R}_R(\theta^*)$ represents the coordinate rotation matrix in roll direction and can be expressed as

$$\mathbf{R}_R(\theta^*) = \begin{bmatrix} \cos\theta^* & -\sin\theta^* & 0 \\ \sin\theta^* & \cos\theta^* & 0 \\ 0 & 0 & 1 \end{bmatrix}. \tag{3.29}$$

Thus, the Cartesian coordinate of the mth antenna element on the receive UCA in $\mathrm{Z}' - \mathrm{X}'\overline{\mathrm{O}}\mathrm{Y}'$ coordinate system, denoted by $(\hat{a}_m, \hat{b}_m, \hat{c}_m)$, can be expressed as $[\hat{a}_m, \hat{b}_m, \hat{c}_m]^T = \mathbf{R}_Y(\bar{\gamma})\mathbf{R}_P(\bar{\psi})[\check{a}_m, \check{b}_m, \check{c}_m]^T$, where $\mathbf{R}_P(\bar{\psi})$ and $\mathbf{R}_Y(\bar{\gamma})$ represent the axis rotation matrixes corresponding to the pitch direction and the yaw direction after the mechanical rotation in roll direction, respectively. Hence, after roll rotation, the transmission distance $\hat{d}_{m,n}$ from the nth $(1 \leq n \leq N)$ transmit antenna element to the mth $(1 \leq m \leq N)$ receive antenna element can be calculated as

$$\begin{aligned}\hat{d}_{m,n} = r &- \frac{R_r R_t}{r}\sin\tilde{\theta}\cos\phi\sin\bar{\psi}\sin\bar{\gamma} \\ &- \frac{R_r R_t}{r}\left(\cos\tilde{\theta}\cos\phi\cos\bar{\gamma} + \sin\tilde{\theta}\sin\phi\cos\bar{\psi}\right) \\ &+ R_r\left(\sin\tilde{\theta}\sin\bar{\psi}\cos\bar{\gamma} - \cos\tilde{\theta}\sin\bar{\gamma}\right),\end{aligned} \tag{3.30}$$

where $\tilde{\theta} = \theta + \theta^*$. Correspondingly, the channel matrix can be expressed as $\mathbf{H}_{\mathcal{F}}(w) = [\hat{h}_{m,n}(w)]_{N\times N}$, where the mth-row nth-column element $\hat{h}_{m,n}(w)$ can be written as

$$\begin{aligned}\hat{h}_{m,n}(w) &= \frac{\beta}{2k_w\hat{d}_{m,n}}\exp\left(-ik_w\hat{d}_{m,n}\right) \\ &\overset{(a)}{\approx} \frac{\beta}{2k_w r}\exp\Big(iS_{k_w}\sin\tilde{\theta}\cos\phi\sin\bar{\psi}\sin\bar{\gamma} + iS_{k_w}\cos\tilde{\theta}\cos\phi\cos\bar{\gamma} \\ &\quad + iS_{k_w}\sin\tilde{\theta}\sin\phi\cos\bar{\psi} - ik_w r - ik_w R_r\left(\sin\tilde{\theta}\sin\bar{\psi}\cos\bar{\gamma} - \cos\tilde{\theta}\sin\bar{\gamma}\right)\Big),\end{aligned} \tag{3.31}$$

(a) neglects a few small terms in the denominator and thus only $2k_w r$ is left. Thus, the effective multi-mode OAM channel matrix after the roll rotation can be written as

$$\begin{aligned}\mathbf{H}^{\mathrm{OAM}}_{\mathcal{F}_2 E_1 \mathcal{F}_1}(w) &= \mathbf{H}^{\mathrm{OAM}}_{E_1 \mathcal{F}_2 \mathcal{F}_1}(w) \\ &= (\mathbf{F}_U \odot \mathbf{B}_{E_1}(w))\mathbf{H}_{\mathcal{F}}(w)\mathbf{F}_U^H,\end{aligned} \tag{3.32}$$

where $h^{\mathrm{OAM}}_{\mathcal{F}_2 E_1 \mathcal{F}_1, w}(u, v) = \left(\mathfrak{b}_{E_1}(w) \odot \mathbf{f}(\ell_u)\right)\mathbf{H}_{\mathcal{F}}(w)\mathbf{f}^H(\ell_v)$ is the uth-row and vth-column element of $\mathbf{H}^{\mathrm{OAM}}_{\mathcal{F}_2 E_1 \mathcal{F}_1}(w)$ and can be calculated as

$$\begin{aligned}h^{\mathrm{OAM}}_{\mathcal{F}_2 E_1 \mathcal{F}_1, w}(u, v) = \eta(w)\sum_{m=1}^{N}\sum_{n=1}^{N}\exp\Big(&-i\ell_u\tilde{\theta} + i\ell_v\phi + iS_{k_w}\sin\tilde{\theta}\cos\phi\sin\bar{\psi}\sin\bar{\gamma} \\ + iS_{k_w}(\cos\tilde{\theta}\cos\phi\cos\bar{\gamma} + \sin\tilde{\theta}\sin\phi\cos\bar{\psi}) &- ik_w R_r\left(\sin\tilde{\theta}\sin\bar{\psi}\cos\bar{\gamma} - \cos\tilde{\theta}\sin\bar{\gamma}\right)\end{aligned}$$

$$+ik_wR_r\left(\sin\theta\sin\bar{\psi}\cos\bar{\gamma}-\cos\theta\sin\bar{\gamma}\right)\Big)$$

$$\overset{(a)}{\approx}\eta(w)\sum_{m=1}^{N}\sum_{n=1}^{N}\exp\Bigg(-i\tilde{\theta}t+ik_wR_r\cos\tilde{\theta}\sin\bar{\gamma}$$

$$-ik_wR_r\sin\tilde{\theta}\sin\bar{\psi}\cos\bar{\gamma}+ik_wR_r\sin\theta\sin\bar{\psi}\cos\bar{\gamma}-ik_wR_r\cos\theta\sin\bar{\gamma}\Bigg)$$

$$\times\sum_{\delta=1}^{N}\exp\left(i\left(\frac{2\pi\delta}{N}-\theta^*\right)\ell_v+iS_{k_w}\cos\left(\frac{2\pi\delta}{N}-\theta^*\right)\right), \tag{3.33}$$

(a) applies the approximation $\cos a\approx 1-\frac{a^2}{2}$ for $\cos\bar{\psi}$ and $\cos\bar{\gamma}$ in the case that $\bar{\psi}$ and $\bar{\gamma}$ are relatively small and neglects a few small terms under the condition $r\gg R_r$.

3.4.4 Adjusting Electronic Beam in Three Directions

It can be seen from (3.33) that the effective multi-mode OAM channel matrix after roll rotation $\mathbf{H}^{\mathrm{OAM}}_{\mathcal{F}_2E_1\mathcal{F}_1}(w)$ is not a diagonal matrix any more, because the transmit and receive beam directions are not aligned after the roll rotation. In order to solve the reemerging inter-mode interferences, the electronic beam has to be adjusted in pitch, yaw, and roll directions. The beam adjustment matrix $\mathbf{B}_{E_2}(w)$ could be designed as $\mathbf{B}_{E_2}(w)=\mathbf{1}\otimes\mathfrak{b}_{E_2}(w)$, where $\mathfrak{b}_{E_2}(w)=[e^{i\widehat{W}_1(w)},e^{i\widehat{W}_2(w)},\cdots,e^{i\widehat{W}_N(w)}]$ and $\widehat{W}_m(w)$ has the form

$$\widehat{W}_m(w)=2k_wR_r\sin\frac{\theta^*}{2}\times\left(\cos\bar{\gamma}\cos\left(\frac{\theta^*}{2}+\theta\right)\sin\bar{\psi}+\sin\bar{\gamma}\sin\left(\frac{\theta^*}{2}+\theta\right)\right), \tag{3.34}$$

$m=1,\cdots,N,w=1,\cdots,W$. After involving these phases into the original phases in $\mathbf{F}_U$ at the receive UCA, the effective multi-mode OAM channel matrix at the wth subcarrier becomes

$$\begin{aligned}\mathbf{H}^{\mathrm{OAM}}_{E_2\mathcal{F}_2E_1\mathcal{F}_1}(w)&=\mathbf{H}^{\mathrm{OAM}}_{E_2E_1\mathcal{F}_2\mathcal{F}_1}(w)\\&=\left(\mathbf{F}_U\odot\mathbf{B}_{E_2}(w)\odot\mathbf{B}_{E_1}(w)\right)\mathbf{H}_{\mathcal{F}}(w)\mathbf{F}_U^H\\&=\left(\mathbf{F}_U\odot\mathbf{B}_E(w)\right)\mathbf{H}_{\mathcal{F}}(w)\mathbf{F}_U^H=\mathbf{H}_H^{\mathrm{OAM}}(w).\end{aligned} \tag{3.35}$$

Proposition 3 *The successive implementation of the electronic beam steering matrices $\mathbf{B}_{E_1}(w)$ and $\mathbf{B}_{E_2}(w)$ can eliminate the inter-mode interferences in the channel matrix after mechanical beam steering $\mathbf{H}_{\mathcal{F}}$.*

Proof Successive implementing $\mathbf{B}_{E_1}(w)$ and $\mathbf{B}_{E_2}(w)$ to $\mathbf{H}_{\mathcal{F}}$ leads to the uth-row and vth-column element of $\mathbf{H}_H^{\mathrm{OAM}}(w)h^{\mathrm{OAM}}_{E_2E_1\mathcal{F}_2\mathcal{F}_1,w}(u,v) = h^{\mathrm{OAM}}_{H,w}(u,v) = \left(\mathfrak{b}_{E_2}(w)\odot \mathfrak{b}_{E_1}(w)\odot \mathbf{f}(\ell_u)\right)\mathbf{H}_{\mathcal{F}}(w)\mathbf{f}^H(\ell_v)$ being expressed as

$$
\begin{aligned}
h^{\mathrm{OAM}}_{H,w}(u,v) &= \eta(w)\sum_{m=1}^{N}\sum_{n=1}^{N}\exp\Bigg(-i\ell_u\tilde{\theta}+i\ell_v\phi+iS_{k_w}\sin\tilde{\theta}\cos\phi\sin\bar{\psi}\sin\bar{\gamma}\\
&\quad +iS_{k_w}(\cos\tilde{\theta}\cos\phi\cos\bar{\gamma}+\sin\tilde{\theta}\sin\phi\cos\bar{\psi})\\
&\quad -ik_wR_r\left(\sin\tilde{\theta}\sin\bar{\psi}\cos\bar{\gamma}-\cos\tilde{\theta}\sin\bar{\gamma}\right)\\
&\quad +ik_wR_r\left(\sin\theta\sin\bar{\psi}\cos\bar{\gamma}-\cos\theta\sin\bar{\gamma}\right)\\
&\quad +2ik_wR_r\sin\frac{\theta^*}{2}\left(\cos\bar{\gamma}\cos\left(\frac{\theta^*}{2}+\theta\right)\sin\bar{\psi}+\sin\bar{\gamma}\sin\left(\frac{\theta^*}{2}+\theta\right)\right)\Bigg)\\
&\overset{(a)}{\approx}\eta(w)\sum_{m=1}^{N}\exp\left(-i\tilde{\omega}t\right)\times\sum_{\delta=1}^{N}\exp\left(i\left(\frac{2\pi\delta}{N}-\theta^*\right)\ell_v+iS_{k_w}\cos\left(\frac{2\pi\delta}{N}-\theta^*\right)\right),
\end{aligned}
\tag{3.36}
$$

where (a) applies the approximation $\cos a\approx 1-\frac{a^2}{2}$ for $\cos\bar{\psi}$ and $\cos\bar{\gamma}$ in the case that $\bar{\psi}$ and $\bar{\gamma}$ are relatively small and neglects a few small terms under the condition $r\gg R_r$. According to (3.25), it is apparent that the uth diagonal element $h^{\mathrm{OAM}}_{H,w}(u,u)$ of $\mathbf{H}_H^{\mathrm{OAM}}(w)$ can be expressed as

$$
h^{\mathrm{OAM}}_{H,w}(u,u) = N\eta(w)\times\sum_{\delta=1}^{N}\exp\left(i\left(\frac{2\pi\delta}{N}-\theta^*\right)\ell_u+iS_{k_w}\cos\left(\frac{2\pi\delta}{N}-\theta^*\right)\right),
\tag{3.37}
$$

and the non-diagonal elements of $\mathbf{H}_H^{\mathrm{OAM}}(w)$ approximately equal to zero. It follows that the successive implementation of the electronic beam steering matrices $\mathbf{B}_{E_1}(w)$ and $\mathbf{B}_{E_2}(w)$ can eliminate the inter-mode interferences.

Remark 2 According to Proposition 3, the procedure of the hybrid mechanical and electronic beam steering for the LoS MCMM-OAM receiver can be simplified from the four steps, $\mathcal{F}_1$, E_1, $\mathcal{F}_2$, and E_2, illustrated in Fig. 3.6 to only the two steps: mechanical beam steering in pitch, yaw, and roll directions $\mathcal{F}=\mathcal{F}_2(\mathcal{F}_1(\cdot))$ and electronic beam steering in pitch, yaw, and roll directions $\mathbf{B}_E(w)=\mathbf{B}_{E_2}(w)\odot\mathbf{B}_{E_1}(w)$.

3.4.5 Complexity Analysis

The specific computational complexity of the proposed hybrid mechanical and electronic beam steering scheme is compared with that of traditional electronic beam steering in the Table 3.1. For LoS MCMM-OAM communication systems, whether the coarse distance and AoA estimation method or the refined distance and AoA estimation method, their complexities are both determined by the complexity of the EVD in (16) of reference [4]. However, the complexity of the coarse distance and AoA estimation method is affected by the values of $\overline{W}$ and $\overline{U}$, and factors that affect the complexity of the refined distance and AoA estimation method are the values of $\widetilde{W}$ and $\widetilde{U}$. Moreover, the complexity of the electronic beam steering is determined by the complexity of the multiplying the received signal vector by $(\mathbf{F}_U \odot \mathbf{B}_E(w))$. It follows that the values of W, U, and N decide the complexity of the electronic beam steering. The complexity of $\mathcal{F}_1$ is determined by the complexity of the mechanical rotation in pitch and yaw directions, and the value of $\hat{\psi}$, $\hat{\gamma}$, and ν decide the complexity of the $\mathcal{F}_1$. Meanwhile, the complexity of the $\mathcal{F}_2$ is determined by the complexity of the mechanical rotation in roll direction corresponding to the value of θ^* and ν. The complexity of the optimization algorithm is determined by the values of J, ζ, T_{init}, and $T_{\min}$. The total computational complexity comparison between the hybrid mechanical and electronic beam steering and electronic beam steering scheme is shown in Fig. 3.10. It can be seen from the Fig. 3.10 that due to the coarse AoA estimation using fewer OAM modes and subcarriers, mechanical beam steering and optimization having relatively low complexity, the total complexity of the proposed hybrid beam steering scheme is only slightly higher than that of traditional electronic beam steering. Furthermore, we calculate

Table 3.1 The complexity comparison between hybrid mechanical and electronic beam steering and electronic beam steering

Scheme	Complexity	
Hybrid BS	Coarse AoA estimation	$O\left(\overline{W}^3\overline{U}^3\right)$
	Mechanical beam steering $\mathcal{F}_1$	$O\left(\frac{\hat{\psi}+\hat{\gamma}}{\nu}\right)$
	Optimization algorithm	$O\left(J\log_{\zeta}\frac{T_{\min}}{T_{\max}}\right)$
	Mechanical beam steering $\mathcal{F}_2$	$O\left(\frac{\theta^*}{\nu}\right)$
	Refined AoA estimation	$O(\widetilde{W}^3\widetilde{U}^3)$
	Electronic beam steering	$O(WUN^2)$
Electronic BS	AoA estimation	$O(\widetilde{W}^3\widetilde{U}^3)$
	Electronic beam steering	$O(WUN^2)$

BS Beam steering

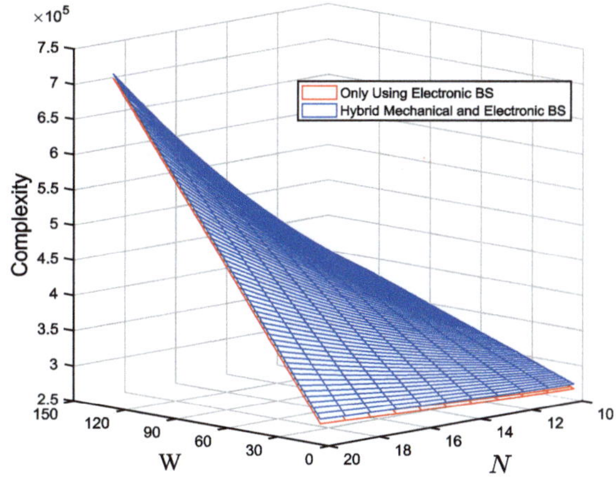

Fig. 3.10 The complexities of the hybrid mechanical and electronic beam steering and electronic beam steering. vs. N and W at $\overline{W}=4$, $\widetilde{W}=8$, $U=9$, $\overline{U}=4$, $\widetilde{U}=8$, $J=20$, $\zeta=0.9$, $T_{\min}=0.001$ and $T_{\text{init}}=100$. BS: Beam steering

the relative complexity from the proposed hybrid beam steering to electronic beam steering as $1+O\left(\frac{\overline{W}^3\overline{U}^3}{\widetilde{W}^3\widetilde{U}^3}\right)$. To show their difference, we select parameters as $\overline{W}=4$, $\widetilde{W}=8$, $U=9$, $\overline{U}=4$, $\widetilde{U}=8$, $J=20$, $\zeta=0.9$, $T_{\min}=0.001$, $T_{\text{init}}=100$, $\nu=0.3°$, $\psi=\gamma=60°$, $N=10$, and $\theta^*=10°$. With these parameters, the complexity of the hybrid beam steering is 1.009 to 1.026 times that of the electronic beam steering.

3.5 Numerical Simulations and Results

In this section we show the performance of the proposed hybrid mechanical and electronic beam steering method by numerical simulations. We first verify the effectiveness of SA algorithm in solving the optimal solution θ^* and show the convergence rate of the algorithm. Then, we compare the channel capacities of the UCA-based LoS MCMM-OAM system with the proposed hybrid beam steering and that with only electronic beam steering. Unless otherwise stated, the SNRs in all the figures are defined as the ratio of the received signal power versus the noise power. In the simulations, we choose $W=8$ subcarriers from 3.9982 GHz to 4.2387 GHz, $\overline{W}=4$, $\widetilde{W}=8$, $U=9$ with $\ell_v\in[-4,4]$, $\overline{U}=4$, $\widetilde{U}=8$, $R_r=R_t=20\lambda_1$, $r=450\lambda_1$, $\lambda_1=2\pi/k_1$, $N=10$, $\nu=0.3°$.

Figure 3.11a shows the solution of the optimization problem (3.28). It can be seen from the figure that the channel capacity with obtained rotation angle θ by Algorithm 1 reaches its optimum. In Fig. 3.11b, we show the convergence rate of Algorithm 1 with the number of outer-layer iterations, where $T_{\text{init}}=100$, $T_{\min}=10^{-3}$ and $\zeta=0.9$. It can be clearly seen that through reasonably setting the parameters of SA algorithm, the target function has a fast convergence rate.

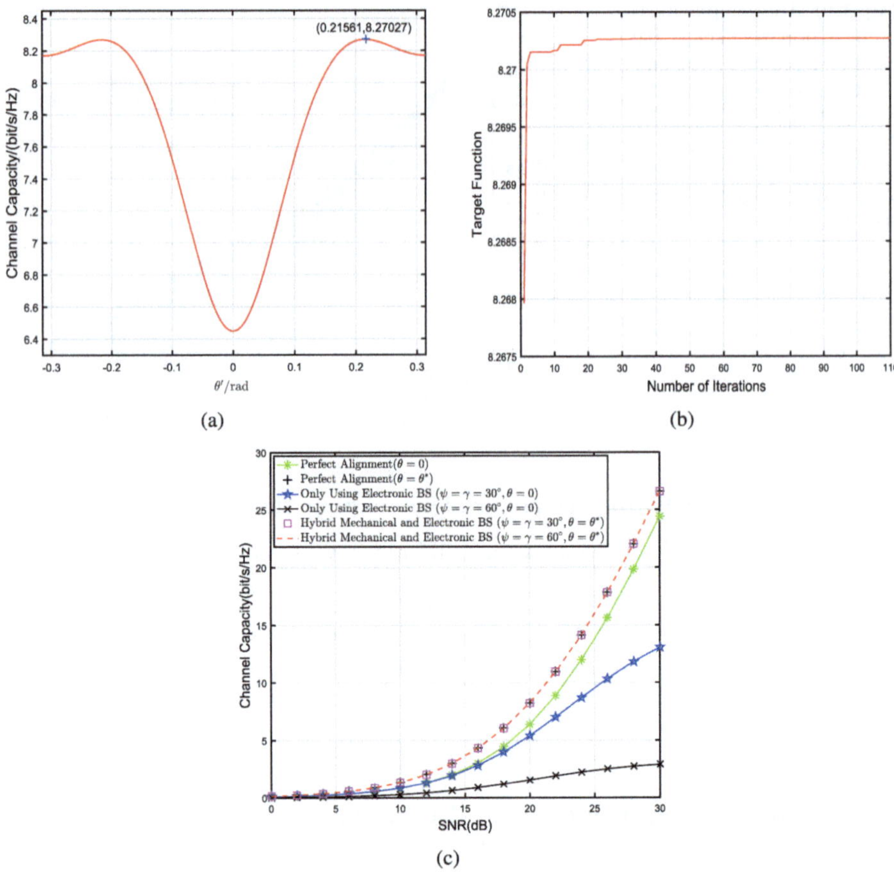

Fig. 3.11 (**a**) The obtained optimal rotation angle θ^* by Algorithm 1 in the range of $[-\pi/10, \pi/10]$. (**b**) The convergence rate of Algorithm 1. (**c**) The channel capacity of the UCA-based LoS MCMM-OAM system. BS: Beam steering

In Fig. 3.11c, we illustrate the channel capacities of the proposed hybrid mechanical and electronic beam steering scheme and the channel capacities of only using electronic beam steering method under different pitch, yaw, and roll rotation angles. Also, the channel capacities in the case of perfect alignment are given for clearer comparison. It can be seen from the results that for the electronic beam steering method, the larger the oblique angles, the lower the channel capacity, while the proposed hybrid mechanical and electronic beam steering scheme can effectively eliminate the effect of large misalignment errors of any practical OAM channel and approaches the performance of perfect aligned OAM channel.

Although we performed simulations at 4 GHz, the proposed hybrid mechanical and electronic beam steering scheme is applicable to all frequency bands because the channel capacity in (3.27) is dependent on the two terms $\eta(w)$ and S_{k_w} related to frequency in (3.26). Recall that $\eta(w) = \frac{\beta}{2k_w rN}\exp(-ik_w r)$, $S_{k_w} = k_w\frac{R_r R_t}{r}$, and $k_w = \frac{2\pi}{\lambda_w}$; we set R_t, R_r, and r to be different times of λ_1 in the simulations. The difference of λ_1 and λ_w is so small within a certain bandwidth that the values of $\eta(w)$ and S_{k_w} do not vary much with frequency. The simulation results in higher frequencies used in 5G and 6G are similar to those at 4 GHz, which are omitted here due to limited space.

3.6 Summary

In this chapter, a hybrid mechanical and electronic beam steering scheme is proposed for LoS multi-carrier and multi-mode OAM communication systems to deal with the capacity loss encountered by electronic beam steering in large misalignment condition. To intuitively reveal the motivation of the proposed hybrid beam steering scheme, we show from both mathematical proof and numerical simulations that only using the electronic beam steering method, the larger the oblique angles, the lower the receive SINR and OAM channel capacity. In our proposed hybrid beam steering scheme, mechanical rotating devices controlled by PWM signals are utilized to eliminate the large misalignment angle, while electronic beam steering is in charge of the remaining small misalignment angle caused by perturbations. The whole procedure is designed as four consecutive steps including mechanical beam steering in pitch and yaw directions, electronic beam steering in pitch and yaw directions, mechanical rotation in roll direction, and adjusting electronic beam in pitch, yaw, and roll directions. As the physical beam steering in mechanical way is nonlinear operation for channel matrix, while electronic beam steering in digital way is linear operation, we show that the four steps of hybrid beam steering can be simplified to only two steps, i.e., mechanical beam steering in pitch, yaw, and roll directions and electronic beam steering in these three directions. Numerical simulations show that searching for the optimal roll rotation angle only requires no more than 30 iterations with the SA algorithm, and the proposed hybrid beam steering scheme can effectively eliminate the effect of large misalignment errors of any practical OAM channel approaching the performance of perfect aligned OAM channel.

It is worth noting that for the OAM communication system in the large off-axis and more general misalignment cases, the proposed hybrid mechanical and electronic beam steering scheme is still applicable with the same idea in [2] that beam steering is performed at both receiver and transmitter. However, the specific procedure needs to be further developed, which can also be used in mmWave massive MIMO systems. Moreover, the directivity of the antenna units will affect the rotation angle θ. Although in this manuscript the antenna elements are assumed to be ideally omnidirectional, we believe that it is of practical significance to study how the directivity of the antenna units affects the rotation angle θ, which is left for our future work.

Appendix

On one hand, when $\psi = 0$ and $\gamma \neq 0$, the uth diagonal element $\tilde{h}_{\mathrm{OAM},w}(u,u)$ of $\widetilde{\mathbf{H}}_{\mathrm{OAM}}(w)$ can be obtained as

$$\tilde{h}_{\mathrm{OAM},w}(u,u) = \eta(w) \times \sum_{\delta=1}^{N} \exp\left(i\frac{2\pi\delta}{N}\ell_u + \frac{1}{2}iS_{k_w}(1+\cos\gamma)\cos\frac{2\pi\delta}{N}\right)\varsigma(m), \tag{3.38}$$

where $\varsigma(m)$ is defined as

$$\varsigma(m) = \sum_{m=1}^{N} \exp\left(\frac{1}{2}iS_{k_w}(\cos\gamma - 1)\cos\frac{2\pi(2m+\delta-2)}{N}\right). \tag{3.39}$$

When S_{k_w} goes to zero, $\varsigma(m) \approx N$. Therefore, $\tilde{h}_{\mathrm{OAM},w}(u,u)$ can be approximated as

$$\tilde{h}_{\mathrm{OAM},w}(u,u) \approx N\eta(w) \sum_{\delta=1}^{N} \exp\left(i\frac{2\pi\delta}{N}\ell_u + \frac{1}{2}iS_{k_w}(1+\cos\gamma)\cos\frac{2\pi\delta}{N}\right). \tag{3.40}$$

According to [4], we can substitute the Taylor series expansion of the exponential function $e^x = \sum_{g=0}^{+\infty}(x^g)/g!$ into (3.40) and eventually have $\tilde{h}_{\mathrm{OAM},w}(u,u) \approx \eta(w)\frac{N^2}{2^\tau}\cdot\frac{i^\tau}{\tau!}\left(\frac{1}{2}S_{k_w}(1+\cos\gamma)\right)^\tau$, where $\tau = \min\{|\ell_u|, N-|\ell_u|\}$.

$$\begin{aligned}
\bar{\varsigma}(m) &= \sum_{m=1}^{N} \exp\left(-i\frac{2\pi(m-1)}{N}t\right)\exp\left(\frac{1}{2}iS_{k_w}(\cos\gamma-1)\cos\frac{2\pi(2m+\delta-2)}{N}\right) \\
&= \sum_{g=0}^{+\infty}\frac{i^g}{2^g g!}\sum_{l=0}^{g}\binom{g}{l}\exp\left(i\frac{2\pi t}{N} + i(2l-g)\frac{2\pi\,(\delta-2)}{N}\right) \\
&\quad \times \sum_{m=1}^{N}\exp\left(i(2l-g-\frac{t}{2})\frac{2\pi(2m)}{N}\right)\left(\frac{1}{2}S_{k_w}(\cos\gamma-1)\right)^g.
\end{aligned} \tag{3.41}$$

On the other hand, when $\psi = 0$ and $\gamma \neq 0$, the uth-row and vth-column element $\tilde{h}_{\mathrm{OAM},w}(u,v)$ of $\widetilde{\mathbf{H}}_{\mathrm{OAM}}(w)$ can be obtained as

$$\tilde{h}_{\mathrm{OAM},w}(u,v) = \eta(w) \sum_{\delta=1}^{N} \exp\left(i \frac{2\pi\delta}{N}\ell_v + \frac{1}{2} i S_{k_w}(1+\cos\gamma)\cos\frac{2\pi\delta}{N} \right) \bar{\varsigma}(m), \tag{3.42}$$

where $\bar{\varsigma}(m)$ is defined as

$$\bar{\varsigma}(m) = \sum_{m=1}^{N} \exp\left(\frac{1}{2} i S_{k_w} \cos\gamma \cos\frac{2\pi(2m+\delta-2)}{N} - \frac{1}{2} i S_{k_w} \cos\frac{2\pi(2m+\delta-2)}{N} - i\frac{2\pi(m-1)t}{N} \right). \tag{3.43}$$

We substitute $e^x = \sum_{g=0}^{+\infty}(x^g)/g!$ into $\bar{\varsigma}(m)$ and obtain (3.41). For the last summation in (3.41), it is easy to verify that

$$\sum_{m=1}^{N} \exp\left(i(2l - g - \frac{t}{2})\frac{4\pi m}{N} \right) = \begin{cases} 0, & \mathrm{Mod}_N(2l-g-\frac{t}{2}) \neq 0; \\ N, & \mathrm{Mod}_N(2l-g-\frac{t}{2}) = 0. \end{cases} \tag{3.44}$$

When $0 \leq g \leq \min\left\{|\frac{t}{2}|, N-|\frac{t}{2}|\right\}$, we always have (recall that $0 \leq l \leq g$) $-N < 2l - g - \frac{t}{2} < N$ according to (34) of [4]. When $0 \leq g < \min\left\{|\frac{t}{2}|, N-|\frac{t}{2}|\right\}$, (3.44) is always zero, i.e., the Taylor series of $\tilde{h}_{\mathrm{OAM},w}(u,v)$ does not contain the terms $S_{k_w}^g$, and when $g = \min\left\{|\frac{t}{2}|, N-|\frac{t}{2}|\right\}$, (3.44) equals to N if and only if $l = g (t \geq 0)$; $l = 0 (t < 0)$. Furthermore, when $g > \min\left\{|\frac{t}{2}|, N-|\frac{t}{2}|\right\}$, the coefficient of $S_{k_w}^g$ is much smaller due to the term $\frac{1}{2^g g!}$, and hence the term $\{S_{k_w}^g | g > \min\left\{|\frac{t}{2}|, N-|\frac{t}{2}|\right\}$ is counted into the higher-order infinitesimal of $S_{k_w}^g$ denoted as $o\{S_{k_w}^g\}$ as S_{k_w} goes to zero. Thus, (3.43) can be approximated as

$$\bar{\varsigma}(m) \approx \frac{N}{2^{\bar{\tau}}} \cdot \frac{i^{\bar{\tau}}}{\bar{\tau}!} \exp\left(i\frac{\pi\delta t}{N} \right) \left(\frac{1}{2} S_{k_w}(\cos\gamma - 1) \right)^{\bar{\tau}}, \tag{3.45}$$

where $\bar{\tau} = \min\left\{|\frac{t}{2}|, N-|\frac{t}{2}|\right\}$. Therefore, (3.42) can be approximated as

$$\tilde{h}_{\mathrm{OAM},w}(u,v) \approx \eta(w) \frac{N}{2^{\bar{\tau}}} \cdot \frac{i^{\bar{\tau}}}{\bar{\tau}!} \left(\frac{1}{2} S_{k_w}(\cos\gamma - 1) \right)^{\bar{\tau}} \times \sum_{\delta=1}^{N} \exp\left(i\frac{2\pi\delta}{N}(\ell_v + \frac{t}{2}) + \frac{1}{2} i S_{k_w} \cos\frac{2\pi\delta}{N} + \frac{1}{2} i S_{k_w} \cos\gamma \cos\frac{2\pi\delta}{N} \right). \tag{3.46}$$

Then, similar to the approximation of the summation term with Taylor series expansion in (3.40), (3.46) can be further approximated as

$$\begin{aligned}\tilde{h}_{\mathrm{OAM},w}(u,v) \approx& \eta(w)(-1)^{\bar{\tau}} \frac{N^2}{4^{(\bar{\tau}+\chi)}} \cdot \frac{i^{(\bar{\tau}+\chi)}}{\bar{\tau}!\chi!} \cdot S_{k_w}{}^{(\bar{\tau}+\chi)} \\ &\times (1-\cos\gamma)^{\bar{\tau}}(1+\cos\gamma)^{\chi},\end{aligned} \tag{3.47}$$

where $\chi = \min\left\{|\ell_v + \frac{t}{2}|, N - |\ell_v + \frac{t}{2}|\right\}$.

In order to analyze how $\mathrm{SIR}_{EO}(w,u)$ changes with the increase of $|\gamma|$ when $\psi = 0$, we first express the reciprocal of $\mathrm{SIR}_{EO}(w,u)$ when $\ell_u \neq 0$ as

$$\begin{aligned}\frac{1}{\mathrm{SIR}_{EO}(w,u)} =& \frac{\sum\limits_{u\neq v}\left|\tilde{h}_{\mathrm{OAM},w}(u,v)\right|^2}{\left|\tilde{h}_{\mathrm{OAM},w}(u,u)\right|^2} \\ =& \left.\frac{\left|\tilde{h}_{\mathrm{OAM},w}(u,v)\right|^2}{\left|\tilde{h}_{\mathrm{OAM},w}(u,u)\right|^2}\right|_{\ell_v=0} + \sum_{\ell_u\ell_v<0}\frac{\left|\tilde{h}_{\mathrm{OAM},w}(u,v)\right|^2}{\left|\tilde{h}_{\mathrm{OAM},w}(u,u)\right|^2} \\ &+ \sum_{\substack{\ell_u\ell_v>0\\ u\neq v}}\frac{\left|\tilde{h}_{\mathrm{OAM},w}(u,v)\right|^2}{\left|\tilde{h}_{\mathrm{OAM},w}(u,u)\right|^2}.\end{aligned} \tag{3.48}$$

It is obvious that (3.48) consists of three terms and the derivative of (3.48) is the sum of their derivatives. Moreover, the derivative of $\frac{\left|\tilde{h}_{\mathrm{OAM},w}(u,v)\right|^2}{\left|\tilde{h}_{\mathrm{OAM},w}(u,u)\right|^2}$ is given as

$$\frac{\partial\left(\frac{\left|\tilde{h}_{\mathrm{OAM},w}(u,v)\right|^2}{\left|\tilde{h}_{\mathrm{OAM},w}(u,u)\right|^2}\right)}{\partial\gamma} = T(u,v,\gamma)\left(\kappa(u,v,\gamma) + \bar{\kappa}(u,v,\gamma)\right), \tag{3.49}$$

where

$$T(u,v,\gamma) = \frac{\frac{1}{4^{2(\bar{\tau}+\chi)}} \cdot \frac{1}{(\bar{\tau}!\chi!)^2} \cdot S_{k_w}{}^{2(\bar{\tau}+\chi)} \sin\gamma}{\frac{1}{4^{2\tau}} \cdot \frac{S_{k_w}{}^{2\tau}}{(\tau!)^2}(1+\cos\gamma)^{4\tau}}, \tag{3.50}$$

$$\begin{aligned}\kappa(u,v,\gamma) =& (1-\cos\gamma)^{2\bar{\tau}-1}(1+\cos\gamma)^{2\chi+2\tau-1} \\ &\times \left(2(\bar{\tau}-\chi) + 2(\bar{\tau}+\chi)\cos\gamma\right),\end{aligned} \tag{3.51}$$

$$\bar{\kappa}(u, v, \gamma) = 2\tau(1 - \cos\gamma)^{2\bar{\tau}}(1 + \cos\gamma)^{2\chi+2\tau-1}, \tag{3.52}$$

$\bar{\tau} - \chi = \frac{|\ell_u - \ell_v|}{2} - \frac{|\ell_u + \ell_v|}{2}$ and $\bar{\tau} + \chi = \frac{|\ell_u - \ell_v|}{2} + \frac{|\ell_u + \ell_v|}{2}$.

When $0 < \gamma \le \frac{\pi}{2}$, $T(u, v, \gamma) > 0$, thus the sign of the derivative of $\frac{\left|\tilde{h}_{\mathrm{OAM},w}(u,v)\right|^2}{\left|\tilde{h}_{\mathrm{OAM},w}(u,u)\right|^2}$ depends on $\kappa(u, v, \gamma)$ and $\bar{\kappa}(u, v, \gamma)$. We can find that when $\ell_v = 0$, $\bar{\tau} - \chi = 0$ and $\bar{\tau} + \chi = |\ell_u|$, thus $\kappa(u, v, \gamma) + \bar{\kappa}(u, v, \gamma)$ can be written as

$$\begin{aligned}\kappa(u, v, \gamma) + \bar{\kappa}(u, v, \gamma) =& (1 - \cos\gamma)^{2\bar{\tau}-1}(1 + \cos\gamma)^{2\chi+2\tau-1} \\ & \times \big(2(|\ell_u| - \tau)\cos\gamma + 2\tau\big).\end{aligned} \tag{3.53}$$

Because $|\ell_u| \ge \tau$, $\kappa(u, v, \gamma) + \bar{\kappa}(u, v, \gamma) > 0$. Hence, the derivative of $\left.\frac{\left|\tilde{h}_{\mathrm{OAM},w}(u,v)\right|^2}{\left|\tilde{h}_{\mathrm{OAM},w}(u,u)\right|^2}\right|_{\ell_v=0}$ is greater than zero so that $\left.\frac{\left|\tilde{h}_{\mathrm{OAM},w}(u,v)\right|^2}{\left|\tilde{h}_{\mathrm{OAM},w}(u,u)\right|^2}\right|_{\ell_v=0}$ is increasing with γ; when $\ell_u\ell_v < 0$, $\bar{\tau} - \chi = \min\{|\ell_u|, |\ell_v|\}$ and $\bar{\tau} + \chi = \max\{|\ell_u|, |\ell_v|\}$, thus $\kappa(u, v, \gamma) + \bar{\kappa}(u, v, \gamma)$ can be written as

$$\begin{aligned}&\kappa(u, v, \gamma) + \bar{\kappa}(u, v, \gamma) \\ &= (1 - \cos\gamma)^{2\bar{\tau}-1}(1 + \cos\gamma)^{2\chi+2\tau-1} \times \Bigg(2(\min\{|\ell_u|, |\ell_v|\}) \\ &\quad + 2(\max\{|\ell_u|, |\ell_v|\})\cos\gamma + 2\tau(1 - \cos\gamma)\Bigg).\end{aligned} \tag{3.54}$$

It can be seen from (3.54) that $\kappa(u, v, \gamma) + \bar{\kappa}(u, v, \gamma) > 0$. Therefore, the derivatives of every term of $\sum_{\ell_u\ell_v<0} \frac{\left|\tilde{h}_{\mathrm{OAM},w}(u,v)\right|^2}{\left|\tilde{h}_{\mathrm{OAM},w}(u,u)\right|^2}$ are greater than zero so that the term $\sum_{\ell_u\ell_v<0} \frac{\left|\tilde{h}_{\mathrm{OAM},w}(u,v)\right|^2}{\left|\tilde{h}_{\mathrm{OAM},w}(u,u)\right|^2}$ is increasing with γ; when $\ell_u\ell_v > 0$, $\bar{\tau} - \chi = -\min\{|\ell_u|, |\ell_v|\}$ and $\bar{\tau} + \chi = \max\{|\ell_u|, |\ell_v|\}$. We can find that when $|\ell_u| < |\ell_v|$, $\kappa(u, v, \gamma) + \bar{\kappa}(u, v, \gamma)$ can be expressed as

$$\begin{aligned}&\kappa(u, v, \gamma) + \bar{\kappa}(u, v, \gamma) \\ &= 2(1 - \cos\gamma)^{2\bar{\tau}-1}(1 + \cos\gamma)^{2\chi+2\tau-1}\cos\gamma(|\ell_v| - |\ell_u|),\end{aligned} \tag{3.55}$$

when $|\ell_u| > |\ell_v|$, $\kappa(u, v, \gamma) + \bar{\kappa}(u, v, \gamma)$ can be expressed as

$$\begin{aligned}&\kappa(u, v, \gamma) + \bar{\kappa}(u, v, \gamma) \\ &= 2(1 - \cos\gamma)^{2\bar{\tau}-1}(1 + \cos\gamma)^{2\chi+2\tau-1}(|\ell_u| - |\ell_v|).\end{aligned} \tag{3.56}$$

Thus, when $\ell_u\ell_v > 0$, $\kappa(u, v, \gamma) + \bar{\kappa}(u, v, \gamma) > 0$. Therefore, the derivatives of every term of $\sum_{\substack{\ell_u\ell_v>0\\u\neq v}} \frac{\left|\tilde{h}_{\mathrm{OAM},w}(u,v)\right|^2}{\left|\tilde{h}_{\mathrm{OAM},w}(u,u)\right|^2}$ are greater than zero so that $\sum_{\substack{\ell_u\ell_v>0\\u\neq v}} \frac{\left|\tilde{h}_{\mathrm{OAM},w}(u,v)\right|^2}{\left|\tilde{h}_{\mathrm{OAM},w}(u,u)\right|^2}$ is increasing with γ.

When $\ell_u = 0$, the reciprocal of $\mathrm{SIR}_{EO}(w, u)$ can be written as

$$\frac{1}{\mathrm{SIR}_{EO}(w, u)} = \sum_{\ell_v<0} \frac{\left|\tilde{h}_{\mathrm{OAM},w}(u, v)\right|^2}{\left|\tilde{h}_{\mathrm{OAM},w}(u, u)\right|^2} + \sum_{\ell_v>0} \frac{\left|\tilde{h}_{\mathrm{OAM},w}(u, v)\right|^2}{\left|\tilde{h}_{\mathrm{OAM},w}(u, u)\right|^2}. \tag{3.57}$$

It is obvious that (3.57) consists of two terms and the derivative of (3.57) is the sum of their derivatives. Moreover, when $\ell_u = 0$, $\bar{\kappa}(u, v, \gamma) = 0$. When $\ell_v < 0$ or $\ell_v > 0$, $\bar{\tau} - \chi = 0$ and $\bar{\tau} + \chi = |\ell_v|$, thus

$$\begin{aligned}\kappa(u, v, \gamma) + \bar{\kappa}(u, v, \gamma) &= \kappa(u, v, \gamma)\\ &= 2|\ell_v| \cos\gamma (1 - \cos\gamma)^{2\bar{\tau}-1}(1 + \cos\gamma)^{2\chi+2\tau-1}.\end{aligned} \tag{3.58}$$

Hence, the derivatives of every term of $\sum_{\ell_v<0} \frac{\left|\tilde{h}_{\mathrm{OAM},w}(u,v)\right|^2}{\left|\tilde{h}_{\mathrm{OAM},w}(u,u)\right|^2}$ and $\sum_{\ell_v>0} \frac{\left|\tilde{h}_{\mathrm{OAM},w}(u,v)\right|^2}{\left|\tilde{h}_{\mathrm{OAM},w}(u,u)\right|^2}$ are greater than zero so that $\sum_{\ell_v<0} \frac{\left|\tilde{h}_{\mathrm{OAM},w}(u,v)\right|^{2\,2}}{\left|\tilde{h}_{\mathrm{OAM},w}(u,u)\right|}$ and $\sum_{\ell_v>0} \frac{\left|\tilde{h}_{\mathrm{OAM},w}(u,v)\right|^2}{\left|\tilde{h}_{\mathrm{OAM},w}(u,u)\right|^2}$ are increasing with γ.

Therefore, we get the conclusion that $\mathrm{SIR}_{EO}(w, u)$ is decreasing with γ when $0 < \gamma \leq \frac{\pi}{2}$. On the contrary, we can prove in the same way that $\mathrm{SIR}_{EO}(w, u)$ is increasing with γ when $-\frac{\pi}{2} \leq \gamma < 0$. Overall, Proposition 1 that $\mathrm{SIR}_{EO}(w, u)$ is decreasing with $|\gamma|$ is proved.

References

1. M.A. Abido, Robust design of multimachine power system stabilizers using simulated annealing. IEEE Trans. Energy conversion **15**(3), 297–304 (2000)
2. R. Chen, H. Xu, M. Moretti, J. Li, Beam steering for the misalignment in UCA-based OAM communication systems. IEEE Wireless Commun. Lett. **7**(4), 582–585 (2018)
3. R. Chen, W. Yang, H. Xu, J. Li, A 2-D FFT-based transceiver architecture for OAM-OFDM systems with UCA antennas. IEEE Trans. Veh. Technol. **67**(6), 5481–5485 (2018)
4. R. Chen, W.-X. Long, X. Wang, J. Li, Multi-mode OAM radio waves: Generation, angle of arrival estimation and reception with UCAs. IEEE Trans. Wirel. Commun. **19**(10), 6932–6947 (2020)

5. R. Chen, H. Zhou, M. Moretti, X. Wang, J. Li, Orbital angular momentum waves: Generation, detection and emerging applications. IEEE Commun. Surveys Tuts. **22**(2), 840–868, 2nd Quart. (2020)
6. M. Jian, Y. Chen, G. Yu, Non-coaxial OAM: Precoding design, misaligned parameter estimation and capacity compensation. IEEE Access **9**, 37726–37738 (2021)
7. W.-X. Long, R. Chen, M. Moretti, J. Li, AoA estimation for OAM communication systems with mode-frequency multi-time ESPRIT method. IEEE Trans. Veh. Technol. **99**(1), 5094–5098 (2021)
8. N. Pinckney, Pulse-width modulation for microcontroller servo control. IEEE Potentials **25**(1), 27–29 (2006)
9. Y. Rahmat-Samii, Useful coordinate transformations for antenna applications. IEEE Trans. Antennas Propagat. **27**(4), 571–574 (1979)
10. M.F.N. Tajuddin, N.H. Ghazali, M.F. Mohammed, B. Ismail, Z.M. Isa, T.C. Siong, N. Ghazali, TMS320F2812 digital signal processor (DSP) implementation of DPWM, in *Proceedings of the IEEE Student Conference on Research and Development (SCOReD)* (2009), pp. 142–145
11. Z. Tian, R. Chen, W.-X. Long, H. Zhou, M. Moretti, Broadband beam steering for misaligned multi-mode OAM communication systems. J. Syst. Eng. Electron. **32**(4), 779–788 (2021)
12. S. Xavier-de Souza, J.A.K. Suykens, J. Vandewalle, D. Bolle, Coupled simulated annealing. IEEE Trans. Syst. Man Cybern. B **40**(2), 320–335 (2010)
13. Y. Yagi, H. Sasaki, T. Yamada, D. Lee, 200 Gb/s wireless transmission using dual-polarized OAM-MIMO multiplexing with uniform circular array on 28 GHz band. IEEE Antennas and Wireless Propag. Lett. **20**(5), 833–837 (2021)
14. W. Zhang, S. Zheng, X. Hui, R. Dong, X. Jin, H. Chi, X. Zhang, Mode division multiplexing communication using microwave orbital angular momentum: An experimental study. IEEE Trans. Wirel. Commun. **16**(2), 1308–1318 (2017)

Chapter 4
Hybrid Circular Array and Luneberg Lens for Point-to-Point Long-Distance OAM Wireless Communications

4.1 Introduction

The highest data rate of OAM wireless communications has been continuously broken. In 2021, a dual-polarized OAM-MIMO system developed by NTT has realized 200 Gbps wireless data rate at 28 GHz band [32]. In 2022, Tsinghua University successfully completed the world's first 1 Tbps high-speed OAM relay transmission demonstration and verification experiment at W-band [2]. So far, there are a number of approaches to generate radio OAM beams [10], e.g., SPP [27], helicoidal parabolic antenna, [19] and uniform circular array (UCA) [20]. Since antenna arrays are popular in modern wireless communication systems (e.g., 4G, 5G, and 6G), UCA is the most commonly used method to generate OAM beams [9].

However, there are still big challenges for achieving the expected extremely high data rate with orthogonal multiplexed OAM beams in a long-distance communication link. On the one hand, the integer-order OAM beams are divergent, and the larger the OAM mode is, the larger the beam divergence angle becomes. Thus, after a long-distance propagation, to ensure that the enough energy of divergent OAM beams is received, OAM beams can only be received through UCA with very large aperture. On the other hand, different integer OAM modes result in the large difference in the received signal energies. Especially, high-order integer OAM modes result in little received signal energy and low SNR at the receive UCA with a limited aperture. Therefore, reducing the divergence angle of the generated OAM beams is very important to increase the wireless transmission distance. Currently, there are a number of ways to realize OAM beam convergence, such as using metasurface [11], time-modulated array [30], and various dielectric lenses [5, 14, 34]. The Luneberg lens is recognized as one of the key technologies of 5G antennas [25], and it has been intensively studied [17, 23, 31] and commercially applied in 5G base station (BS) antenna arrays [18] as shown in Fig. 4.1. Currently, the Luneberg lens has been proposed to converge non-0-mode OAM beams generated by the UCA for decreasing the divergence

R. Chen et al., *Circular Array-Based Radio Frequency OAM Communications*,
Wireless Networks, https://doi.org/10.1007/978-3-031-37589-7_4

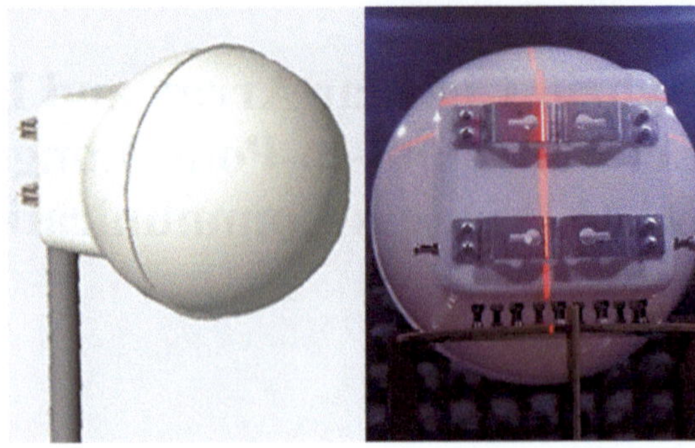

Fig. 4.1 The 5G Luneberg lens antenna [18]

angles [34]. However, the UCA aperture relative to the Luneberg lens is non-negligible. Thus, the 0-mode OAM beam becomes more divergent after passing through Luneberg lens, which is verified by EM simulation in this chapter. In other words, the Luneberg lens can converge the non-0-mode OAM beams but would diverge the 0-mode OAM beam generated by the UCA, which would reduce the capacity of long-distance multi-mode OAM wireless communication system with high proportion of the 0-mode OAM channel capacity.

In this chapter, to deal with the aforementioned challenges, we propose the hybrid circular array with a center element (CAC) and spherical Luneberg lens (HCCL) structure for OAM communication systems, and both the HCCL-based OAM channel model and corresponding system performance are discussed in detail. The novelty and major contributions of this chapter are summarized as follows:

- First, the divergence of OAM beams with different modes generated by the UCA and the convergence performance of Luneberg lens for non-0-mode OAM beams are verified by EM simulation. Besides, EM simulation shows that the 0-mode OAM beam becomes more divergent after passing through Luneberg lens when the UCA aperture relative to the Luneberg lens is non-negligible.
- Second, we propose the novel HCCL structure for OAM communication systems to converge OAM beams and improve the OAM channel capacity. EM simulations prove the proposed HCCL structure can converge not only non-0-mode OAM beams but also the 0-mode OAM beam.
- Third, we propose the Tx/Rx/Tx&Rx HCCL-based OAM communication systems, in which the HCCL structure is applied at the transmitter, receiver, and both at transmitter and receiver, respectively. Then, we derive their corresponding channel models respectively, based on which the multiplexed OAM modes in the Tx/Rx/Tx&Rx HCCL-based OAM communication systems keep orthogonal. Furthermore, the capacities of Tx/Rx/Tx&Rx HCCL-based OAM communication systems are quantitatively calculated and compared. Numerical simulations show the capacities of Tx/Rx/Tx&Rx HCCL-based OAM communication systems are significantly improved, which outperforms the UCA/CAC-based OAM communication systems in long-distance communication link. Meanwhile, the capacity of the OAM communication system equipped with HCCL structure both

at transmitter and receiver greatly outperforms that with only HCCL structure at transmitter or receiver.

The remainder of this chapter is organized as follows. In Sect. 4.2, we show the divergence of OAM beams with different modes generated by the UCA and verify the convergence performance of Luneberg lens for OAM beams, and we propose the HCCL structure to improve OAM beam transmission performance. In Sect. 4.3, we propose the Tx/Rx/Tx&Rx HCCL-based OAM communication system model, respectively. In Sect. 4.4, to further analyze the convergence effect of HCCL structure, the Tx/Rx/Tx&Rx HCCL-based OAM channels are first modeled, then corresponding channel capacities are calculated respectively, and finally a performance comparison is given.

Notations Bold lowercase letters denote vectors and bold uppercase letters denote matrices. $(\cdot)^T$ and $(\cdot)^H$ denote transpose and conjugate transpose, respectively. $\mathbf{I}$ is the identity matrix, and $\mathbf{0}$ is the all-zero column vector.

4.2 Convergence Effect of Luneberg Lens on OAM Beams Generated by Uniform Circular Array

In this section, we first utilize the high-frequency structure simulator (HFSS) to show the divergence of different mode OAM beams generated by the UCA and verify the convergence effect of Luneberg lens for OAM beams. Then, the HCCL structure is proposed to converge OAM beams.

4.2.1 Effect of Luneberg Lens on the Phase Structure of OAM Beams

The feasibility of utilizing UCA to generate OAM waves has been verified in theory [20] and in practice [13]. Specifically, the antenna elements in a UCA are fed with the same input signal but with a successive phase shift from element to element, such that after a full turn, the phase has the increment of $2\pi\ell$, where ℓ is an unbounded integer termed as *topological charge or OAM mode number* [9]. To figure out the convergence performance of Luneberg lens for the OAM beams, we first simulate the phase distributions of the OAM beams with different modes generated by the UCA. The radius of UCA cannot be set too small. According to [20], the divergence angle of an OAM beam is related to the OAM mode and the radius of the UCA. Specifically, for a given OAM mode, the divergence angle of the OAM beam increases as the radius of the UCA decreases. Therefore, in order to ensure the transmission quality of long-distance OAM wireless communication, the divergence angle of OAM beam has to be limited to less than a certain degree so that the receiving end can receive enough energy. Moreover, the radius of UCA is

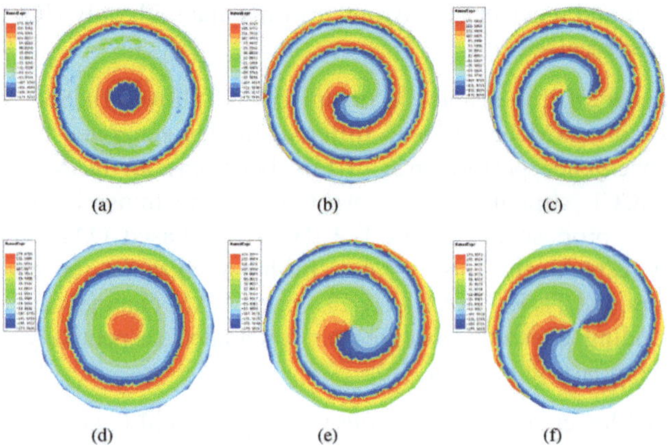

Fig. 4.2 The phase distributions of OAM beams without Luneberg lens (above) and with Luneberg lens (below): **(a)** **(d)** $\ell = 0$, **(b)** **(e)** $\ell = +1$, **(c)** **(f)** $\ell = +2$

also constrained by the half-wavelength limitation between array elements and the practical physical size of antenna elements. Therefore, the UCA in the simulation is set with the radius 0.75λ and $N = 8$ linearly polarised patch antennas working at the frequency $f = 2.45$ GHz. Then, the phase distributions of OAM beams with modes $\ell = 0, +1, +2$ generated by the UCA are shown in Fig. 4.2a, b, and c.

After that, we simulate the phase distributions of the OAM beams with different modes generated by the UCA and passing through a Luneberg lens. Each element in the UCA is placed at the focus of the Luneberg lens as a feed. The size of spherical Luneberg lens cannot be assumed too large. Theoretically, the size of the Luneberg lens can be infinite, but considering the limited cost and size constraints in practical applications, the minimum-size Luneberg lens that can meet the performance requirements is normally selected. At present, the fabrication methods of spherical Luneberg lens mainly include layering [15, 22], 3D printing [8, 16], and artificial dielectric material stacking [36], of which costs and time consumptions rise sharply with the increasing radius of the spherical lens. Therefore, the diameter of commercial Luneberg lens on the market working in the S-band is about several hundred millimeters [1, 3]. Therefore, the radius of Luneberg lens in the simulation is set to 2.5λ. The UCA aperture relative to the Luneberg lens cannot be negligible. As the spherical Luneberg lens can be approximated by incremental steps of constant index material [34], the simulated Luneberg lens with a radius of 2.5λ is divided into $K = 10$ layers and the relative dielectric constant varying with the layer number k is $\varepsilon_r = 2 - (\frac{2k-1}{2K})^2$. Then, the phase distributions of OAM beams with modes $\ell = 0, +1, +2$ are shown in Fig. 4.2d, e, and f. It is obvious to see from Fig. 4.2 that the phase structure of OAM beams would not be changed after passing through the lens. It means the Luneberg lens does not change the mode of the OAM beam, which provides a prerequisite for utilizing the Luneberg lens to converge OAM beams.

4.2.2 Convergence Effects of Luneberg Lens for Different OAM Modes

The radiation patterns of OAM beams with different modes generated by the UCA without Luneberg lens and with Luneberg lens are shown in Fig. 4.3. It can be seen from the figure that the divergence angles of the OAM beams with modes $\ell = +1, +2$ are obviously reduced after passing the Luneberg lens, but the 0-mode OAM beam becomes more divergent after passing through Luneberg lens. The counterintuitive phenomenon for the 0-mode OAM beam is due to the fact that the aperture of the UCA relative to the Luneberg lens cannot be ignored in practice; thus, the array elements in UCA cannot be deemed as they are at the focus of the Luneberg lens in the central axis transmission direction. To observe this phenomenon carefully, we show in Fig. 4.4a and b that the lens collimates the beam generated by a single element on one side of the UCA into the plane wave deviating from the central axis and toward the center of the UCA. Similarly, the beams generated by different elements of the UCA, after passing through the Luneberg lens, all deviate toward the center of the UCA and tend to cross above the center of the UCA. The fact that the superposition of waves does not change their own propagation directions makes the waves coming out of the lens appear divergent, which result in the divergence of 0-mode OAM beam. Different from 0-mode OAM beam having all the element with the same phase, the non-0-mode OAM beams have the phase difference between the elements of the UCA, not allowing the plane waves from every element to cross and the superposed wave appearing convergence in contrast to the direct superposition without the Luneberg lens. Overall, the Luneberg

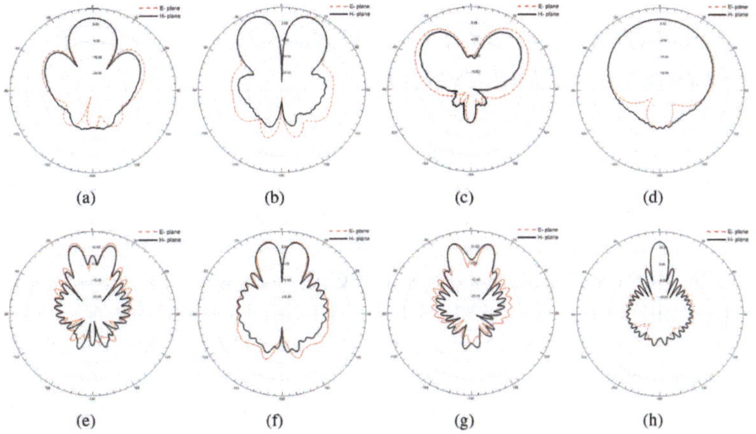

Fig. 4.3 The 2D radiation patterns of OAM beams transmitted by UCA without Luneberg lens (above) and with Luneberg lens (below): (**a**) (**e**) $\ell = 0$, (**b**) (**f**) $\ell = +1$, (**c**) (**g**) $\ell = +2$. 2D radiation patterns of the 0-mode OAM beam transmitted by CAC: (**d**) without Luneberg lens, (**h**) with Luneberg lens

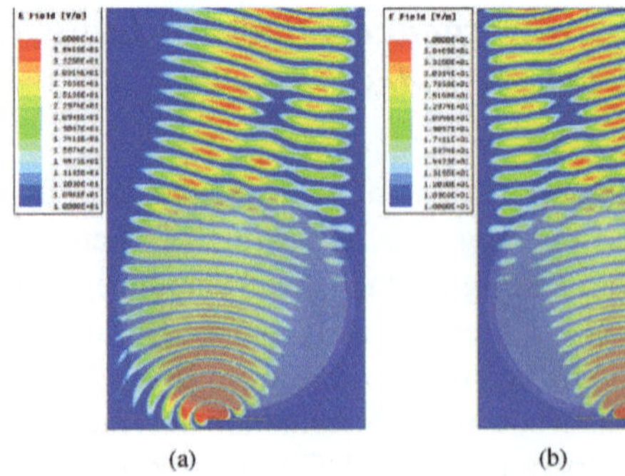

Fig. 4.4 The electric field intensity distributions of the single antenna changed by the Luneberg lens: (**a**) leftmost single antenna of UCA, (**b**) rightmost single antenna of UCA

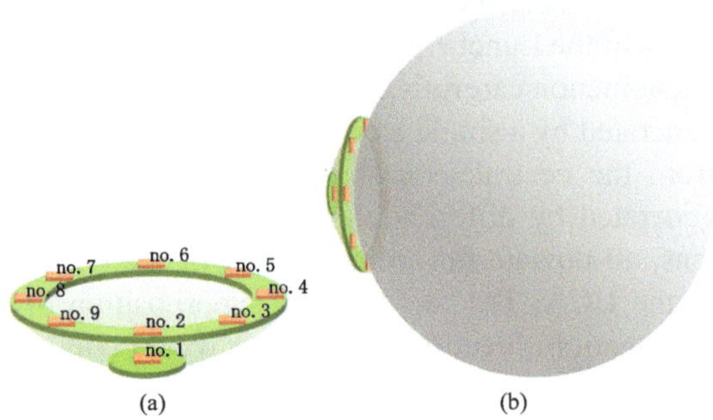

Fig. 4.5 The proposed CAC and HCCL structure: (**a**) CAC, (**b**) the HCCL structure composed of a CAC and a Luneberg lens

lens can converge the non-0-mode OAM beams generated by the UCA but would diverge the 0-mode OAM beam generated by the UCA.

4.2.3 Novel HCCL Structure for Multi-mode OAM Beam Transmission and Reception

Despite the reason for the divergence of the 0-mode OAM beam lying in the non-negligible UCA aperture, it is not feasible to reduce the UCA radius due to the half-wavelength limitation between elements. Meanwhile, reducing the UCA radius will enlarge the divergence angles of other non-0-mode OAM beams. In order to deal with the dilemma, we consider using CAC as shown in Fig. 4.5a instead of UCA to transmit and receive OAM beams, which is composed of a N-element outer ring with radius R to transmit/receive non-0-mode OAM beams and an additional center element designed to transmit/receive the 0-mode OAM beam. Meanwhile, the Luneberg lens is placed at the front end of CAC to converge OAM beams with different modes. This structure as shown in Fig. 4.5b is named HCCL structure. To facilitate the description, the antenna elements located in the outer ring of the CAC are numbered 2 to $N + 1$, and the center element is numbered 1. Figure 4.3d and h show that convergence effect of the Luneberg lens on the 0-mode OAM beam generated by the CAC. It is obvious to see from Fig. 4.3 that the HCCL structure can improve not only the transmission performance of non-0-mode OAM beams but also the transmission performance of 0-mode OAM beam. Furthermore, the characteristic of HCCL structure is not affected by the operating frequency, which has been verified in 40 GHz simulations. However, since the simulation results of 40 GHz showing the same characteristic of HCCL structure in 2.45 GHz, the result figures are omitted due to limited space. Thus, we can conclude that the HCCL structure is applicable to not only several GHz frequency bands but also higher-millimeter wave frequency bands.

4.3 HCCL-Based OAM Communication Systems

In this section, for an in-depth study, the HCCL structure is assumed at the transmitter, receiver, and both at transmitter and receiver, respectively. Then, the Tx/Rx/Tx&Rx HCCL-based orthogonal OAM communication systems are proposed.

4.3.1 Tx HCCL-Based OAM Communication System

In the Tx HCCL-based OAM communication system as shown in Fig. 4.6, the coordinate system Y–ZOX is established at the transmitter using the transmit outer ring plane as the ZOX plane and the axis through the point O and perpendicular to the transmit outer ring plane as the Y-axis, and the coordinate system Y–Z′O′X′ is established at the receiver by the similar approach. The elements in the outer ring of the transmit/receive CAC are numbered clockwise from 2 to $N+1$, and the azimuth angles of the elements in the outer ring of the transmit/receive CAC are expressed

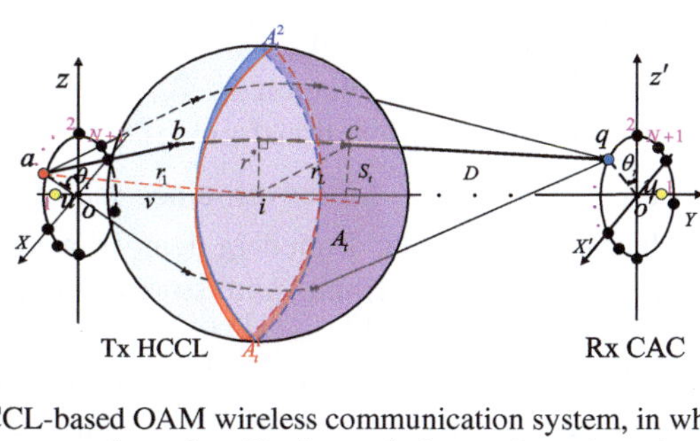

Fig. 4.6 The Tx HCCL-based OAM wireless communication system, in which the purple area on the Luneberg lens corresponds to the effective emission surface A_t, red area corresponds to the hemisphere surface A_t^1 without overlap, and dark blue area corresponds to the hemisphere surface A_t^2 without overlap

as $\theta_t(n) = \frac{2\pi}{N}(n-2)(2 \le n \le N+1)$ and $\theta_r(m) = \frac{2\pi}{N}(m-2)(2 \le m \le N+1)$, respectively. Denote by R the radii of transmit and receive outer rings and the distance between the ZOX plane and the Z′O′X′ plane is D. Moreover, the transmit center element numbered 1 is located at the negative half of Y-axis with the distance u from O, and the receive center element numbered 1 is located at the positive half of Y-axis with the distance $D+u$ from O. The center of the Tx spherical Luneberg lens with radius r_L is placed at $i = (0, v, 0)$ and r_1 away from the elements in transmit CAC.

The non-0-mode OAM beams are transmitted and received by the elements in the outer ring, and the 0-mode OAM beam is transmitted and received by the center element. Since for the N-element outer ring, at most N OAM modes can be generated and received, i.e., $|\ell| < N/2$ [20], our proposed Tx HCCL-based OAM communication system supports at most N OAM modes. All the antenna elements in CAC are assumed to be isotropic. In this case, the only way for radio waves to propagate from the transmit CAC to the receive CAC is to pass through the Tx Luneberg lens. Then, the channel matrix of the Tx HCCL-based communication system can be written as

$$\mathbf{H}^{\text{Tx HCCL}}_{(N+1)\times(N+1)} = \left[\begin{array}{c:c} h^{\text{Tx HCCL}}_{1,1} & \bar{\mathbf{e}}^{\text{Tx HCCL}}_{1\times N} \\ \hdashline \bar{\mathbf{t}}^{\text{Tx HCCL}}_{N\times 1} & \bar{\mathbf{H}}^{\text{Tx HCCL}}_{N\times N} \end{array}\right], \tag{4.1}$$

where $h^{\text{Tx HCCL}}_{1,1}$ denotes the channel from the center element in the transmit CAC to the center element in the receive CAC; $\bar{\mathbf{t}}^{\text{Tx HCCL}}$ and $\bar{\mathbf{e}}^{\text{Tx HCCL}}$ denote the channel vector from the center element in the transmit CAC to the elements in the outer ring of the receive CAC and that from the elements in the outer ring of the transmit CAC to the center element in the receive CAC, respectively; and $\bar{\mathbf{H}}^{\text{Tx HCCL}} = [h^{\text{Tx HCCL}}_{m,n}]_{N\times N}$ denotes the $N \times N$ channel matrix from the elements in the outer ring of the transmit CAC to the elements in the outer ring of the receive CAC, $2 \le m \le N+1, 2 \le n \le N+1$.

In LoS propagation environment, when the transmit CAC and receive CAC are aligned, $\bar{\mathbf{H}}^{\text{Tx HCCL}}$ is a circulant matrix. It can be decomposed as $\bar{\mathbf{H}}^{\text{Tx HCCL}} = \mathbf{F}^H_N \boldsymbol{\Lambda}^{\text{Tx HCCL}}_N \mathbf{F}_N$, where $\boldsymbol{\Lambda}^{\text{Tx HCCL}}_N$ is a diagonal matrix with N eigenvalues of $\bar{\mathbf{H}}^{\text{Tx HCCL}}$ on its diagonal, $\mathbf{F}_N = [\mathbf{f}^H_{\ell_1}, \mathbf{f}^H_{\ell_2}, \cdots, \mathbf{f}^H_{\ell_N}]^H$ is the N-dimensional discrete Fourier transform (DFT) matrix, and $\mathbf{f}_{\ell_l} = \frac{1}{\sqrt{N}}[1, e^{-j2\pi\frac{\ell_l}{N}}, \cdots, e^{-j2\pi\frac{\ell_l(N-1)}{N}}]$ [7]. Moreover, since the elements in the outer ring are uniformly distributed and the Luneberg lens is a symmetrical sphere, the elements in vector $\bar{\mathbf{t}}^{\text{Tx HCCL}}$ are equal, and so are the elements in vector $\bar{\mathbf{e}}^{\text{Tx HCCL}}$, which can be denoted as $\bar{\mathbf{t}}^{\text{Tx HCCL}} = [t^{\text{Tx HCCL}}, t^{\text{Tx HCCL}}, \cdots, t^{\text{Tx HCCL}}]^T$ and $\bar{\mathbf{e}}^{\text{Tx HCCL}} = [e^{\text{Tx HCCL}}, e^{\text{Tx HCCL}}, \cdots, e^{\text{Tx HCCL}}]$, respectively. Then, due to $\sum_{n=2}^{N+1} e^{j2\pi\frac{\ell_l(n-2)}{N}} = 0(\ell_l \ne 0)$ and $\sum_{n=2}^{N+1} e^{-j2\pi\frac{\ell_l(n-2)}{N}} = 0(\ell_l \ne 0)$, the two terms $\mathbf{F}_{U-1}\bar{\mathbf{t}}^{\text{Tx HCCL}}$ and $\bar{\mathbf{e}}^{\text{Tx HCCL}}\mathbf{F}^H_{U-1}$ can be written as

$$\mathbf{F}_{U-1}\bar{\mathbf{t}}^{\text{Tx HCCL}} = \frac{1}{\sqrt{N}} \begin{bmatrix} t^{\text{Tx HCCL}} \sum_{n=2}^{N+1} e^{-j2\pi \frac{\ell_1(n-2)}{N}} \\ t^{\text{Tx HCCL}} \sum_{n=2}^{N+1} e^{-j2\pi \frac{\ell_2(n-2)}{N}} \\ \vdots \\ t^{\text{Tx HCCL}} \sum_{n=2}^{N+1} e^{-j2\pi \frac{\ell_{U-1}(n-2)}{N}} \end{bmatrix} = \mathbf{0}, \tag{4.2}$$

$$\bar{\mathbf{e}}^{\text{Tx HCCL}}\mathbf{F}_{U-1}^H = \frac{1}{\sqrt{N}} \begin{bmatrix} e^{\text{Tx HCCL}} \sum_{n=2}^{N+1} e^{j2\pi \frac{\ell_1(n-2)}{N}} \\ e^{\text{Tx HCCL}} \sum_{n=2}^{N+1} e^{j2\pi \frac{\ell_2(n-2)}{N}} \\ \vdots \\ e^{\text{Tx HCCL}} \sum_{n=2}^{N+1} e^{j2\pi \frac{\ell_{U-1}(n-2)}{N}} \end{bmatrix}^T = \mathbf{0}^T, \tag{4.3}$$

where $\mathbf{F}_{U-1} = [\mathbf{f}_{\ell_1}^H, \mathbf{f}_{\ell_2}^H, \cdots, \mathbf{f}_{\ell_{U-1}}^H]^H$.

Thus, for the Tx HCCL-based OAM communication system multiplexed $L(1 < U \leq N)$ modes, the received information signal vector $\mathbf{y}^{\text{Tx HCCL}}$ can be expressed as

$$\begin{aligned} \mathbf{y}^{\text{Tx HCCL}} &= \mathbf{F}_U(\mathbf{H}^{\text{Tx HCCL}}\mathbf{F}_U^H\mathbf{x} + \mathbf{n}) \\ &= \left[\begin{array}{c:c} Nh_{1,1}^{\text{Tx HCCL}} & \sqrt{N}\bar{\mathbf{e}}^{\text{Tx HCCL}}\mathbf{F}_{U-1}^H \\ \hdashline \sqrt{N}\mathbf{F}_{U-1}\bar{\mathbf{t}}^{\text{Tx HCCL}} & \mathbf{F}_{U-1}\bar{\mathbf{H}}^{\text{Tx HCCL}}\mathbf{F}_{U-1}^H \end{array}\right]\mathbf{x} + \mathbf{F}_U\mathbf{n} \\ &= \left[\begin{array}{c:c} Nh_{1,1}^{\text{Tx HCCL}} & \mathbf{0}^T \\ \hdashline \mathbf{0} & \boldsymbol{\Lambda}_{U-1}^{\text{Tx HCCL}} \end{array}\right]\mathbf{x} + \mathbf{F}_U\mathbf{n} \\ &= \mathbf{H}_{\text{OAM}}^{\text{Tx HCCL}}\mathbf{x} + \mathbf{F}_U\mathbf{n}, \end{aligned} \tag{4.4}$$

where $\mathbf{x} = [x_0, x_{\ell_1}, \cdots, x_{\ell_{U-1}}]^T$, $\mathbf{y}^{\text{HCCL}} = [y_0, y_{\ell_1}, \cdots, y_{\ell_{U-1}}]^T$, x_{ℓ_l} and $y_{\ell_l}(l = 1, 2, \cdots, U-1, \ell_l \neq 0)$ are, respectively, the information symbol transmitted and received on the ℓ-mode OAM by the elements in the outer ring; x_0 and y_0 are, respectively, the information symbol transmitted and received on the 0-mode OAM by the center element; $\mathbf{F}_U = \text{diag}\{\sqrt{N}, \mathbf{F}_{U-1}\}$ is a diagonal matrix; and $\mathbf{n}$ is the complex Gaussian noise vector with zero mean and covariance matrix $\sigma_n^2\mathbf{I}_N$. In this case, we rank the modes from high to low, e.g., mode ℓ_1 is higher than mode ℓ_{U-1}. Then, one obtains $\mathbf{F}_{U-1}\bar{\mathbf{H}}^{\text{Tx HCCL}}\mathbf{F}_{U-1}^H = \mathbf{F}_{U-1}\mathbf{F}_N^H\boldsymbol{\Lambda}_N^{\text{Tx HCCL}}\mathbf{F}_N\mathbf{F}_{U-1}^H = \boldsymbol{\Lambda}_{U-1}^{\text{Tx HCCL}}$ and $\boldsymbol{\Lambda}_{U-1}^{\text{Tx HCCL}} = \text{diag}\{\gamma_{1,1}^{\text{Tx HCCL}}(\ell_1), \gamma_{2,2}^{\text{Tx HCCL}}(\ell_2), \cdots, \gamma_{U-1,U-1}^{\text{Tx HCCL}}(\ell_{U-1})\}$ is a diagonal matrix composed of $U-1$ eigenvalues of $\bar{\mathbf{H}}^{\text{Tx HCCL}}$. $\mathbf{H}_{\text{OAM}}^{\text{Tx HCCL}} = \text{diag}\{Nh_{1,1}^{\text{Tx HCCL}}, \boldsymbol{\Lambda}_{U-1}^{\text{Tx HCCL}}\}$ is termed as the Tx HCCL-based OAM channel, which shows the U different OAM modes keep orthogonal in the proposed Tx HCCL-based OAM communication system.

4.3.2 Rx HCCL-Based OAM Communication System

In the Rx HCCL-based OAM communication system as shown in Fig. 4.7, the coordinate system Y–ZOX, the coordinate system Y–Z′O′X′, and the elements in the transmit/receive CAC are established by the same approach in Tx HCCL-based OAM communication system. The center of the Rx spherical Luneberg lens with radius r_L is placed at $j = (0, D - v, 0)$ and r_1 away from the elements in receive CAC. In this case, the only way for radio waves to propagate from the transmit CAC to the receive CAC is to pass through the Rx Luneberg lens. Then, the channel matrix of the Rx HCCL-based communication system can be similarly written as

$$\mathbf{H}_{(N+1)\times(N+1)}^{\text{Rx HCCL}} = \left[\begin{array}{c:c} h_{1,1}^{\text{Rx HCCL}} & \bar{\mathbf{e}}_{1\times N}^{\text{Rx HCCL}} \\ \hdashline \bar{\mathbf{t}}_{N\times 1}^{\text{Rx HCCL}} & \bar{\mathbf{H}}_{N\times N}^{\text{Rx HCCL}} \end{array}\right], \tag{4.5}$$

where $h_{1,1}^{\text{Rx HCCL}}$ denotes the channel from the center element in the transmit CAC to the center element in the receive CAC; $\bar{\mathbf{t}}^{\text{Rx HCCL}}$ and $\bar{\mathbf{e}}^{\text{Rx HCCL}}$ denote the channel vector from the center element in the transmit CAC to the elements in the outer ring of the receive CAC and that from the elements in the outer ring of the transmit CAC to the center element in the receive CAC, respectively; and $\bar{\mathbf{H}}^{\text{Rx HCCL}} = [h_{m,n}^{\text{Rx HCCL}}]_{N\times N}$ denotes the $N \times N$ channel matrix from the elements in the outer ring of the transmit CAC to the elements in the outer ring of the receive CAC, $2 \leq m \leq N+1, 2 \leq n \leq N+1$.

When the transmit CAC and receive CAC are aligned, $\bar{\mathbf{H}}^{\text{Rx HCCL}}$ can be decomposed as $\bar{\mathbf{H}}^{\text{Rx HCCL}} = \mathbf{F}_N^H \boldsymbol{\Lambda}_N^{\text{Rx HCCL}} \mathbf{F}_N$. Similarly, we have $\bar{\mathbf{t}}^{\text{Rx HCCL}} = [t^{\text{Rx HCCL}}, t^{\text{Rx HCCL}}, \cdots, t^{\text{Rx HCCL}}]^T$ and $\bar{\mathbf{e}}^{\text{Rx HCCL}} = [e^{\text{Rx HCCL}}, e^{\text{Rx HCCL}}, \cdots,$

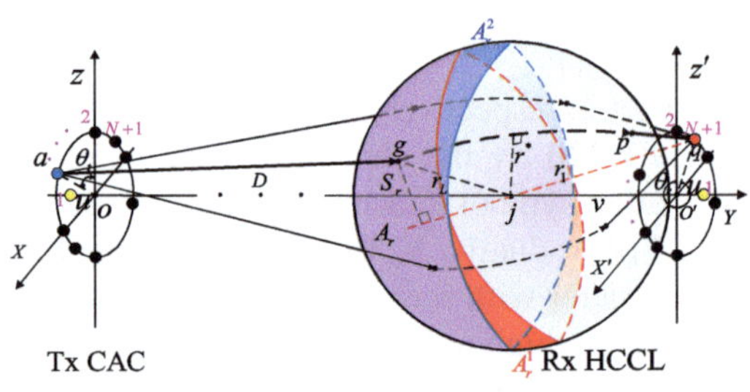

Fig. 4.7 The Rx HCCL-based OAM wireless communication system, in which the purple area on the Luneberg lens corresponds to the effective receiving surface A_r, red area corresponds to the hemisphere surface A_r^1 without overlap, and dark blue area corresponds to the hemisphere surface A_r^2 without overlap

$e^{\text{Rx HCCL}}]$. Then, the two terms $\mathbf{F}_{U-1}\bar{\mathbf{t}}^{\text{Rx HCCL}}$ and $\bar{\mathbf{e}}^{\text{Rx HCCL}}\mathbf{F}_{U-1}^{H}$ can be written as $\mathbf{F}_{U-1}\bar{\mathbf{t}}^{\text{Rx HCCL}} = \mathbf{0}$ and $\bar{\mathbf{e}}^{\text{Rx HCCL}}\mathbf{F}_{U-1}^{H} = \mathbf{0}^{T}$.

Correspondingly, for the Rx HCCL-based OAM communication system multiplexed U modes, the received information signal vector $\mathbf{y}^{\text{Rx HCCL}}$ can be expressed as

$$\begin{aligned}
\mathbf{y}^{\text{Rx HCCL}} &= \mathbf{F}_U(\mathbf{H}^{\text{Rx HCCL}}\mathbf{F}_U^H\mathbf{x} + \mathbf{n}) \\
&= \left[\begin{array}{c|c} Nh_{1,1}^{\text{Rx HCCL}} & \sqrt{N}\bar{\mathbf{e}}^{\text{Rx HCCL}}\mathbf{F}_{U-1}^{H} \\ \hline \sqrt{N}\mathbf{F}_{U-1}\bar{\mathbf{t}}^{\text{Rx HCCL}} & \mathbf{F}_{U-1}\bar{\mathbf{H}}^{\text{Rx HCCL}}\mathbf{F}_{U-1}^{H} \end{array}\right]\mathbf{x} + \mathbf{F}_U\mathbf{n} \\
&= \left[\begin{array}{c|c} Nh_{1,1}^{\text{Rx HCCL}} & \mathbf{0}^T \\ \hline \mathbf{0} & \boldsymbol{\Lambda}_{U-1}^{\text{Rx HCCL}} \end{array}\right]\mathbf{x} + \mathbf{F}_U\mathbf{n} \\
&= \mathbf{H}_{\text{OAM}}^{\text{Rx HCCL}}\mathbf{x} + \mathbf{F}_U\mathbf{n},
\end{aligned} \tag{4.6}$$

where $\mathbf{F}_{U-1}\bar{\mathbf{H}}^{\text{Rx HCCL}}\mathbf{F}_{U-1}^{H} = \mathbf{F}_{U-1}\mathbf{F}_{N}^{H}\boldsymbol{\Lambda}_{N}^{\text{Rx HCCL}}\mathbf{F}_{N}\mathbf{F}_{U-1}^{H} = \boldsymbol{\Lambda}_{U-1}^{\text{Rx HCCL}}$ and $\boldsymbol{\Lambda}_{U-1}^{\text{Rx HCCL}} = \text{diag}\{\gamma_{1,1}^{\text{Rx HCCL}}(\ell_1), \gamma_{2,2}^{\text{Rx HCCL}}(\ell_2), \cdots, \gamma_{U-1,U-1}^{\text{Rx HCCL}}(\ell_{U-1})\}$ is a diagonal matrix composed of $U-1$ eigenvalues of $\bar{\mathbf{H}}^{\text{Rx HCCL}}$. $\mathbf{H}_{\text{OAM}}^{\text{Rx HCCL}} = \text{diag}\{Nh_{1,1}^{\text{Rx HCCL}}, \boldsymbol{\Lambda}_{U-1}^{\text{Rx HCCL}}\}$ is termed as the Rx HCCL-based OAM channel, which shows the U different OAM modes keep orthogonal in the proposed Rx HCCL-based OAM communication system.

4.3.3 Tx&Rx HCCL-Based OAM Communication System

The Tx&Rx HCCL-based OAM communication system is shown in Fig. 4.8, which is the combination of Tx HCCL-based and Rx HCCL-based OAM communication system. In this case, the only way for radio waves to propagate from the transmit CAC to the receive CAC is to pass through both the Tx Luneberg lens and the Rx Luneberg lens. Then, the channel matrix of the Tx&Rx HCCL-based communication system can be similarly written as

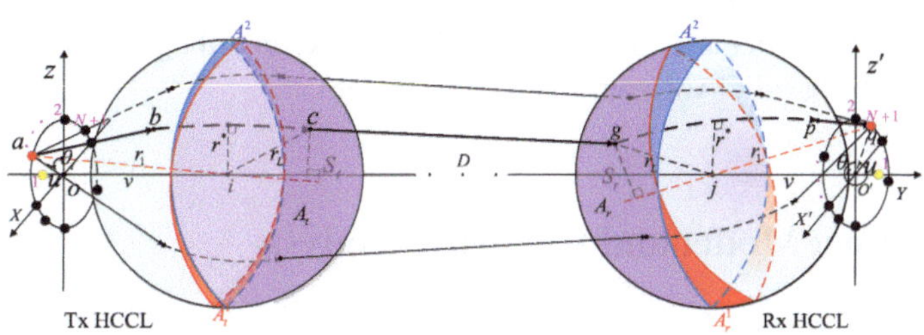

Fig. 4.8 The Tx&Rx HCCL-based OAM wireless communication system, in which the purple areas on the Luneberg lenses correspond to the effective emission surface A_t and effective receiving surface A_r, red areas correspond to hemisphere surfaces A_t^1 and A_r^1 without overlap, and dark blue areas correspond to hemisphere surfaces A_t^2 and A_r^2 without overlap

$$\mathbf{H}_{(N+1)\times(N+1)}^{\text{Tx\&Rx HCCL}} = \left[\begin{array}{c|c} h_{1,1}^{\text{Tx\&Rx HCCL}} & \bar{\mathbf{e}}_{1\times N}^{\text{Tx\&Rx HCCL}} \\ \hline \bar{\mathbf{t}}_{N\times 1}^{\text{Tx\&Rx HCCL}} & \bar{\mathbf{H}}_{N\times N}^{\text{Tx\&Rx HCCL}} \end{array}\right], \tag{4.7}$$

where $h_{1,1}^{\text{Tx\&Rx HCCL}}$ denotes the channel from the center element in the transmit CAC to the center element in the receive CAC; $\bar{\mathbf{t}}^{\text{Tx\&Rx HCCL}}$ and $\bar{\mathbf{e}}^{\text{Tx\&Rx HCCL}}$ denote the channel vector from the center element in the transmit CAC to the elements in the outer ring of the receive CAC and that from the elements in the outer ring of the transmit CAC to the center element in the receive CAC, respectively; and $\bar{\mathbf{H}}^{\text{Tx\&Rx HCCL}} = [h_{m,n}^{\text{Tx\&Rx HCCL}}]_{N\times N}$ denotes the $N \times N$ channel matrix from the elements in the outer ring of the transmit CAC to the elements in the outer ring of the receive CAC, $2 \le m \le N+1$, $2 \le n \le N+1$.

When transmit CAC and receive CAC are aligned, $\bar{\mathbf{H}}^{\text{Tx\&Rx HCCL}}$ can be decomposed as $\bar{\mathbf{H}}^{\text{Tx\&Rx HCCL}} = \mathbf{F}_N^H \boldsymbol{\Lambda}_N^{\text{Tx\&Rx HCCL}} \mathbf{F}_N$, and $\bar{\mathbf{t}}^{\text{Tx\&Rx HCCL}}$, and $\bar{\mathbf{e}}^{\text{Tx\&Rx HCCL}}$ can be similarly written as $\bar{\mathbf{t}}^{\text{Tx\&Rx HCCL}} = [t^{\text{Tx\&Rx HCCL}}, t^{\text{Tx\&Rx HCCL}}, \cdots, t^{\text{Tx\&Rx HCCL}}]^T$, and $\bar{\mathbf{e}}^{\text{Tx\&Rx HCCL}} = [e^{\text{Tx\&Rx HCCL}}, e^{\text{Tx\&Rx HCCL}}, \cdots, e^{\text{Tx\&Rx HCCL}}]$. Then, the two terms $\mathbf{F}_{U-1}\bar{\mathbf{t}}^{\text{Tx\&Rx HCCL}}$ and $\bar{\mathbf{e}}^{\text{Tx\&Rx HCCL}}\mathbf{F}_{U-1}^H$ can be written as $\mathbf{F}_{U-1}\bar{\mathbf{t}}^{\text{Tx\&Rx HCCL}} = \mathbf{0}$ and $\bar{\mathbf{e}}^{\text{Tx\&Rx HCCL}}\mathbf{F}_{U-1}^H = \mathbf{0}^T$.

Correspondingly, for the Tx&Rx-based OAM communication system multiplexed U modes, the received information signal vector $\mathbf{y}^{\text{Tx\&Rx HCCL}}$ can be expressed as

$$\begin{aligned} \mathbf{y}^{\text{Tx\&Rx HCCL}} &= \mathbf{F}_U(\mathbf{H}^{\text{Tx\&Rx HCCL}}\mathbf{F}_U^H\mathbf{x} + \mathbf{n}) \\ &= \left[\begin{array}{c|c} N h_{1,1}^{\text{Tx\&Rx HCCL}} & \sqrt{N}\bar{\mathbf{e}}^{\text{Tx\&Rx HCCL}}\mathbf{F}_{U-1}^H \\ \hline \sqrt{N}\mathbf{F}_{U-1}\bar{\mathbf{t}}^{\text{Tx\&Rx HCCL}} & \mathbf{F}_{U-1}\bar{\mathbf{H}}^{\text{Tx\&Rx HCCL}}\mathbf{F}_{U-1}^H \end{array}\right]\mathbf{x} + \mathbf{F}_U\mathbf{n} \\ &= \left[\begin{array}{c|c} N h_{1,1}^{\text{Tx\&Rx HCCL}} & \mathbf{0}^T \\ \hline \mathbf{0} & \boldsymbol{\Lambda}_{U-1}^{\text{Tx\&Rx HCCL}} \end{array}\right]\mathbf{x} + \mathbf{F}_U\mathbf{n} \\ &= \mathbf{H}_{\text{OAM}}^{\text{Tx\&Rx HCCL}}\mathbf{x} + \mathbf{F}_U\mathbf{n}, \end{aligned} \tag{4.8}$$

where $\mathbf{F}_{U-1}\bar{\mathbf{H}}^{\text{Tx\&Rx HCCL}}\mathbf{F}_{U-1}^H = \boldsymbol{\Lambda}_{U-1}^{\text{Tx\&Rx HCCL}}$ and $\boldsymbol{\Lambda}_{U-1}^{\text{Tx\&Rx HCCL}} = \text{diag}\{\gamma_{1,1}^{\text{Tx\&Rx HCCL}}(\ell_1), \gamma_{2,2}^{\text{Tx\&Rx HCCL}}(\ell_2), \cdots, \gamma_{U-1,U-1}^{\text{Tx\&Rx HCCL}}(\ell_{U-1})\}$. $\mathbf{H}_{\text{OAM}}^{\text{Tx\&Rx HCCL}} = \text{diag}\{N h_{1,1}^{\text{Tx\&Rx HCCL}}, \boldsymbol{\Lambda}_{U-1}^{\text{Tx\&Rx HCCL}}\}$ is termed as the Tx&Rx HCCL-based OAM channel, which shows the U different OAM modes keep orthogonal in the proposed Tx&Rx HCCL-based OAM communication system.

4.4 Channel Model and Capacity Analysis

In this section, to further quantitatively analyze the convergence effect of HCCL structure, the Tx/Rx/Tx&Rx HCCL-based OAM channels are modeled, and their corresponding channel capacities are calculated respectively. Then, we perform the numerical analysis for the effective channel coefficients of CAC-based and

Tx/Rx/Tx&Rx HCCL-based OAM communication systems to reveal why the HCCL-based OAM channels have larger capacities. Finally, the improvement of channel capacity by the proposed HCCL structure is verified through numerical simulations.

4.4.1 Tx HCCL-Based OAM Channel Capacity

Compared with the LoS CAC-based OAM communication system, the channel matrix $\mathbf{H}^{\mathrm{Tx\ HCCL}}$ is changed due to the refraction of the Luneberg lens as shown in Fig. 4.6. First, we model the channel matrix $\bar{\mathbf{H}}^{\mathrm{Tx\ HCCL}} = [h_{m,n}^{\mathrm{Tx\ HCCL}}]_{N\times N}$. For illustration, we draw only three paths for the point a which is the position of the n-th element in the outer ring of the transmit CAC and select a path to illustrate. The beam from the point $a = (R\sin\theta_t, 0, R\cos\theta_t)$, which is r_1 away from the spherical center $i = (0, v, 0)$ of the Luneberg lens and forms an angle of θ_t with the Z-axis, passes through the incident point b and point $c = (x_t, y_t, z_t)$ on the Luneberg lens effective emission surface. It finally arrives at the point $q = (R\sin\theta_r, D, R\cos\theta_r)$ which is the position of the m-th element in the outer ring of the receive CAC. Due to that the aperture of the outer ring relative to the Luneberg lens cannot be ignored in Tx HCCL structure, and the directions of transmit red hemispherical surfaces A_t^1 on the lens determined by different elements in the outer ring deviate from the Y-axis and face different directions which are all parallel to the lines between point i and elements in the transmit outer ring. In LoS propagation environment, the beams emitted from the transmit dark blue hemispherical surface A_t^2 on the lens facing the Y-axis can be received by the element in the receive CAC when $D \gg R$. Thus, the effective emission surface A_t formed by overlapping the two hemispheres mentioned above is presented as a purple area on the Luneberg lens.

Then, the effective propagation distances in the Tx HCCL structure can be respectively written as [4, 21]

$$d_{ab} = \int_{r_L}^{r_1} \frac{r}{\sqrt{r^2 - S_t)^2}} \mathrm{d}r, \tag{4.9}$$

$$d_{bc} = \frac{\psi_t(S_t)}{k} = \frac{4\pi}{k\lambda} \int_{r^*}^{r_L} \frac{rn(r)^2}{\sqrt{n^2(r)r^2 - (S_t)^2}} \mathrm{d}r, \tag{4.10}$$

where ψ_t is the phase shift in the lens; k is the wave number in vacuum; $n(r)$ is the refractive index of the Luneberg lens varying along the radial length r; r_1 and r_L are the lens' focus length and radius, respectively; r^* is referred to as a turning point's radial coordinate; and $\varphi = \pm\pi/2$ at the turning point where φ denotes the angle from the positive direction along the ray to the direction of increasing r. S_t is the distance from the point c on the effective emission surface to the straight line in space formed by the point a and spherical center i of Tx Luneberg lens and is written as (4.11).

$$
\begin{aligned}
S_t &= \frac{|\overrightarrow{ic} \times \overrightarrow{ai}|}{|\overrightarrow{ai}|} \\
&= \frac{\left|(x_t, y_t - v, z_t) \times \left(-R\sin\left[\frac{2\pi}{N}(n-2)\right], v, -R\cos\left[\frac{2\pi}{N}(n-2)\right]\right)\right|}{\left|\left(-R\sin\left[\frac{2\pi}{N}(n-2)\right], v, -R\cos\left[\frac{2\pi}{N}(n-2)\right]\right)\right|},
\end{aligned}
\tag{4.11}
$$

Therefore, the signal y_{A_t} arrived at any point c on the Luneberg lens' effective emission surface A_t can be expressed as [33]

$$
y_{A_t} = \frac{\beta_t \lambda}{4\pi d_{ab}} \frac{\lambda}{4\pi d_{bc}} e^{-j\frac{2\pi}{\lambda}(d_{ab}+d_{bc})} x', \tag{4.12}
$$

where x' is the signal transmitted from the point a, d_{ab} is the effective propagation distance outside the lens between a and b, d_{bc} is the effective propagation distance in the lens between b and c, λ is the wavelength of carrier, β_t model all the constants relative to transmit elements, $\lambda/4\pi d$ denotes the degradation of amplitude, and the complex exponential term is the phase difference due to the propagation distance.

Due to the linear superposition principle, the received signal $y_q^{\text{Tx HCCL}}$ at the point q can be expressed as

$$
y_q^{\text{Tx HCCL}} = \iint\limits_{A_t} \frac{\beta_t \beta_r \lambda^3}{(4\pi)^3 d_{ab} d_{bc} d_{cq}} e^{-j\frac{2\pi}{\lambda}(d_{ab}+d_{bc}+d_{cq})} x' \mathrm{d}\sigma_t, \tag{4.13}
$$

where $\mathrm{d}\sigma_t$ is the surface area element of the effective emission surface A_t, β_r model all the constants relative to receive elements, and $d_{cq} = \sqrt{[R\sin\theta_r(m) - x_t]^2 + (D - y_t)^2 + [R\cos\theta_r(m) - z_t]^2}$.

Thus, the channel coefficient $h_{m,n}^{\text{Tx HCCL}}$ from the n-th element in the transmit outer ring to the m-th element in the receive outer ring in $\bar{\mathbf{H}}^{\text{Tx HCCL}}$ for non-0-mode OAM beam transmission can be written as (4.14) [33].

$$
\begin{aligned}
h_{m,n}^{\text{Tx HCCL}} &= \iint\limits_{A_t} \frac{\beta_t \beta_r \lambda^3 e^{-j\frac{2\pi}{\lambda}[d_{ab}(x_t,y_t,z_t,n)+d_{bc}(x_t,y_t,z_t,n)+d_{cq}(x_t,y_t,z_t,m)]}}{(4\pi)^3 \left[d_{ab}(x_t,y_t,z_t,n) d_{bc}(x_t,y_t,z_t,n) d_{cq}(x_t,y_t,z_t,m)\right]} \mathrm{d}\sigma_t, \\
&\quad \forall m, n \in \{2, \cdots, N+1\}.
\end{aligned}
\tag{4.14}
$$

Then, the 0-mode OAM channel $h_{1,1}^{\text{Tx HCCL}}$ from the center element in the transmit CAC to the center element in the receive CAC can be written as [14, 21]

$$
h_{1,1}^{\text{Tx HCCL}} = \frac{\beta_t' \beta_r \lambda}{4\pi(D+2u)} e^{-j\frac{2\pi}{\lambda}(D+2u)}, \tag{4.15}
$$

where $\beta_t' = \frac{\beta_t}{r_1\sqrt{T}}$ is the antenna gain of transmit center element enhanced by lens and T is the loss in lens and can be written as [28]

$$T = 10^{\int_0^{r_L} \frac{54.6 \tan\delta n(r)}{10\lambda} dr}, \tag{4.16}$$

where $\tan\delta$ is the loss tangent of the Luneberg lens medium.

Similarly, with (4.14) and (4.15), we can substitute the expression of $\mathbf{H}^{\text{Tx HCCL}}$ into the following expression (4.17) to obtain the Tx HCCL-based OAM channel capacity:

$$\begin{aligned} C^{\text{Tx HCCL}} &= \log_2 \left| \mathbf{I} + \frac{P}{U\sigma_n^2} \mathbf{H}_{\text{OAM}}^{\text{Tx HCCL}} \mathbf{H}_{\text{OAM}}^{\text{Tx HCCL}^H} \right| \\ &= \log_2 \left| \mathbf{I} + \frac{P}{U\sigma_n^2} \mathbf{F}_U \mathbf{H}^{\text{Tx HCCL}} \mathbf{H}^{\text{Tx HCCL}^H} \mathbf{F}_U^H \right|, \end{aligned} \tag{4.17}$$

where P is the transmit power and P/σ_n^2 is the transmit SNR.

4.4.2 Rx HCCL-Based OAM Channel Capacity

First, we model the channel matrix $\bar{\mathbf{H}}^{\text{Rx HCCL}} = [h_{m,n}^{\text{Rx HCCL}}]_{N\times N}$. In the LoS Rx HCCL-based OAM communication system as shown in Fig. 4.7, the beam from the point a passes through the incident point $g = (x_r, y_r, z_r)$ at the effective receiving surface A_r and exit point p in turn and finally arrives at the point q. Similarly, the effective receiving surface A_r formed by overlapping the two hemispheres are presented as a purple area on the Luneberg lens.

The signal y_{A_r} arrived at any point g on the effective receiving surface A_r can be expressed as

$$y_{A_r} = \frac{\beta_t \lambda}{4\pi d_{ag}} e^{-j\frac{2\pi}{\lambda} d_{ag}} x', \tag{4.18}$$

where d_{ag} is the distance from the point a to the any point g on the effective receiving surface A_r and $d_{ag} = \sqrt{([x_r - R\sin\theta_t(n)]^2 + y_r^2 + [z_r - R\cos\theta_t(n)]^2)}$.

The effective propagation distances in the Rx HCCL structure can be respectively written as

$$d_{pq} = \int_{r_L}^{r_1} \frac{r}{\sqrt{r^2 - (S_r)^2}} dr, \tag{4.19}$$

$$d_{gp} = \frac{4\pi}{k\lambda} \int_{r*}^{r_L} \frac{r n(r)^2}{\sqrt{n^2(r) r^2 - (S_r)^2}} dr, \tag{4.20}$$

where $d\sigma_r$ is the surface area element of the effective receiving surface A_r, $j = (0, D - v, 0)$ is the spherical center of Rx Luneberg lens, and S_r is written as

$$\begin{aligned} S_r &= \frac{|\overrightarrow{jg} \times \overrightarrow{qj}|}{|\overrightarrow{qj}|} \\ &= \frac{\left|(x_r, y_r - D + v, z_r) \times \left(-R\sin\left[\frac{2\pi}{N}(m-2)\right], -v, -R\cos\left[\frac{2\pi}{N}(m-2)\right]\right)\right|}{\left|\left(-R\sin\left[\frac{2\pi}{N}(m-2)\right], -v, -R\cos\left[\frac{2\pi}{N}(m-2)\right]\right)\right|}, \end{aligned} \tag{4.21}$$

Therefore, the received signal $y_q^{\text{Rx HCCL}}$ at the point q can be expressed as

$$y_q^{\text{Rx HCCL}} = \iint\limits_{A_r} \frac{\beta_t \beta_r \lambda^3}{(4\pi)^3 d_{ag} d_{gp} d_{pq}} e^{-j\frac{2\pi}{\lambda}(d_{ag}+d_{gp}+d_{pq})} x' \mathrm{d}\sigma_r, \tag{4.22}$$

Thus, the channel coefficient $h_{m,n}^{\text{Rx HCCL}}$ in channel matrix $\bar{\mathbf{H}}^{\text{Rx HCCL}}$ for non-0-mode OAM beam transmission is given by

$$\begin{aligned} h_{m,n}^{\text{Rx HCCL}} &= \iint\limits_{A_t} \frac{\beta_t \beta_r \lambda^3 e^{-j\frac{2\pi}{\lambda}\left[d_{ag}(x_r,y_r,z_r,n)+d_{gp}(x_r,y_r,z_r,m)+d_{pq}(x_r,y_r,z_r,m)\right]}}{(4\pi)^3 \left[d_{ag}(x_r, y_r, z_r, n) d_{gp}(x_r, y_r, z_r, m) d_{pq}(x_r, y_r, z_r, m)\right]} \mathrm{d}\sigma_r, \\ &\forall m, n \in \{2, \cdots, N+1\}. \end{aligned} \tag{4.23}$$

Then, the 0-mode OAM channel $h_{1,1}^{\text{Rx HCCL}}$ can be written as

$$h_{1,1}^{\text{Rx HCCL}} = \frac{\beta_t \beta_r' \lambda}{4\pi(D+2u)} e^{-j\frac{2\pi}{\lambda}(D+2u)}, \tag{4.24}$$

where $\beta_r' = \beta_r'$ due to the antenna reciprocity theorem.

Correspondingly, with (4.23) and (4.24), we can substitute the expression of $\mathbf{H}^{\text{Rx HCCL}}$ into the following expression (4.25) to obtain the Rx HCCL-based OAM channel capacity:

$$\begin{aligned} C^{\text{Rx HCCL}} &= \log_2 \left| \mathbf{I} + \frac{P}{U\sigma_n^2} \mathbf{H}_{\text{OAM}}^{\text{Rx HCCL}} \mathbf{H}_{\text{OAM}}^{\text{Rx HCCL}^H} \right| \\ &= \log_2 \left| \mathbf{I} + \frac{P}{U\sigma_n^2} \mathbf{F}_U \mathbf{H}^{\text{Rx HCCL}} \mathbf{H}^{\text{Rx HCCL}^H} \mathbf{F}_U^H \right|. \end{aligned} \tag{4.25}$$

4.4.3 Tx&Rx HCCL-Based OAM Channel Capacity

Similarly, we first model the channel matrix $\bar{\mathbf{H}}^{\text{Tx\&Rx HCCL}} = [h_{m,n}^{\text{Tx\&Rx HCCL}}]_{N\times N}$ from the elements in the outer ring of the transmit CAC to the elements in the outer ring of the receive CAC. In the LoS Tx&Rx HCCL-based OAM communication system as shown in the Fig. 4.8, the beam from the point a on the Z-axis passes through the incident point b, exit point c at the effective emission surface A_t, incident point g at the effective receiving surface A_r and exit point p in turn and finally arrives at the point q. Then, the signal y'_{A_t} arrived at any point c on the effective emission surface A_t is the same as the y_{A_t}. The signal y'_{A_r} arrived at any point g on the effective receiving surface A_r is given by

$$y'_{A_r} = \iint\limits_{A_t} \frac{\beta_t \lambda^3}{(4\pi)^3 d_{ab} d_{bc} d_{cg}} e^{-j\frac{2\pi}{\lambda}(d_{ab}+d_{bc}+d_{cg})} x' \mathrm{d}\sigma_r, \tag{4.26}$$

where $d_{cg} = \sqrt{(x_r - x_t)^2 + (y_r - y_t)^2 + (z_r - z_t)^2}$.

For convenience of expression, we define $d_1 = d_{ab} + d_{bc} + d_{cg} + d_{gp} + d_{pq}$ and $d_2 = d_{ab} d_{bc} d_{cg} d_{gp} d_{pq}$. Then, the received signal $y_q^{\text{Tx\&Rx HCCL}}$ can be expressed as

$$y_q^{\text{Tx\&Rx HCCL}} = \iint\limits_{A_t} \iint\limits_{A_r} \frac{\beta_t \beta_r \lambda^5 e^{-j\frac{2\pi}{\lambda} d_2(x_t, y_t, z_t, x_r, y_r, z_r, m, n)}}{(4\pi)^5 d_1(x_t, y_t, z_t, x_r, y_r, z_r, m, n)} x' \mathrm{d}\sigma_r \mathrm{d}\sigma_t. \tag{4.27}$$

Thus, the channel coefficient $h_{m,n}^{\text{Tx\&Rx HCCL}}$ in channel matrix $\bar{\mathbf{H}}^{\text{Tx\&Rx HCCL}}$ for non-0-mode OAM beam transmission can be written as

$$h_{m,n}^{\text{Tx\&Rx HCCL}} = \iint\limits_{A_t} \iint\limits_{A_r} \frac{\beta_t \beta_r \lambda^5 e^{-j\frac{2\pi}{\lambda} d_1(x_t, y_t, z_t, x_r, y_r, z_r, m, n)}}{(4\pi)^5 d_2(x_t, y_t, z_t, x_r, y_r, z_r, m, n)} \mathrm{d}\sigma_r \mathrm{d}\sigma_t,$$

$$\forall m, n \in \{2, \cdots, N+1\}. \tag{4.28}$$

Then, the 0-mode OAM channel $h_{1,1}^{\text{Tx\&Rx HCCL}}$ can be written as

$$h_{1,1}^{\text{Tx\&Rx HCCL}} = \frac{\beta'_t \beta'_r \lambda}{4\pi(D+2u)} e^{-j\frac{2\pi}{\lambda}(D+2u)}. \tag{4.29}$$

Correspondingly, with (4.28) and (4.29), we can substitute the expression of $\mathbf{H}^{\text{Tx\&Rx HCCL}}$ into the following expression (4.30) to obtain the Tx&Rx HCCL-based OAM channel capacity:

$$
\begin{aligned}
C^{\text{Tx\&Rx HCCL}} &= \log_2 \left| \mathbf{I} + \frac{P}{U\sigma_n^2} \mathbf{H}_{\text{OAM}}^{\text{Tx\&Rx HCCL}} \mathbf{H}_{\text{OAM}}^{\text{Tx\&Rx HCCL}^H} \right| \\
&= \log_2 \left| \mathbf{I} + \frac{P}{U\sigma_n^2} \mathbf{F}_U \mathbf{H}^{\text{Tx\&Rx HCCL}} \mathbf{H}^{\text{Tx\&Rx HCCL}^H} \mathbf{F}_U^H \right|.
\end{aligned} \tag{4.30}
$$

4.4.4 Performance Comparison

(1) Numerical Analysis:

Considering $e^{j2\pi \frac{\ell_l x}{N}} = e^{j2\pi \frac{\ell_l (x+N)}{N}}$, the channel matrix $\bar{\mathbf{H}}$ is a circulant matrix, and all the channel coefficients in $\bar{\mathbf{H}}$ can be represented by the channel coefficients in a row of $\bar{\mathbf{H}}$, the modulus of the lth diagonal element in the effective OAM channel matrix $\boldsymbol{\Lambda}_{U-1}$ can be written as

$$
\begin{aligned}
|\gamma_{l,l}(\ell_l)| &= |\mathbf{f}_{\ell_l} \bar{\mathbf{H}} \mathbf{f}_{\ell_l}^H| \\
&= \frac{1}{N} \left| \sum_{m=2}^{N+1} \sum_{n=2}^{N+1} h_{m,n} e^{[-j\ell_l \theta_r(m) + j\ell_l \theta_t(n)]} \right| \\
&= \frac{1}{N} \left| \sum_{m=2}^{N+1} \sum_{n=2}^{N+1} h_{m,n} e^{j2\pi \frac{\ell_l (n-m)}{N}} \right| \\
&= \frac{1}{N} \left| \sum_{n=2}^{N+1} h_{2,n} e^{j2\pi \frac{\ell_l (n-2)}{N}} + \sum_{n=2}^{N+1} h_{3,n} e^{j2\pi \frac{\ell_l (n-3)}{N}} + \cdots \right. \\
&\quad \left. + \sum_{n=2}^{N+1} h_{N+1,n} e^{j2\pi \frac{\ell_l (n-N-1)}{N}} \right| \\
&= \left| \sum_{n=2}^{N+1} h_{2,n} e^{j2\pi \frac{\ell_l (n-2)}{N}} \right|.
\end{aligned} \tag{4.31}
$$

Due to $\sum_{n=2}^{N+1} e^{j2\pi \frac{\ell_l (n-2)}{N}} = 0$, when $h_{2,n}$, $\forall n \in \{2, \cdots, N+1\}$ are all the same, the ℓ_l-mode effective OAM channel $\gamma_{l,l}(\ell_l)$ can be regarded as the sum of vectors uniformly scattered at various points on a circle with radius $h_{2,n}$ and $|\gamma_{l,l}(\ell_l) = 0|$. Thus, in the case of long-distance transmission, the $h_{2,n}$ of all systems becomes small, and the greater the difference between channel coefficients $h_{2,n}$, $\forall n \in \{2, \cdots, N+1\}$, the greater the modulus of the ℓ_l-mode effective OAM channel $\gamma_{l,l}(\ell_l)$.

The channel coefficient $h_{2,n}^{\text{CAC}}$ of the CAC-based OAM communication system shown in Fig. 4.9 can be written as [12]

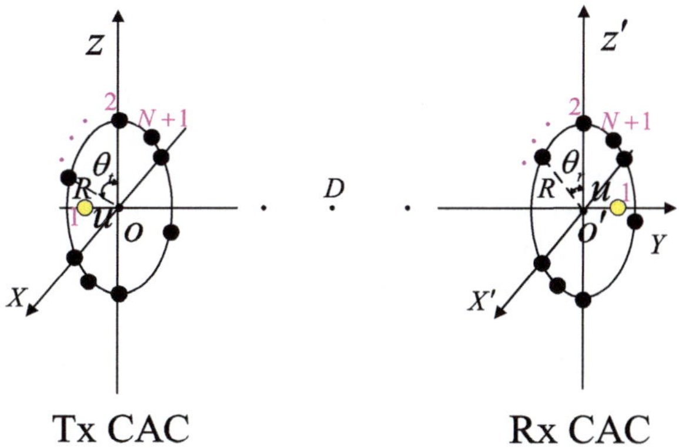

Fig. 4.9 The CAC-based OAM wireless communication system

$$h_{m,n}^{\mathrm{CAC}} = \frac{\beta_t \beta_r \lambda}{4\pi d_{m,n}^{\mathrm{CAC}}} e^{-j\frac{2\pi}{\lambda} d_{m,n}^{\mathrm{CAC}}}, \tag{4.32}$$

where $d_{m,n}^{\mathrm{CAC}}$ is the distance between the n-th transmit element and the m-th receive element. In the geometrical model shown in Fig. 4.9, $d_{1,1}^{\mathrm{CAC}} = D + 2u$, and the $d_{m,n}^{\mathrm{CAC}}$ can be written as [29]

$$d_{m,n}^{\mathrm{CAC}} = \sqrt{D^2 + 2R^2 - 2R^2 \cos\left[2\pi(n-m)/N\right]}, \quad \forall m, n \in \{2, \cdots, N+1\}. \tag{4.33}$$

Similarly, with (4.32) and (4.33), we can substitute the expression of $\mathbf{H}^{\mathrm{CAC}}$ into the following expression (4.34) to obtain the CAC-based OAM channel capacity:

$$\begin{aligned} C^{\mathrm{CAC}} &= \log_2 \left| \mathbf{I} + \frac{P}{U\sigma_n^2} \mathbf{H}_{\mathrm{OAM}}^{\mathrm{CAC}} \mathbf{H}_{\mathrm{OAM}}^{\mathrm{CAC}\,H} \right| \\ &= \log_2 \left| \mathbf{I} + \frac{P}{U\sigma_n^2} \mathbf{F}_U \mathbf{H}^{\mathrm{CAC}} \mathbf{H}^{\mathrm{CAC}\,H} \mathbf{F}_U^H \right|. \end{aligned} \tag{4.34}$$

The channel coefficient $h_{2,n}^{\mathrm{CAC}}$ is determined by the distance between the n-th element in the outer ring of the transmit CAC and the 2-th element in the outer ring of the receive CAC. The channel coefficient $h_{2,n}^{\mathrm{Tx\ HCCL}}$, $\forall n \in \{2, \cdots, N+1\}$ of the Tx HCCL-based OAM communication system in (4.14) is the surface integration. The integral surface of the surface integration is the effective surface A_t. The integrand function is the channel coefficients of the sub path passing through the effective surface A_t from the n-th element in the outer ring of the transmit CAC to the 2-th element in the outer ring of the receive CAC, and similar are the $h_{2,n}^{\mathrm{Rx\ HCCL}}$ and $h_{2,n}^{\mathrm{Tx\&Rx\ HCCL}}$. Since the radius of the integral surface is greater than that of the outer ring of the CAC, the difference between the different channel coefficients in the Tx/Rx/Tx&Rx HCCL-based OAM communication system is higher than that in the CAC-based OAM communication system. Therefore, the moduli of non-0-mode effective OAM channels of the Tx/Rx/Tx&Rx HCCL-based OAM communication systems are higher than those of the CAC-based OAM communication system. Due to the antenna reciprocity theorem, the moduli of non-0-mode effective OAM channels of the Tx HCCL-based OAM communication system and those of the Rx HCCL-based OAM communication system are equal. Thus, the non-0-mode OAM channel capacities of Tx/Rx/Tx&Rx HCCL-based OAM communication systems are higher than those of the CAC-based OAM communication system, and the non-0-mode OAM channel capacities of Tx HCCL-based OAM communication system are equal to the non-0-mode OAM channel capacities of Rx HCCL-based OAM communication system.

In the far field, i.e., $D \gg R$, compared with the Tx&Rx HCCL-based OAM communication system, the elements in the outer ring of the transmitter or the receiver without lens in the CAC/Tx HCCL/Rx HCCL-based OAM communication systems can be regarded as on the Y-axis. Therefore, in the CAC-based OAM communication system, the path losses from different transmit elements to the receive element on the Y-axis are equal, leading to the difference between the different channel coefficients of approximately 0 and so are in the Tx/Rx HCCL-based OAM communication systems. Conversely, in the Tx&Rx HCCL-based OAM communication system, due to the non-negligible outer ring's aperture relative to the Luneberg lens and the asymmetry of effective receiving surface A_r, the elements in the outer ring of the transmitter or the receiver cannot be regarded as on the Y-axis. The path losses from different elements in the outer ring of the transmit CAC to the 2-th element in the outer ring of the receive CAC are not equal. Therefore, the difference between the different channel coefficients in the Tx&Rx HCCL-based OAM communication system is not approximately 0. It is much larger than the difference between the different channel coefficients in the CAC-based OAM communication system, the difference between different channel coefficients in the Tx HCCL-based OAM communication system, and the difference between different channel coefficients in the Rx HCCL-based OAM communication system. The moduli of non-0-mode effective OAM channels of the Tx&Rx HCCL-based OAM communication systems are much higher than that of the CAC/Tx HCCL/Rx HCCL-based OAM communication systems. Thus, the non-0-mode OAM channel capacities of the Tx&Rx HCCL-based OAM communication system are much higher than that of the CAC/Tx HCCL/Rx HCCL-based OAM communication systems.

Moreover, the Luneberg lens collimates the beam emitted by the center element in CAC into the plane wave, which enhances the energy in the central axis transmission direction. Therefore, the antenna gain of the hybrid center element and Luneberg lens structure in the central axis transmission direction is higher than that of center element, improving 0-mode effective OAM channel and bringing higher 0-mode OAM channel capacity. Thus, the 0-mode OAM channel capacity of the Tx&Rx HCCL-based OAM communication system is higher than that of the CAC/Tx HCCL/Rx HCCL-based OAM communication systems, and the 0-mode OAM channel capacities of the Tx/Rx HCCL-based OAM communication systems

Table 4.1 The moduli of the effective OAM channels

Parameter	Value
$\lvert\gamma_{1,1}^{\text{CAC}}(\ell_1 = +3)\rvert$	5.19×10^{-7}
$\lvert\gamma_{1,1}^{\text{Tx HCCL}}(\ell_1 = +3)\rvert$	8.95×10^{-6}
$\lvert\gamma_{1,1}^{\text{Rx HCCL}}(\ell_1 = +3)\rvert$	8.95×10^{-6}
$\lvert\gamma_{1,1}^{\text{Tx\&Rx HCCL}}(\ell_1 = +3)\rvert$	2.30×10^{-3}
$\lvert Nh_{1,1}^{\text{CAC}}\rvert$	6.35×10^{-1}
$\lvert Nh_{1,1}^{\text{Tx HCCL}}\rvert$	1.94
$\lvert Nh_{1,1}^{\text{Rx HCCL}}\rvert$	1.94
$\lvert Nh_{1,1}^{\text{Tx\&Rx HCCL}}\rvert$	5.95

are equal and are all higher than that of the CAC-based OAM communication system.

(2) Numerical Simulations:

We compare the capacities of existing UCA-based communication system [12], CAC-based OAM communication system, Tx HCCL-based OAM communication system, Rx HCCL-based OAM communication system, and Tx&Rx HCCL-based OAM communication system with multiple modes. We choose $U = 7$, $R = 0.75\lambda$, $r_L = r_1 = 2.5\lambda$, $u = 0.12\lambda$ at $f = 2.45$ GHz and the multiplexed modes of all systems are $(\ell = -3, -2, \cdots, 0, \cdots, +3)$. The element numbers in UCA/CAC/Tx HCCL/Rx HCCL/Tx&Rx HCCL-based OAM communication systems are all $N = 9$. We assume that the dielectric loss of the Luneberg lens to be $\tan\delta = 0.001$ [35]. The signal power of the transmitter is 3.981W, and the noise power is 3×10^{-9}mW [26]. Unless otherwise stated, the total transmit power of all systems is the same, and equal power distribution is adopted for different OAM modes in all systems.

Table 4.1 shows that the numerical results of the effective OAM channel moduli with $D = 10^2\lambda$. It is obvious to see from the table that the moduli of the +3-mode effective OAM channel $|\gamma_{1,1}^{\text{Tx HCCL}}(\ell_1 = +3)|$, $|\gamma_{1,1}^{\text{Rx HCCL}}(\ell_1 = +3)|$, and $|\gamma_{1,1}^{\text{Tx\&Rx HCCL}}(\ell_1 = +3)|$ are higher than $|\gamma_{1,1}^{\text{CAC}}(\ell_1 = +3)|$. $|\gamma_{1,1}^{\text{Tx HCCL}}(\ell_1 = +3)| = |\gamma_{1,1}^{\text{Rx HCCL}}(\ell_1 = +3)|$. Meanwhile, $|\gamma_{1,1}^{\text{Tx\&Rx HCCL}}(\ell_1 = +3)|$ is much higher than $|\gamma_{1,1}^{\text{CAC}}(\ell_1 = +3)|$, $|\gamma_{1,1}^{\text{Tx HCCL}}(\ell_1 = +3)|$ and $|\gamma_{1,1}^{\text{Rx HCCL}}(\ell_1 = +3)|$. Besides, the moduli of the 0-mode effective OAM channel $|Nh_{1,1}^{\text{Tx HCCL}}|$, $|Nh_{1,1}^{\text{Rx HCCL}}|$ and $|Nh_{1,1}^{\text{Tx\&Rx HCCL}}|$ are higher than $|Nh_{1,1}^{\text{CAC}}|$, and $|Nh_{1,1}^{\text{Tx\&Rx HCCL}}|$ is higher than $|Nh_{1,1}^{\text{Tx HCCL}}|$ and $|Nh_{1,1}^{\text{Rx HCCL}}|$. $|Nh_{1,1}^{\text{Tx HCCL}}| = |Nh_{1,1}^{\text{Rx HCCL}}|$. The numerical results in Table 4.1 prove that the HCCL structure can effectively improve the modulus of effective OAM channel.

Figure 4.10 shows the capacities of the +3-mode OAM channel of the UCA/CAC/Tx HCCL/Rx HCCL/Tx&Rx HCCL-based OAM communication systems respectively. Figure 4.11 shows the capacities of the 0-mode OAM channel of UCA/CAC/Tx HCCL/Rx HCCL/Tx&Rx HCCL-based OAM communication systems. It is obvious to see from the two figures that when the number of antenna elements and total transmit power are the same, the performance of UCA/CAC-based OAM communication systems is the same, and the performance of Tx/Rx

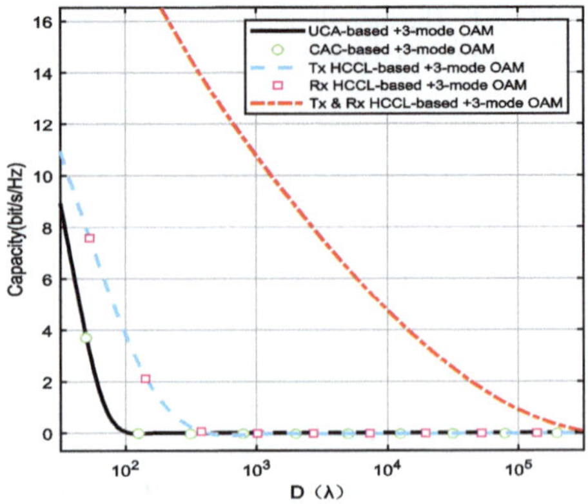

Fig. 4.10 Capacity performance comparison between CAC-based OAM system and HCCL-based OAM systems with +3-mode

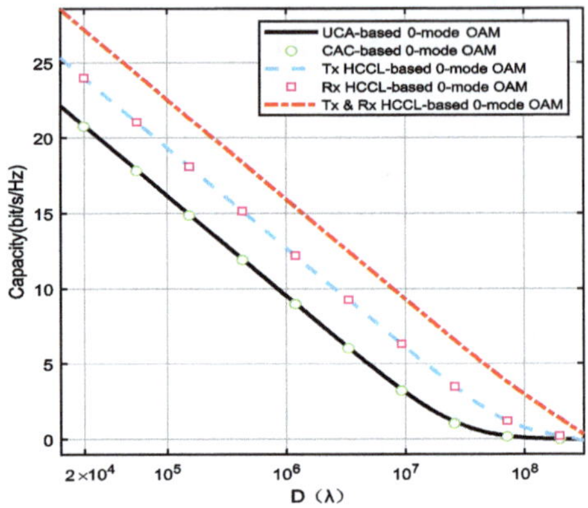

Fig. 4.11 Capacity performance comparison between CAC-based OAM system and HCCL-based OAM systems with 0-mode

HCCL-based OAM communication systems is also the same. Meanwhile, the capacities of Tx/Rx/Tx&Rx HCCL-based OAM communication systems are higher than that of the existing UCA-based OAM communication system, which proves the performance improvement brought by HCCL structure for OAM beam transmission. Besides, the performance of the system equipped with the HCCL structure both at transmitter and receiver is significantly higher than that of the system equipped with the HCCL structure at one side. Moreover, comparing Figs. 4.10 and 4.11, it can be seen that at the same transmission distance, the capacities of the 0-mode OAM channel of different communication systems are higher than those of the non-0-mode OAM channel because the divergence angle reduced by the lens still exists. The capacity comparison of UCA/CAC/Tx HCCL/Rx HCCL/Tx&Rx HCCL-based communication systems with multiple modes are shown in Fig. 4.12. From the figure, one can get the same conclusion as in Figs. 4.10 and 4.11, which verifies that the proposed HCCL structure can indeed improve the capacity of the OAM communication system.

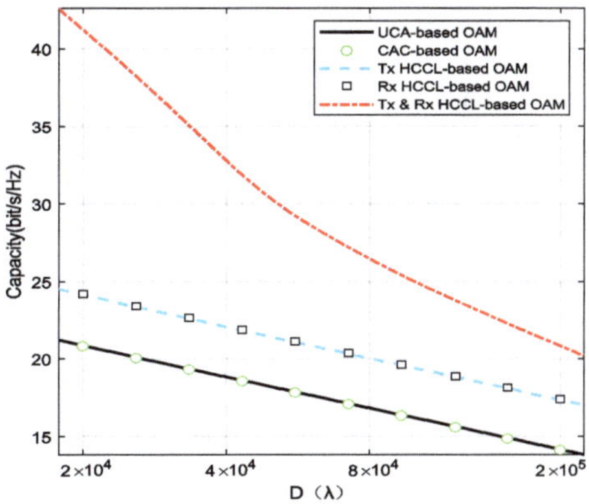

Fig. 4.12 Capacity performance comparison between CAC-based OAM system and HALL-based OAM systems

4.5 Summary

In this chapter, we propose a novel HCCL structure for long-distance OAM wireless communications. Specifically, the Luneberg lens are equipped at the transmitter, receiver, and both at transmitter and receiver of the CAC-based OAM communication system, respectively. Thereafter, the channels of different HCCL-based systems are modeled, and the capacities are quantitatively analyzed and compared. The EM simulations show that the HCCL structure can effectively converge both 0-mode and non-0-mode OAM beams. Meanwhile, both numerical analysis and simulation results validate that the capacities of the HCCL-based systems are significantly improved, which outperforms the existing UCA-based OAM communication systems. Moreover, the capacity of Tx HCCL-based communication system is equal to the capacity of Rx HCCL-based communication system. The capacity of the OAM system equipped with the HCCL structure both at transmitter and receiver is significantly better than that of the system equipped with the HCCL structure at one side. It is hopeful to further enhance the space utilization efficiency and channel capacity of our proposed HCCL-based OAM communication system by applying OAM-MIMO multiplexing [24] to our system. Moreover, for dealing with the non-orthogonality of the OAM channels when there is Tx-Rx misalignment, electronic beam steering-based Tx-Rx beam alignment [6] can be adopted. However, since the lens has nonlinear refraction effect on the electronic beam steering, the OAM beam steering and beam alignment with the lens are still an open problem, which is a very challenging but interesting future work.

References

1. 6-port Three Beam Lens Antenna (2021). http://www.eahison.com/cp_31.html
2. 6G, World's First Vortex Electromagnetic Wave Experiment Manages to Send 1 TB of Data in Only 1 Second (2022). https://optocrypto.com/6g-worlds-first-vortex-electromagnetic-wave-experiment-manages-to-send-1-tb-of-data-in-only-1-second/

3. 20 dbi gain, 2.2 ghz to 3.3 ghz, 20 db Return Loss, Luneburg Lens Antenna (2022). https://www.rfecho.com/product/20-dbi-gain-2-2-ghz-to-3-3-ghz-20-db-return-loss-luneburg-lens-antenna/
4. M. Arai, K. Hiraga, D. Uchida, T. Kaho, T. Seki, A beamforming network using a 3-dimensional dielectric lens for SDMA systems, in *Proceedings of the Asia Pacific Microwave Conference (APMC)* (2017), pp. 65–68
5. Y. Chen, S. Zheng, X. Hui, A flat-lensed spiral phase plate based on phase-shifting surface for generation of millimeter-wave OAM beam. IEEE Antennas Wireless Propag. Lett. **15**, 1156–1158 (2016)
6. R. Chen, H. Xu, M. Moretti, J. Li, Beam steering for the misalignment in UCA-based OAM communication systems. IEEE Wireless Commun. Lett. **7**(4), 582–585 (2018)
7. R. Chen, H. Xu, X. Wang, J. Li, On the performance of OAM in keyhole channels. IEEE Wireless Commun. Lett. **8**(1), 313–316 (2019)
8. M. Chen, Z. Zhang, Z. Li, Y. Li, L. Ge, J. Wang, Multibeam 3-d-printed luneburg lens fed by magnetoelectric dipole antennas for millimeter-wave mimo applications. IEEE Trans. Antennas Propag. **67**(5), 2923–2933 (2019)
9. R. Chen, W.X. Long, X. Wang, J. Li, Multi-mode OAM radio waves: Generation, angle of arrival estimation and reception with UCAs. IEEE Trans. Wirel. Commun. **19**(10), 6932–6947 (2020)
10. R. Chen, H. Zhou, M. Moretti, X. Wang, J. Li, Orbital angular momentum waves: Generation, detection, and emerging applications. IEEE Commun. Surveys Tuts. **22**(2), 840–868, 2nd Quart. (2020)
11. X. Dong, H. Sun, C. Gu, L. Zhuo, B. Xu, Reflective metasurface for vortex wave generating and divergence reducing in X-Band. Prog. Electromagn. Res. **87**, 97–106 (2018)
12. O. Edfors, A.J. Johansson, Is orbital angular momentum (OAM) based radio communication an unexploited area? IEEE Trans. Antennas Propag. **60**(2), 1126–1131 (2012)
13. R. Gaffoglio, A. Cagliero, A. De Vita, B. Sacco, OAM multiple transmission using uniform circular arrays: numerical modeling and experimental verification with two digital television signals. Radio Sci. **51**(6), 645–658 (2016)
14. S. Gao, W. Cheng, H. Zhang, Bifocal-lens converging based OAM wireless communications. J. Commun. Inf. Netw. **4**(2), 89–100 (2019)
15. O. Lafond S. Rondineau B. Fuchs, L.L. Coq, M. Himdi, Design optimization of multishell luneburg lenses. IEEE Trans. Antennas Propag. **55**(2), 283–289 (2007)
16. X. Li S. Lei, K. Han, G. Wei, A design of broadband 3-d-printed circularly polarized spherical luneburg lens antenna for x-band. IEEE Antennas Wirel. Propag. Lett. **20**(4), 528–532 (2021)
17. J. Liu, H. Lu, Z. Dong, Z. Liu, Y. Liu, X. Lv, Fully metallic dual-polarized Luneburg lens antenna based on gradient parallel plate waveguide loaded with non-uniform nails. IEEE Trans. Antennas Propa. **70**(1), 697–701 (2022)
18. LuneburgLens (2 $\sim$ 48 Beams) (2019). http://www.eahison.com/Luneburg-lens.html
19. E. Mari, F. Spinello, M. Oldoni, R.A. Ravanelli, F. Romanato, G. Parisi, Near-field experimental verification of separation of OAM channels. IEEE Antennas Wireless Propag. Lett. **14**, 556–558 (2015)
20. S.M. Mohammadi, et al., Orbital angular momentum in radio—A system study. IEEE Trans. Antennas Propag. **58**(2), 565–572 (2010)
21. S.P. Morgan, General solution of the luneberg lens problem. J. Appl. Phys. **29**(9), 1358–1368 (1958)
22. H. Mosallaei, Y. Rahmat-Samii, Nonuniform luneburg and two-shell lens antennas: radiation characteristics and design optimization. IEEE Trans. Antennas Propag. **49**(1), 60–69 (2001)
23. O. Quevedo-Teruel, J. Miao, M. Mattsson, A. Algaba-Brazalez, M. Johansson, L. Manholm, Glide-symmetric fully metallic luneburg lens for 5G communications at K-Band. IEEE Antennas Wireless Propag. Lett. **17**(9), 1588–1592 (2018)
24. H. Sasaki, et al., Experiment on over-100-gbps wireless transmission with oam-mimo multiplexing system in 28-ghz band, in Proceedings of the IEEE GlobeCom 2018 (2018), pp. 1–6

25. Scope of Luneburg lens Antenna Industry 2021–2026: Market Analysis with Trends and Opportunities (2021). https://themarketgossip.com/2021/08/10/scope-of-luneburg-lens-antenna-industry-2021-2026-market-analysis-with-trends-and-opportunities-tongyu-communication-ec-microwave-eahison-communication-rfecho-lunewave-and-more-affluence/
26. W.L. Stutzman, G.A. Thiele, *Antenna Theory and Design* (Wiley, New York, 2012)
27. G.A. Turnbull, D.A. Robertson, G.M. Smith, L. Allen, M.J. Padgett, The generation of free-space Laguerre-Gaussian modes at millimetre-wave frequencies by use of a spiral phaseplate. Opt. Commun. **127**(4), 183–188 (1996)
28. J. Volakis, *Antenna Engineering Handbook* (McGraw-Hill, New York, 2007)
29. P. Wang, Y. Li, B. Vucetic, Millimeter wave communications with symmetric uniform circular antenna arrays. IEEE Commun. Lett. **18**(8), 1307–1310 (2014)
30. Y. Wang, J. Liu, T. Hu, W. Jie, D. Yang, A. Tennant, Convergence of OAM beams using time-modulated concentric circular arrays, in *Proceedings of the 14th European Conference on Antennas and Propagation (EuCAP)* (2020), pp. 1–3
31. X. Wang, Y. Pan, Y. Dong, An E-plane-focused triple-layer multi-beam Luneburg lens antenna for 5G millimeter-wave applications. IEEE Antennas Wirel. Propag. Lett. **21**(2), 227–231 (2022)
32. Y. Yagi, H. Sasaki, T. Yamada, D. Lee, 200 Gb/s wireless transmission using dual-polarized OAM-MIMO multiplexing with uniform circular array on 28 GHz band. IEEE Antennas Wireless Propag. Lett. **20**(5), 833–837 (2021)
33. J. Yang, Y. Zeng, S. Jin, C. Wen, P. Xu, Communication and localization with extremely large lens antenna array. IEEE Trans. Wireless Commun. **20**(5), 3031–3048 (2021)
34. Y. Yao, X. Liang, W. Zhu, J. Li, J. Geng, R. Jin, K. Zhuang, Realizing orbital angular momentum (OAM) beam with small divergence angle by luneberg lens, in *Proceedings of the IEEE International Symposium on Antennas and Propagation & USNC/URSI National Radio Science Meeting* (2017), pp. 367–368
35. M. Zhong, S. Yang, Z. Nie, Optimization of a luneberg lens antenna using the differential evolution algorithm, in *Proceedings of the IEEE Antennas and Propagation Society International Symposium* (2008), pp. 1–4
36. H. Zhu N. Shariati M. Ansari, B. Jones, Y.J. Guo, A highly efficient spherical luneburg lens for low microwave frequencies realized with a metal-based artificial medium. IEEE Trans. Antennas Propag. **69**(7), 3758–3770 (2021)

Chapter 5
Circular Array-Based Radio Frequency Point-to-Multipoint OAM Wireless Backhaul

5.1 Introduction

The multi-tier heterogeneous network (HetNet), incorporated with flexible backhaul connections, distributed caching, and MIMO technologies [1, 7, 8, 15, 19, 20], is a promising solution for next-generation wireless communications. Typically, HetNet consists of a macro base station (MBS) tier and a small base station (SBS) tier as shown in Fig. 5.1. Backhaul links between MBS and SBSs exploit wired and/or wireless connectivity. In 6G systems, considering the massive number of small cells and the implementation costs, wireless backhaul [16, 18] will become a mainstream approach to handle the backhaul connectivity of small cells. One potential approach for 6G wireless backhaul is the OAM-based wireless backhaul scheme. The phase front of an EM wave carrying OAM rotates with azimuth exhibiting a helical structure $e^{j\ell\varphi}$ in space, as shown in Fig. 2.3. Due to inherent orthogonality among different OAM modes, the OAM-based wireless communication enables a novel coaxial multiplexing approach as shown in Fig. 5.2a, which utilizes a set of information-bearing modes on the same frequency channel to achieve a high SE [2, 4, 6, 11, 14, 17, 22, 24–26].

Up to now, a significant progress in the research of OAM-based wireless communications has been made [4]. In [17], the OAM-based wireless transmission experiment is realized for the first time, which successfully multiplexes two radio signals at the same frequency. Moreover, a 4 Gbps uncompressed video transmission link over a 60 GHz OAM radio channel is implemented in [11]. It is shown in [22] that multiplexing 4 dual-polarized OAM modes (mode number $\ell = \pm 1, \pm 3$) can achieve 32 Gbit/s data rate in a wireless communication link at 28 GHz. In [14], a 2×2 antenna aperture architecture, where each aperture multiplexes two OAM modes, is implemented in the 28 GHz band achieving a 16 Gbit/s transmission rate. Recently, a communication link with a transmission rate of over 200 Gbit/s is obtained by multiplexing five dual-polarized OAM modes (mode number $\ell = 0, \pm 1, \pm 2$) in the 28 GHz band [21].

R. Chen et al., *Circular Array-Based Radio Frequency OAM Communications*, Wireless Networks, https://doi.org/10.1007/978-3-031-37589-7_5

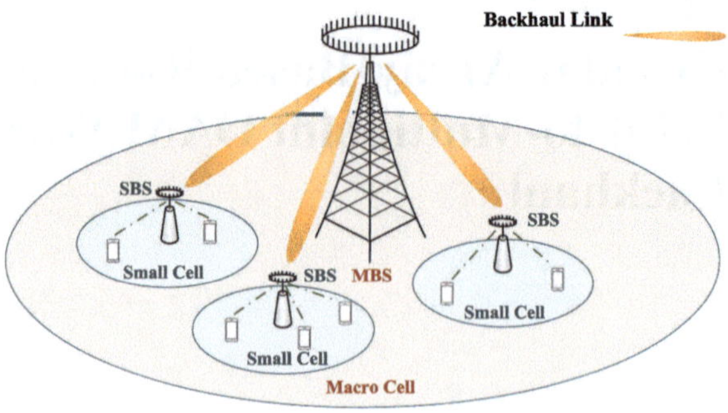

Fig. 5.1 The UCA-based downlink MU-OAM wireless backhaul system

However, the existing research on OAM wireless communications is mainly focused on the point-to-point coaxial transmission in the LoS scenario [5, 6, 11, 17, 22, 23, 25, 26], i.e., the single-user OAM (SU-OAM) communications, few paper considers the multi-user OAM (MU-OAM) transmission. To the best of our knowledge, only an OAM combined with a MIMO multi-user communication scheme is proposed in [14], shown in Fig. 5.2b. However, due to the harsh precondition that each pair of the transmit and receive antenna apertures need perfect alignment, the practical application of such an OAM+MIMO system is constrained by the application scenario. To get around that bottleneck, we consider a downlink MU-OAM wireless communication scheme for HetNet wireless backhaul systems with the positions of MBS and each SBS as shown in Fig. 5.1, which is illustrated in Fig. 5.2c for comparison with the existing two OAM multiplexing schemes. Furthermore, we propose an overall communication scheme for the downlink MU-OAM and MU-OAM-MIMO wireless backhaul system including the OAM-based multi-user distance and AoA estimation and the preprocessing of MU-OAM signals. The novelty and major contributions of this chapter are summarized as follows:

- First, we propose a downlink MU-OAM wireless backhaul system based on UCAs, which enables joint spatial division and coaxial multiplexing (JSDCM). Compared with the space division multiplexing in traditional downlink multi-user MIMO (MU-MIMO) wireless backhaul system, JSDCM in MU-OAM system can simultaneously multiplex the axis space between the transmit UCA center, and each receive UCA center and the radial space perpendicular to the axis space due to the inherent orthogonality among different OAM modes. Therefore, in contrast to the traditional MU-MIMO system, the proposed MU-OAM system is expected to provide a higher SE.
- Second, we formulate the signal models of training and data transmission phases for the proposed MU-OAM system. It's worth noting that the channel matrix of the proposed system is completely characterized by the position of the UCA center in each SBS. According to this character, we propose an effective MU-OAM distance and AoA estimation method based on a novel recursive ESPRIT algorithm, estimating the positions of all the SBSs simultaneously during the training phase. Compared with the traditional MU-MIMO system, the proposed

Fig. 5.2 (**a**) The coaxial multiplexing of OAM wireless communications, (**b**) the OAM and MIMO combined communication scheme in [14], (**c**) the proposed joint spatial division and coaxial multiplexing scheme for the downlink MU-OAM wireless backhaul

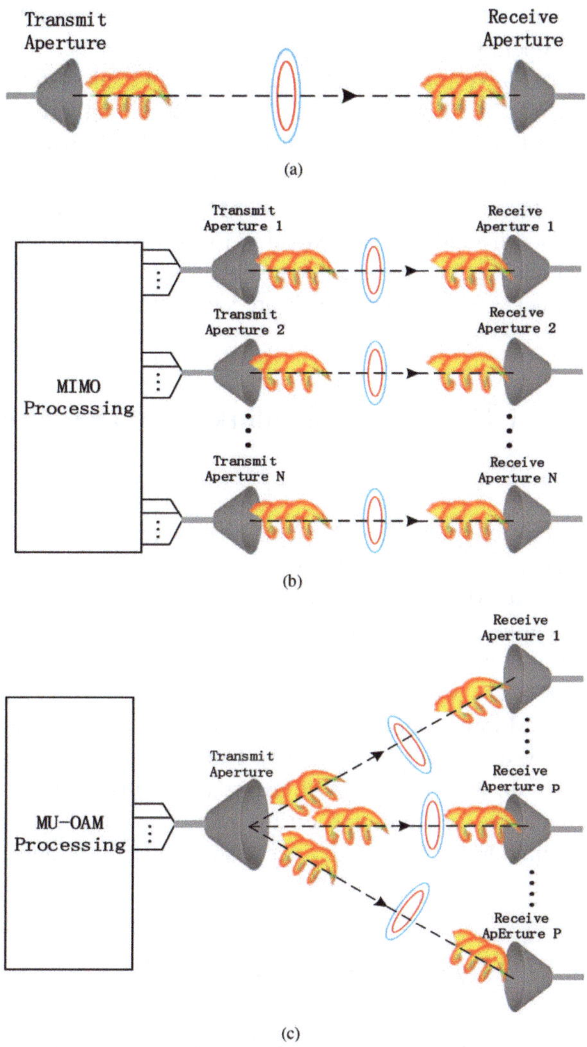

MU-OAM system does not need to estimate a large channel matrix, which can greatly reduce the required training overhead.

- Third, we apply a MU-OAM preprocessing scheme to eliminate the co-mode and the inter-mode interferences in the downlink MU-OAM channel. Since interferences in the system are eliminated through preprocessing at MBS, each SBS only needs to separate and despiralize its own received OAM signals, which thus reduces the complexity of SBSs. Furthermore, we extend the proposed scheme to the downlink MU-OAM-MIMO wireless backhaul system based on UCCAs, for which much higher SE and energy efficiency (EE) than traditional MU-MIMO systems could be achieved.

The remainder of this chapter is organized as follows. Based on the feasibility of generating and receiving multi-mode OAM beams by UCA, we model the UCA-based downlink MU-OAM wireless backhaul system in Sect. 3.2. In Sect. 3.3, the MU-OAM distance and AoA estimation method based on the recursive ESPRI algorithm is proposed. With the estimated distances and AoAs, a MU-OAM preprocessing scheme is applied to effectively eliminate co-mode and inter-mode interferences in the downlink MU-OAM channel in Sect. 3.4. In Sect. 3.5, the proposed methods are extended to the UCCA-based downlink MU-OAM-MIMO wireless backhaul system to achieve higher EE and SE. Simulation results are shown in Sect. 3.6, and conclusions are summarized in Sect. 3.6.

5.2 UCA-Based Downlink MU-OAM Wireless Backhaul System

In this chapter, we also consider a UCA-based downlink MU-OAM wireless backhaul system, where multi-mode OAM beams are generated by an N-element UCA at MBS and received by M-element UCAs at P SBSs separately. In the downlink MU-OAM wireless backhaul system, the UCAs at MBS and P SBSs are off-axis misaligned as shown in Fig. 5.3, where the UCA at MBS and the UCAs at P SBSs are parallel, and the UCAs at P SBSs are not centered with the same axis. Moreover, for the M-element UCAs at SBSs, at most M OAM modes can be resolved, i.e., $|\ell| < M/2$ [13]. Hence, the N-element UCA at MBS is required to generate PM-mode OAM beams for P SBSs in the proposed MU-OAM system. Owing to the limitation on the number of multiplexed OAM modes, $N \geq PM$. Without loss of generality, we assume $N = PM$ here.

5.2.1 Downlink Channel Model

The geometrical model of the UCAs at MBS and the p-th SBS in the off-axis case is illustrated in Fig. 5.3. In the geometrical model, the coordinate system $\mathrm{Z} - \mathrm{XOY}$ is established at MBS using the MBS UCA plane as the XOY plane and the axis through the point O and perpendicular to the UCA plane as the Z-axis, and the coordinate system $\mathrm{Z}_p - \mathrm{X}_p\mathrm{O}_p\mathrm{Y}_p$ is established at the p-th SBS by the similar approach. According to the geometric relationship in Fig. 5.3, the coordinate of the UCA center at the p-th SBS is denoted as $\mathrm{O}_p(r_p, \theta_p, \varphi_p)$ in $\mathrm{Z} - \mathrm{XOY}$, and the coordinate of the UCA center at MBS can be denoted as $\mathrm{O}(r_p, \theta_p, \varphi_p + \pi)$ in $\mathrm{Z}_p - \mathrm{X}_p\mathrm{O}_p\mathrm{Y}_p$, where r_p is the distance between the UCA center of MBS and the UCA center of the p-th SBS; φ_p and θ_p are, respectively, the azimuth angle and the elevation angle of the UCA center at the p-th SBS in $\mathrm{Z} - \mathrm{XOY}$; and θ_p *and* φ_p *are defined as the AoA of OAM beams transmitted by the p-th SBS.*

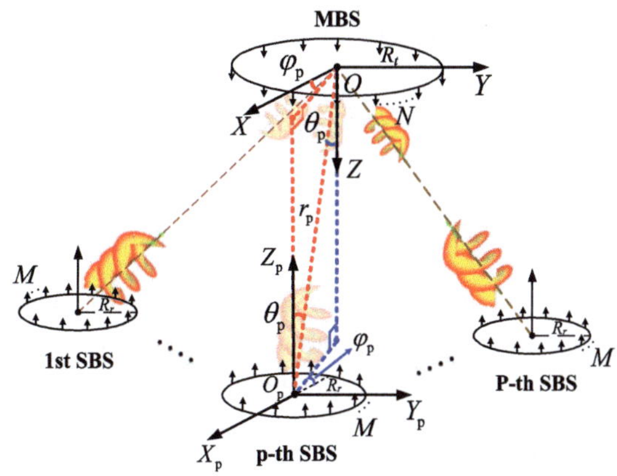

Fig. 5.3 The geometrical model of the transmit and receive UCAs in the downlink MU-OAM wireless backhaul system

Denote ϕ as the azimuth angle of the n-th $(1 \le n \le N)$ UCA element at MBS and α_p as the azimuth angle of the m-th $(1 \le m \le M)$ UCA element at the p-th SBS. According to the geometric relationship in Fig. 5.3, the transmission distance $d_{m,n}^p$ from the n-th antenna element of MBS to the m-th antenna element of the p-th SBS can be calculated as

$$
\begin{aligned}
d_{m,n}^p =& \big[R_t^2 + R_r^2 + r_p^2 + 2r_pR_r \sin\theta_p \cos(\varphi_p - \alpha_p) - 2r_pR_t \sin\theta_p \cos(\varphi_p - \phi) \\
& - 2R_tR_r \cos(\alpha_p - \phi)\big]^{1/2},
\end{aligned} \tag{5.1}
$$

where R_t and R_r are, respectively, the radii of UCAs at MBS and SBSs, $\phi = [\frac{2\pi(n-1)}{N} + \phi_0]$, $n = 1, 2, \cdots, N$, $\alpha_p = [\frac{2\pi(m-1)}{M} + \alpha_0]$, $m = 1, 2, \cdots, M$, ϕ_0 and α_0 are, respectively, the corresponding initial angles of the first reference antenna elements in both UCAs. For easier analysis, we assume $\phi_0 = 0$ and $\alpha_0 = 0$ in this chapter. By assuming that all the P SBSs are located in the far-field distance region of MBS, i.e., $r_p \gg R_t$ and $r_p \gg R_r$, $p = 1, 2, \cdots, P$, we can approximate $d_{m,n}^p$ as

$$
\begin{aligned}
d_{m,n}^p &\overset{(a)}{\approx} \sqrt{R_t^2 + R_r^2 + r_p^2} + \frac{r_pR_r \sin\theta_p \cos(\varphi_p - \alpha_p)}{\sqrt{R_t^2 + R_r^2 + r_p^2}} - \frac{r_pR_t \sin\theta_p \cos(\varphi_p - \phi)}{\sqrt{R_t^2 + R_r^2 + r_p^2}} \\
&\quad - \frac{R_tR_r \cos(\alpha_p - \phi)}{\sqrt{R_t^2 + R_r^2 + r_p^2}} \\
&\overset{(b)}{\approx} r_p + R_r \sin\theta_p \cos(\varphi_p - \alpha_p) - R_t \sin\theta_p \cos(\varphi_p - \phi) - \frac{R_tR_r}{r_p} \cos(\alpha_p - \phi),
\end{aligned} \tag{5.2}
$$

where (a) uses the method of completing a square and the condition $r_p \gg R_t, R_r$ as same as the simple case $\sqrt{a^2 - 2b} \approx a - \frac{b}{a}, a \gg b$; (b) is directly obtained from the condition $r_p \gg R_t, R_r$. Then, substituting (5.2) into (2.5) and abbreviating

$h(k, d^p_{m,n})$ to $h^p_{m,n}(k)$, we thus have

$$h^p_{m,n}(k) \overset{(a)}{\approx} \frac{\beta}{2kr_p} \exp\Bigg(- ikr_p - ikR_r \sin\theta_p \cos(\varphi_p - \alpha_p) + ikR_t \sin\theta_p \cos(\varphi_p - \phi) + i\frac{kR_tR_r}{r_p} \cos(\alpha_p - \phi)\Bigg), \tag{5.3}$$

where (a) neglects a few small terms in the denominator and thus only $2kr_p$ is left.

In this way, the channel matrix from MBS to P SBSs can be expressed as $\mathbf{H}(k) = [h^p_{m,n}(k)]_{PM\times N}$. In the proposed MU-OAM system, MBS is required to simultaneously transmit the P group OAM-based signals to P SBSs in the multi-user downlink channel; thus, the N $(N = PM)$ elements of UCA at MBS should be divided into P groups, and each group contains M elements. Therefore, the channel matrix $\mathbf{H}(k)$ can be rewritten as

$$\mathbf{H}(k) = \begin{bmatrix} \mathbf{H}_{1,1}(k) & \mathbf{H}_{1,2}(k) & \cdots & \mathbf{H}_{1,P}(k) \\ \mathbf{H}_{2,1}(k) & \mathbf{H}_{2,2}(k) & \cdots & \mathbf{H}_{2,P}(k) \\ \vdots & \vdots & \ddots & \vdots \\ \mathbf{H}_{P,1}(k) & \mathbf{H}_{P,2}(k) & \cdots & \mathbf{H}_{P,P}(k) \end{bmatrix}, \tag{5.4}$$

where $\mathbf{H}_{p,q}(k)$ is the $M \times M$ submatrix that denotes the channel from the q-th $(1 \leq q \leq P)$ group elements of MBS to the p-th SBS, which can be written as

$$\mathbf{H}_{p,q}(k) = \begin{bmatrix} h^p_{1,q}(k) & h^p_{1,q+P}(k) & \cdots & h^p_{1,q+P\times(M-1)}(k) \\ h^p_{2,q}(k) & h^p_{2,q+P}(k) & \cdots & h^p_{2,q+P\times(M-1)}(k) \\ \vdots & \vdots & \ddots & \vdots \\ h^p_{M,q}(k) & h^p_{M,q+P}(k) & \cdots & h^p_{M,q+P\times(M-1)}(k) \end{bmatrix}. \tag{5.5}$$

5.2.2 *Signal Model*

Similar to the single-user OAM communications, in the downlink MU-OAM wireless backhaul system, we also assume W subcarriers and U OAM modes for data transmission, i.e., $k_1 \leq k_w \leq k_W$, $-\lfloor U/2 \rfloor \leq \ell_u \leq \lfloor U/2 \rfloor, 0 \leq U \leq M$, and $\widetilde{W}$ subcarriers and $\widetilde{U}$ OAM modes for training, $1 \leq \widetilde{W} \leq W$, $1 \leq \widetilde{U} \leq U$. The transceiver structure of OAM-OFDM for downlink MU-OAM transmission is illustrated in Fig. 5.4.

5.2.2.1 Uplink Training Phase

In the uplink training phase, pilot data is exploited for performing channel estimation. First, MBS broadcast a message to P SBSs. After receiving the broadcast

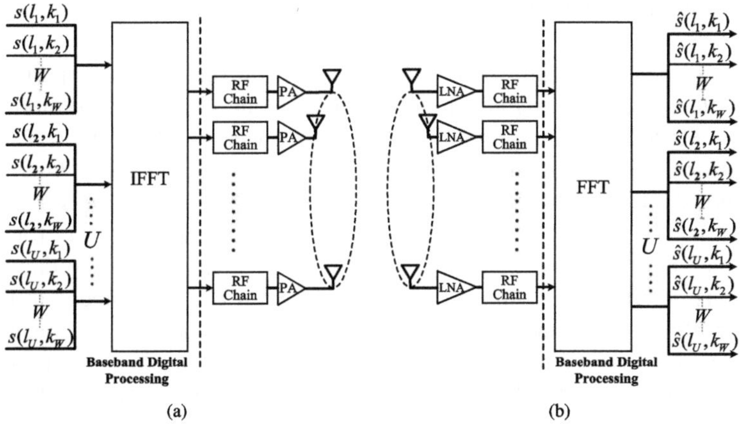

Fig. 5.4 The transceiver structure of OAM-OFDM for the downlink MU-OAM transmission. PA: the power amplifier. LNA: low noise amplifier. (**a**) The transmitter structure. (**b**) The receiver structure

signal, P SBSs simultaneously send OAM-based training symbols to MBS. The equivalent baseband training signal model sent by the p-th SBS can be expressed as $\mathbf{F}_{\widetilde{U}}^H \mathbf{S'}_p(k_w)$, where $\mathbf{F}_{\widetilde{U}}$ is the $\widetilde{U} \times M$ (partial) DFT matrix used to generate OAM beams with different modes and $\mathbf{S}'_p(k_w) = \text{diag}\{s'_p(k_w), s'_p(k_w), \cdots, s'_p(k_w)\}$ is the $\widetilde{U}$-dimensional training symbol matrix transmitted by the p-th SBS on $\widetilde{U}$ OAM modes at the w-th $(1 \leq w \leq \widetilde{W})$ subcarrier. Without loss of generality, the transmitted training symbols $\{\mathbf{S}'_p(k_w)|p = 1, 2, \cdots, P\}$ are assumed to be known by MBS.

After that, the received uplink baseband training symbol matrix $\mathbf{Y'}_t(k_w)$ at MBS can be written as

$$\mathbf{Y'}_t(k_w) = \sum_{p=1}^{P} \mathbf{H}_p^T(k_w)\mathbf{F}_{\widetilde{U}}^H \mathbf{S'}_p(k_w) + \mathbf{Z'}_t(k_w), \tag{5.6}$$

where $\mathbf{Y'}_t(k_w)$ is the $N \times \widetilde{U}$ matrix, $\mathbf{H}_p(k_w) = [h_{m,n}^p(k_w)]_{M\times N}$ is the channel matrix from MBS to the p-th SBS at the w-th subcarrier, and $\mathbf{Z}'_t(k_w)$ is the related additive noise matrix. Then, the received uplink signals $\mathbf{Y'}_t(k_w)$ on all the N elements of MBS are combined, which can be formulated as

$$\mathbf{x'}_t(k_w) = \mathbf{1}^T \mathbf{Y}'_t(k_w) = \mathbf{1}^T \sum_{p=1}^{P} \left(\mathbf{H}_p^T(k_w)\mathbf{F}_{\widetilde{U}}^H \mathbf{S'}_p(k_w) + \mathbf{Z'}_t(k_w) \right), \tag{5.7}$$

where $\mathbf{x}'_t(k_w) = [x'_t(\ell_1, k_w), x'_t(\ell_2, k_w), \cdots, x'_t(\ell_{\widetilde{U}}, k_w)]$ contains all the combined signals. Furthermore, how to perform channel estimation at MBS by the received uplink training symbols will be specified in Sect. 3.3.

5.2.2.2 Downlink Data Transmission Phase

As UCA can generate multi-mode OAM beams with the baseband (partial) DFT, the equivalent baseband signal model of multi-mode OAM data symbols transmitted by MBS can be expressed as $\mathbf{F}^H\mathbf{s}(k_w)$, where $\mathbf{F}^H = \mathbf{I}_P \otimes \mathbf{F}_U^H$ is the P-dimensional block diagonal matrix used to generate multi-mode OAM beams simultaneously, $\mathbf{F}_U$ is a $U \times M$ (partial) DFT matrix, $\otimes$ denotes Kronecker product, $\mathbf{s}^T(k_w) = [\mathbf{s}_1^T(k_w), \mathbf{s}_2^T(k_w), \cdots, \mathbf{s}_P^T(k_w)]$ contains all the data symbols transmitted to P SBSs, and $\mathbf{s}_p^T(k_w) = [s_p(\ell_1, k_w), s_p(\ell_2, k_w), \cdots, s_p(\ell_U, k_w)]$ is the data symbol vector transmitted to the p-th $(1 \le p \le P)$ SBS on U OAM modes at the w-th subcarrier. Besides, the interferences in the MU-OAM channel are considered to be eliminated at MBS; thus, the data symbol vector $\mathbf{s}(k_w)$ needs to be preprocessed before performing baseband (partial) DFT, i.e., $\mathbf{F}^H\mathbf{P}(k_w)\mathbf{s}(k_w)$, where $\mathbf{P}(k_w)$ is the $PU \times PU$ preprocessing matrix at the w-th subcarrier.

Correspondingly, the received downlink signal vector at the p-th SBS can be written as

$$\mathbf{y}_p(k_w) = \mathbf{H}_p(k_w)\mathbf{F}^H\mathbf{P}(k_w)\mathbf{s}(k_w) + \mathbf{z}_p(k_w), \tag{5.8}$$

where $\mathbf{z}_p(k_w)$ is the noise vector. The UCA-based OAM receiver at SBSs has the similar baseband digital structure to the transmitter but with the opposite operations, i.e., separating different OAM modes and despiralizing each mode, as shown in Fig. 5.4b. Thus, the detected OAM data symbol vector $\mathbf{x}_p(k_w)$ at the p-th SBS can be expressed as

$$\mathbf{x}_p(k_w) = \mathbf{F}_U\left(\mathbf{H}_p(k_w)\mathbf{F}^H\mathbf{P}(k_w)\mathbf{s}(k_w) + \mathbf{z}_p(k_w)\right), \tag{5.9}$$

where $\mathbf{x}_p^T(k_w) = [x_p(\ell_1, k_w), x_p(\ell_2, k_w), \cdots, x_p(\ell_U, k_w)]$, $x_p(\ell_u, k_w)$ is the detected data symbol of the p-th SBS on the u-th $(1 \le u \le U)$ OAM mode at the w-th $(1 \le w \le W)$ subcarrier. Furthermore, how to design the preprocessing matrix $\mathbf{P}(k_w)$ will be specified in Sect. 3.4.

5.3 MU-OAM Distance and AoA Estimation

In this section, we propose a MU-OAM distance and AoA estimation method based on the novel recursive ESPRIT algorithm, which forms the basis of the following MU-OAM preprocessing.

5.3.1 Problem Formulation

We can observe from (5.3) and (5.4) that if $k_w (1 \le w \le \widetilde{W} \le W)$, R_r and R_t are known to MBS, the channel matrix from MBS to P SBSs is only characterized by the position of each SBS. Accordingly, pilot data should be exploited for the distance and AoA estimations of P SBSs during the training phase.

According to (5.7), the combined signal $x_t'(\ell_u, k_w)$ can be derived in (5.10), where $\mathbf{h}_n^p(k_w) = [h_{n,1}^p(k_w), h_{n,2}^p(k_w), \cdots, h_{n,M}^p(k_w)]$, $\boldsymbol{k}_w$ is the wave vector, $\boldsymbol{d}_{n,m}^p$ is the position vector from the m-th antenna element of p-th SBS to the n-th antenna element of MBS, $\sigma_{\ell_u,k_w} = \frac{MN\beta i^{-\ell_u}}{2k_w}$, $J_{\ell_u}(\cdot)$ is ℓ_uth-order Bessel function of the first kind, and $z_t'(\ell_u, k_w)$ is the noise. In far field, $|\boldsymbol{r}_p - \boldsymbol{r}_m'^p + \boldsymbol{r}_n'| \approx r_p$ for amplitudes, and $|\boldsymbol{r}_p - \boldsymbol{r}_m'^p + \boldsymbol{r}_n'| \approx r_p - \hat{\boldsymbol{r}}_p \cdot \boldsymbol{r}_m^p + \hat{\boldsymbol{r}}_p \cdot \boldsymbol{r}_n$ for phases, where $\boldsymbol{r}_p$ is the position vector from the UCA center of the p-th SBS to the UCA center of MBS, $\hat{\boldsymbol{r}}_p$ is the unit vector of $\boldsymbol{r}_p$, $\boldsymbol{r}_m^p = R_r(\boldsymbol{x}_p \cos\alpha_p + \boldsymbol{y}_p \sin\alpha_p)$, $\boldsymbol{r}_n = R_t(\boldsymbol{x}\cos\phi + \boldsymbol{y}\sin\phi)$, and $\boldsymbol{x}_p$, $\boldsymbol{y}_p$, $\boldsymbol{x}$, and $\boldsymbol{y}$ are the unit vectors of X$_p$-axis, Y$_p$-axis, X-axis, and Y-axis, respectively.

Overall, the aim of the MU-OAM distance and AoA estimation is to obtain the distance, the azimuthal angle and the elevation angle of each SBS's UCA center from the signals $\{x_t'(\ell_u, k_w) | u = 1, 2, \ldots, \widetilde{U}; w = 1, 2, \cdots, \widetilde{W}\}$.

5.3.2 *Distance and AoA Estimation*

Since the training signals of multiple SBSs have different amplitudes arriving at MBS, it is difficult to directly extract the exponentials containing r_p and φ_p from (5.10). Hence, the conventional ESPRIT-based distance and AoA estimation methods proposed in Sect. 2.4 are not suitable for MU-OAM scenario. From (5.10), we can observe that φ_p and θ_p are both intrinsically related to the OAM mode ℓ_u and r_p is related to the subcarrier k_w. Accordingly, we estimate $\{(\theta_p, \varphi_p) | p = 1, 2, \cdots, P\}$ by processing training signals in the OAM mode domain and estimate $\{r_p, | p = 1, 2, \cdots, P\}$ by processing training signals in the frequency domain.

5.3.2.1 Estimation of θ_p and φ_p Based on Recursive ESPRIT

Neglecting constant terms and performing phase compensation, the combined signal $x_t'(\ell_u, k_w)$ in (5.10) can be simplified as

$$
\begin{aligned}
x_t'(\ell_u, k_w) &= \sum_{n=1}^{N} \left(\sum_{p=1}^{P} \mathbf{h}_n^p(k_w) \mathbf{f}^H(\ell_u) s_p'(k_w) \right) + z_t'(\ell_u, k_w) \\
&= \frac{\beta}{2k_w} \sum_{p=1}^{P} \sum_{n=1}^{N} \sum_{m=1}^{M} \frac{e^{i k_w |\boldsymbol{d}_{n,m}^p|}}{|\boldsymbol{d}_{n,m}^p|} e^{i\ell_u \alpha_p} s_p'(k_w) + z_t'(\ell_u, k_w) \\
&= \frac{\beta}{2k_w} \sum_{p=1}^{P} \sum_{n=1}^{N} \sum_{m=1}^{M} \frac{e^{i k_w |\boldsymbol{r}_p - \boldsymbol{r}_m'^p + \boldsymbol{r}_n'|}}{|\boldsymbol{r}_p - \boldsymbol{r}_m'^p + \boldsymbol{r}_n'|} e^{i\ell_u \alpha_p} s_p'(k_w) + z_t'(\ell_u, k_w)
\end{aligned}
$$

$$\approx \frac{\beta}{2k_w}\sum_{p=1}^{P}\frac{e^{ik_w r_p}}{r_p}\sum_{n=1}^{N}e^{i\boldsymbol{k}_w\cdot\boldsymbol{r}_n}\sum_{m=1}^{M}e^{-i(\boldsymbol{k}_w\cdot\boldsymbol{r}_m^p-\ell_u\alpha_p)}s'_p(k_w)+z'_t(\ell_u,k_w)$$

$$\approx \sigma_{\ell_u,k_w}\sum_{p=1}^{P}s'_p(k_w)\frac{e^{ik_w r_p}}{r_p}e^{i\ell_u\varphi_p}J_{\ell_u}(k_w R_r\sin\theta_p)$$

$$\times J_0(k_w R_t\sin\theta_p)+z'_t(\ell_u,k_w), \quad (5.10)$$

$$\tilde{x}'_t(\ell_u,k_w)=\sum_{p=1}^{P}s'_p(k_w)\frac{e^{ik_w r_p}}{r_p}e^{i\ell_u\varphi_p}J_{\ell_u}(k_w R_r\sin\theta_p)J_0(k_w R_t\sin\theta_p)+\tilde{z}'_t(\ell_u,k_w), \quad (5.11)$$

where $\tilde{z}'_t(\ell_u,k_w)$ is the noise. Then, all the signals received on the $\widetilde{U}$ OAM modes at the $\widetilde{W}$ subcarriers can be collected in the matrix

$$\widetilde{\mathbf{X}}=\begin{bmatrix}\tilde{x}'_t(\ell_1,k_1) & \tilde{x}'_t(\ell_1,k_2) & \cdots & \tilde{x}'_t(\ell_1,k_{\widetilde{W}})\\ \tilde{x}'_t(\ell_2,k_1) & \tilde{x}'_t(\ell_2,k_2) & \cdots & \tilde{x}'_t(\ell_2,k_{\widetilde{W}})\\ \vdots & \vdots & \ddots & \vdots\\ \tilde{x}'_t(\ell_{\widetilde{U}},k_1) & \tilde{x}'_t(\ell_{\widetilde{U}},k_2) & \cdots & \tilde{x}'_t(\ell_{\widetilde{U}},k_{\widetilde{W}})\end{bmatrix}. \quad (5.12)$$

In the estimation of θ_p and φ_p, we first denote the w-th column of $\widetilde{\mathbf{X}}$ as a column vector $\tilde{\mathbf{x}}_d^w$, i.e.,

$$\tilde{\mathbf{x}}_d^w=\left[\tilde{x}'_t(\ell_1,k_w),\cdots,\tilde{x}'_t(\ell_{\widetilde{U}},k_w),)\right]^T. \quad (5.13)$$

Considering the symmetry of OAM modes used in the training phase, $\tilde{\mathbf{x}}_d^w$ can be rewritten as

$$\tilde{\mathbf{x}}_d^w=\left[\tilde{x}'_t(-\lfloor\widetilde{U}/2\rfloor,k_w),\cdots,\tilde{x}'_t(0,k_w),\cdots,\tilde{x}'_t(\lfloor\widetilde{U}/2\rfloor,k_w),)\right]^T. \quad (5.14)$$

Then, the vector $\tilde{\mathbf{x}}_d^w$ can be given in compact form by

$$\tilde{\mathbf{x}}_d^w=\mathbf{A}_w\tilde{\mathbf{s}}(k_w)+\mathbf{z}'(k_w), \quad (5.15)$$

where the $\widetilde{U}\times P$ *direction matrix* takes the form $\mathbf{A}_w=[\mathbf{a}_w(\theta_1,\varphi_1),\cdots,\mathbf{a}_w(\theta_p,\varphi_p),\cdots,\mathbf{a}_w(\theta_P,\varphi_P)]$; the vector $\mathbf{a}_w(\theta_p,\varphi_p)=[e^{-i\lfloor\widetilde{U}/2\rfloor\varphi_p}J_{-\lfloor\widetilde{U}/2\rfloor}(k_w R_r\sin\theta_p),\cdots,J_0(k_w R_r\sin\theta_p),\cdots,e^{i\lfloor\widetilde{U}/2\rfloor\varphi_p}J_{\lfloor\widetilde{U}/2\rfloor}(k_w R_r\sin\theta_p)]^T$ contains the AoA information of the p-th SBS, $\tilde{\mathbf{s}}(k_w)=\boldsymbol{\Upsilon}(k_w)\mathbf{s}(k_w)$, $\boldsymbol{\Upsilon}(k_w)=\mathrm{diag}\{J_0(k_w R_t\sin\theta_1)e^{ik_w r_1}/r_1,\cdots,J_0(k_w R_t\sin\theta_P)e^{ik_w r_P}/r_P\}$, $\mathbf{s}=[s'_1(k_w),\cdots,s'_P(k_w)]^T$; and $\mathbf{z}'(k_w)$ is the noise vector.

By introducing the $(\widetilde{U}-2)\times\widetilde{U}$ matrices $\Delta^{(-1)}=[\mathbf{I}_{\widetilde{U}-2},\mathbf{0}_{(\widetilde{U}-2)\times 2}]$, $\Delta^{(0)}=[\mathbf{0}_{(\widetilde{U}-2)\times 1},\mathbf{I}_{\widetilde{U}-2},\mathbf{0}_{(\widetilde{U}-2)\times 1}]$ and $\Delta^{(1)}=[\mathbf{0}_{(\widetilde{U}-2)\times 2},\mathbf{I}_{\widetilde{U}-2}]$ that select the first, middle, and last $\widetilde{U}-2$ rows of $\mathbf{a}_w(\theta_p,\varphi_p)$, we can choose a subset of the elements of $\mathbf{a}_w(\theta_p,\varphi_p)$ to obtain the three vectors

$$\mathbf{a}_w^{(\gamma)}(\theta_p,\varphi_p)=\Delta^{(\gamma)}\mathbf{a}_w(\theta_p,\varphi_p),\quad \gamma=-1,0,1, \tag{5.16}$$

such that the phases (excluding the Bessel functions) of $\mathbf{a}_w^{(0)}(\theta_p,\varphi_p)$, $e^{i\varphi_p}\mathbf{a}_w^{(-1)}(\theta_p,\varphi_p)$, and $e^{-i\varphi_p}\mathbf{a}_w^{(1)}(\theta_p,\varphi_p)$ are the same. Moreover, considering the recursive relation, the characteristic of Bessel functions,

$$J_{x-1}(\xi)+J_{x+1}(\xi)=\frac{2x}{\xi}J_x(\xi) \tag{5.17}$$

the three vectors $\mathbf{a}_w^{(-1)}(\theta_p,\varphi_p)$, $\mathbf{a}_w^{(0)}(\theta_p,\varphi_p)$, and $\mathbf{a}_w^{(1)}(\theta_p,\varphi_p)$ satisfy the relationship

$$\mathbf{\Gamma}\mathbf{a}_w^{(0)}(\theta_p,\varphi_p)=\mu_p\mathbf{a}_w^{(-1)}(\theta_p,\varphi_p)+\mu_p^*\mathbf{a}_w^{(1)}(\theta_p,\varphi_p), \tag{5.18}$$

where $\mu_p=\sin\theta_p e^{i\varphi_p}$, and $\mathbf{\Gamma}=(2/(kR_r))\mathrm{diag}\{-\lfloor\widetilde{U}/2\rfloor,\cdots,\lfloor\widetilde{U}/2\rfloor\}$. Correspondingly, applying the equivalence (5.18) to all the columns of the direction matrix $\mathbf{A}_w$ yields

$$\mathbf{\Gamma}\mathbf{A}_w^{(0)}=\mathbf{A}_w^{(-1)}\mathbf{\Phi}+\mathbf{A}_w^{(1)}\mathbf{\Phi}^{\mathrm{H}}, \tag{5.19}$$

where $\mathbf{A}_w^{(\gamma)}=\Delta^{(\gamma)}\mathbf{A}_w, \gamma=-1,0,1$, $\mathbf{\Phi}=\mathrm{diag}\{\mu_1,\cdots,\mu_P\}$.

The covariance matrix of $\tilde{\mathbf{x}}_d^w$ can be represented as

where $\mathbf{R}_{\tilde{\mathbf{s}}}^w=\mathbf{\Upsilon}(k_w)\mathbb{E}\left\{\mathbf{s}(k_w)\mathbf{s}^H(k_w)\right\}\mathbf{\Upsilon}^H(k_w)$, $\mathbf{R}_{\mathbf{z}'}^w=\mathbb{E}\left\{\mathbf{z}'(k_w)\mathbf{z}'^H(k_w)\right\}$. Then, the EVD of $\mathbf{R}_d^w$ is written as

$$\mathbf{R}_d^w=\sum_{u=1}^{\widetilde{U}}\lambda_u\mathbf{u}_u\mathbf{u}_u^H=\mathbf{U}\mathbf{\Lambda}\mathbf{U}^H, \tag{5.20}$$

where $\{\lambda_u|u=1,2,\cdots,\widetilde{U}\}$ are the eigenvalues of $\mathbf{R}_d^w$, $\mathbf{\Lambda}=\mathrm{diag}\{\lambda_1,\lambda_2,\cdots,\lambda_{\widetilde{U}}\}$, $\mathbf{u}_u$ is the eigenvector corresponding to the eigenvalue λ_u, $\mathbf{U}=[\mathbf{u}_1,\cdots,\mathbf{u}_{\widetilde{U}}]$. In practice, the training signal and noise are independent of each other; thus, the covariance matrix $\mathbf{R}_d^w$ can be decomposed into two mutually orthogonal parts:

$$\mathbf{R}_d^w=\mathbf{U}_s\mathbf{\Lambda}_s\mathbf{U}_s^{\mathrm{H}}+\mathbf{U}_n\mathbf{\Lambda}_n\mathbf{U}_n^{\mathrm{H}}, \tag{5.21}$$

where $\mathbf{\Lambda}_s$ is the P-dimensional diagonal matrix that contains the P largest eigenvalues of $\mathbf{R}_d^w$, $\mathbf{U}_s$ is the $\widetilde{U}\times P$ signal subspace spanned by the eigenvectors corresponding to the P larger eigenvalues, $\mathbf{\Lambda}_n$ is the $(\widetilde{U}-P)$-dimensional diagonal

Algorithm 1 AoA estimation by recursive ESPRIT

Input: $\tilde{\mathbf{x}}_d^w$

Output: $\{(\hat{\theta}_p^w, \hat{\varphi}_p^w)|p = 1, 2, \cdots, P\}$

procedure

$\mathbf{R}_d^w \leftarrow$

$\mathbf{U}, \mathbf{\Lambda} \leftarrow$ decompose $\mathbf{R}_d^w$ such that $\mathbf{U\Lambda U}^H$

$\mathbf{\Lambda} \leftarrow \mathrm{diag}\{\lambda_1, \cdots, \lambda_{\widetilde{U}}\}, \lambda_1 \geq \cdots \geq \lambda_{\widetilde{U}}$

$\mathbf{\Lambda}_s \leftarrow \mathrm{diag}\{\lambda_1, \cdots, \lambda_P\}$

$\mathbf{U}_s \leftarrow$ the P columns of $\mathbf{U}$ corresponding to $\mathbf{\Lambda}_s$

$\mathbf{U}_s^{(\gamma)} \leftarrow \Delta^{(\gamma)}\mathbf{U}_s, \gamma = -1, 0, 1$

$\mathbf{\Gamma} \leftarrow (2/(kR_r))\mathrm{diag}\{-\lfloor\widetilde{U}/2\rfloor, \cdots, \lfloor\widetilde{U}/2\rfloor\}$

$\widehat{\mathbf{\Psi}} \leftarrow \left[\mathbf{U}_s^{(-1)}, \mathbf{U}_s^{(1)}\right]^{\dagger} \mathbf{\Gamma U}_s^{(0)}$

$\mathbf{\Psi} \leftarrow$ the first P rows of $\widehat{\mathbf{\Psi}}$

$\mathbf{\Psi}' \leftarrow$ the last P rows of $\widehat{\mathbf{\Psi}}$

$\mathbf{\Phi} \leftarrow$ decompose $\mathbf{\Psi}$ such that $\mathbf{T\Phi T}^{-1}$

$\mathbf{\Phi}^{\mathrm{H}} \leftarrow$ decompose $\mathbf{\Psi}'$ such that $\mathbf{T\Phi}^{\mathrm{H}}\mathbf{T}^{-1}$

For $p = 1 \rightarrow P$

$\hat{\varphi}_p^0 \leftarrow \arg(\mathbf{\Phi}(p, p))$

$\hat{\varphi}_p^1 \leftarrow -\arg\left(\mathbf{\Phi}^{\mathrm{H}}(p, p)\right)$

$\hat{\varphi}_p^w \leftarrow (\hat{\varphi}_p^0 + \hat{\varphi}_q^1)/2$

$\hat{\theta}_p^0 \leftarrow \arcsin(|\mathbf{\Phi}(p, p)|)$

$\hat{\theta}_p^1 \leftarrow \arcsin\left(|\mathbf{\Phi}^{\mathrm{H}}(p, p)|\right)$

$\hat{\theta}_p^w \leftarrow (\hat{\theta}_p^0 + \hat{\theta}_p^1)/2$

end for

end procedure

matrix that contains the $\widetilde{U} - P$ smaller eigenvalues of $\mathbf{R}_d^w$, and $\mathbf{U}_n$ is the $\widetilde{U} \times (\widetilde{U} - P)$ noise subspace spanned by the eigenvectors corresponding to the $\widetilde{U} - P$ smaller eigenvalues.

Since the range of $\mathbf{U}_s$ is equal to that of $\mathbf{A}_w$, i.e., $\mathcal{R}\{\mathbf{U}_s\} = \mathcal{R}\{\mathbf{A}_w\}$, there exists a nonsingular matrix $\mathbf{T}$ such that

$$\mathbf{U}_s = \mathbf{A}_w\mathbf{T}. \tag{5.22}$$

Therefore, the signal subspace $\mathbf{U}_s$ can be similarly denoted as

$$\mathbf{\Gamma U}_s^{(0)} = \mathbf{\Gamma A}_w^{(0)}\mathbf{T} = \mathbf{A}_w^{(-1)}\mathbf{TT}^{-1}\mathbf{\Phi T} + \mathbf{A}_w^{(1)}\mathbf{TT}^{-1}\mathbf{\Phi}^{\mathrm{H}}\mathbf{T} = \mathbf{U}_s^{(-1)}\mathbf{\Psi} + \mathbf{U}_s^{(1)}\mathbf{\Psi}', \tag{5.23}$$

where $\mathbf{U}_s^{(\gamma)} = \Delta^{(\gamma)}\mathbf{U}_s, \gamma = -1, 0, 1$, $\mathbf{\Psi} = \mathbf{T}^{-1}\mathbf{\Phi T}$ and $\mathbf{\Psi}' = \mathbf{T}^{-1}\mathbf{\Phi}^{\mathrm{H}}\mathbf{T}$. Based on the (5.23), one obtains

$$\begin{bmatrix} \mathbf{\Psi} \\ \mathbf{\Psi}' \end{bmatrix} = \left[\mathbf{U}_s^{(-1)}, \mathbf{U}_s^{(1)}\right]^{\dagger} \mathbf{\Gamma U}_s^{(0)}, \tag{5.24}$$

and the eigenvalues of $\boldsymbol{\Psi}$ and $\boldsymbol{\Psi}'$ give the diagonal elements of $\boldsymbol{\Phi}$ and $\boldsymbol{\Phi}^{\rm H}$, which are $\{\mu_p = \sin\theta_p e^{i\varphi_p}|p = 1, 2, \cdots, P\}$ and $\{\mu_p^* = \sin\theta_p e^{-i\varphi_p}|p = 1, 2, \cdots, P\}$, respectively. The eigenvalues of $\boldsymbol{\Psi}$ and $\boldsymbol{\Psi}'$ therefore yield automatically paired azimuth angle φ_p and elevation angle θ_p, and the detailed procedure of the proposed recursive ESPRIT algorithm is shown in Algorithm 1.

Hence, there are $\widetilde{W}$ estimates of φ_p and θ_p in total, which can be expressed as

$$\hat{\theta}_p^w = \theta + \varepsilon_\theta^w,\ \hat{\varphi}_p^w = \varphi + \varepsilon_\varphi^w,\ p = 1, 2, \cdots, P,$$

where $\{(\varepsilon_\theta^w, \varepsilon_\varphi^w)|w = 1, 2, \cdots, \widetilde{W}\}$ represent the estimation errors at $\widetilde{W}$ subcarriers. Suppose that $\{(\varepsilon_\theta^w, \varepsilon_\varphi^w)|w = 1, 2, \cdots, \widetilde{W}\}$ have the same average variance $\mathrm{Var}(\varepsilon_\theta)$ and $\mathrm{Var}(\varepsilon_\varphi)$, thus,

$$\begin{aligned} \mathrm{Var}\left(\frac{1}{\widetilde{W}}\sum_{w=1}^{\widetilde{W}}\hat{\theta}_p^w\right) &= \frac{\mathrm{Var}(\varepsilon_\theta)}{\widetilde{W}}, \\ \mathrm{Var}\left(\frac{1}{\widetilde{W}}\sum_{w=1}^{\widetilde{W}}\hat{\varphi}_p^w\right) &= \frac{\mathrm{Var}(\varepsilon_\varphi)}{\widetilde{W}}. \end{aligned} \tag{5.25}$$

Therefore, $\hat{\theta}_p = \frac{1}{\widetilde{W}}\sum_{w=1}^{\widetilde{W}}\hat{\theta}_p^w$ and $\hat{\varphi}_p = \frac{1}{\widetilde{W}}\sum_{w=1}^{\widetilde{W}}\hat{\varphi}_p^w$ are adopted as the estimate of (θ_p, φ_p), $p = 1, 2, \cdots, P$.

5.3.2.2 Estimation of r_p Based on FFT

In the estimation of r_p, we similarly denote the u-th row of $\widetilde{\mathbf{X}}$ as a column vector $\tilde{\mathbf{x}}_e^u$, i.e.,

$$\tilde{\mathbf{x}}_e^u = \left[\tilde{x}_t'(\ell_u, k_1), \cdots, \tilde{x}_t'(\ell_u, k_{\widetilde{W}}),)\right]^T. \tag{5.26}$$

From (5.11), we can find that the subcarrier k_w and the the distance r_p are related through the exponential term $e^{ik_w r_p}$. According to the Fourier transform, k_w and r_p satisfy the dual relationship, which means that the estimates $\{\hat{r}_p^u|p = 1, 2, \cdots, P\}$ can be obtained directly by FFT for (5.26) in frequency domain [10].

Therefore, there are $\widetilde{U}$ estimates of r_p, which can be expressed as

$$\hat{r}_p^u = r + \varepsilon_r^u,\ p = 1, 2, \cdots, P,$$

where $\{\varepsilon_r^u|w = 1, 2, \cdots, \widetilde{W}\}$ are the estimation errors at $\widetilde{U}$ modes. Finally, $\hat{r}_p = \frac{1}{\widetilde{U}}\sum_{u=1}^{\widetilde{U}}\hat{r}_p^u$ is adopted as the estimate of r_p, $p = 1, 2, \cdots, P$. Thus, the MU-OAM distance and AoA estimation is achieved.

5.4 Preprocessing of Downlink MU-OAM System

In this section, we first apply a MU-OAM preprocessing scheme for the proposed system, which can effectively eliminate the co-mode and inter-mode interferences in the downlink MU-OAM channel. After that, we analyze SE and EE performances of the downlink MU-OAM wireless backhaul system.

5.4.1 Problem Formulation

From (5.9), the effective OAM channel matrix from MBS to P SBSs at the w-th subcarrier can be expressed as

$$\mathbf{H}_{\mathrm{OAM}}(k_w) = \mathbf{F}\mathbf{H}(k_w)\mathbf{F}^H = [\mathbf{H}_{\mathrm{OAM}}^{p,q}(k_w)]_{P\times P}, \tag{5.27}$$

where $\mathbf{H}_{\mathrm{OAM}}^{p,q}(k_w) = \mathbf{F}_U\mathbf{H}_{p,q}(k_w)\mathbf{F}_U^H$ is the $U \times U$ effective OAM channel matrix corresponding to $\mathbf{H}_{p,q}(k_w)$, and the expression of the elements in $\mathbf{H}_{\mathrm{OAM}}^{p,q}(k_w)$ can be written as

$$\begin{aligned} h_{\mathrm{OAM}}^{p,q}(u,v) &= \sum_{m=1}^{M}\sum_{\bar{n}=1}^{M} h_{m,q+(\bar{n}-1)\times P}^{p} \exp\left(-i\ell_u\alpha_p + i\ell_v\phi'\right) \\ &= \delta_p(k_w)\sum_{m=1}^{M}\sum_{\bar{n}=1}^{M}\exp\Bigg(-i\ell_u\alpha_p + i\ell_v\phi' - ik_wR_r\sin\theta_p\cos(\varphi_p-\alpha_p) \\ &\quad + i\frac{k_wR_tR_r}{r_p}\cos\left(\alpha_p-\phi''\right) + ik_wR_t\sin\theta_p\cos\left(\varphi_p-\phi''\right)\Bigg), \end{aligned} \tag{5.28}$$

where $\delta_p(k_w) = \frac{\beta}{2k_wr_p}\exp\left(-ik_wr_p\right)$, $\phi' = \frac{2\pi(\bar{n}-1)}{M}$, and $\phi'' = \frac{2\pi[(\bar{n}-1)\times P+q-1]}{N}$. Given that k_w, R_r, R_t, M, and ℓ_u are known to MBS, the effective channel coefficient $h_{\mathrm{OAM}}^{p,q}(u,v)$ is only the function of r_p, φ_p, and θ_p. Hence, MBS can obtain the full information of $\mathbf{H}_{\mathrm{OAM}}(k_w)$ after the MU-OAM distance and AoA estimation. Naturally, the preprocessing process should be performed at MBS to eliminate interferences in the downlink MU-OAM channel.

Denote the effective OAM channel matrix from MBS to the p-th SBS as $\mathbf{H}_{\mathrm{OAM}}^{p}(k_w) = \mathbf{F}_U\mathbf{H}_p(k_w)\mathbf{F}^H$ and the associated preprocessing matrix as $\mathbf{P}_p(k_w)$. Since the role of $\mathbf{P}_p(k_w)$ is to eliminate the co-mode interference from other users and the inter-mode interference in $\mathbf{H}_{\mathrm{OAM}}^{p}(k_w)$, $\mathbf{H}_{\mathrm{OAM}}^{p}(k_w)\mathbf{P}_q(k_w) = 0 (p \neq q)$ and $\mathbf{H}_{\mathrm{OAM}}^{p}(k_w)\mathbf{P}_p(k_w)$ should be the diagonal matrix with $\mathbf{H}_{\mathrm{OAM}}^{p}(k_w)$ being the $U \times PU$ matrix and $\mathbf{P}_p(k_w)$ being the $PU \times U$ matrix.

5.4.2 *Preprocessing for Interference Elimination in Downlink MU-OAM System*

5.4.2.1 Co-mode Interference Elimination

We first construct the co-mode interference elimination matrix as

$$\mathbf{E}(k_w) = [\mathbf{E}_1(k_w), \cdots, \mathbf{E}_p(k_w), \cdots, \mathbf{E}_P(k_w)], \tag{5.29}$$

where $\mathbf{E}_p(k_w)$ is the $PU \times U$ submatrix for the p-th SBS at the w-th subcarrier. To eliminate the co-mode interference from other users, the constraint $\mathbf{H}^p_{\mathrm{OAM}}(k_w)\mathbf{E}_q(k_w) = 0$ for $p \neq q$ should be satisfied. That is, $\mathbf{E}_p(k_w)$ should lie within the null space of $\widehat{\mathbf{H}}^p_{\mathrm{OAM}}(k_w)$, where $\widehat{\mathbf{H}}^p_{\mathrm{OAM}}(k_w)$ is defined as

$$\widehat{\mathbf{H}}^p_{\mathrm{OAM}}(k_w) = \begin{bmatrix} \mathbf{H}^1_{\mathrm{OAM}}(k_w) \\ \vdots \\ \mathbf{H}^{p-1}_{\mathrm{OAM}}(k_w) \\ \mathbf{H}^{p+1}_{\mathrm{OAM}}(k_w) \\ \vdots \\ \mathbf{H}^P_{\mathrm{OAM}}(k_w) \end{bmatrix}. \tag{5.30}$$

Fortunately, the null space of $\widehat{\mathbf{H}}^p_{\mathrm{OAM}}(k_w)$ can be computed by the SVD of $\widehat{\mathbf{H}}^p_{\mathrm{OAM}}(k_w)$, being written as

$$\widehat{\mathbf{H}}^p_{\mathrm{OAM}}(k_w) = \widehat{\mathbf{U}}_{p,w}\widehat{\mathbf{\Sigma}}_{p,w}\widehat{\mathbf{V}}^H_{p,w} = \widehat{\mathbf{U}}_{p,w}\widehat{\mathbf{\Sigma}}_{p,w}\left[\widehat{\mathbf{V}}^{(1)}_{p,w}, \widehat{\mathbf{V}}^{(2)}_{p,w}\right], \tag{5.31}$$

$\widehat{\mathbf{U}}_{p,w}$ is the $(P-1)U \times (P-1)U$ left singular matrix, $\widehat{\mathbf{\Sigma}}_{p,w}$ is the $(P-1)U \times PU$ matrix including all singular values of $\widehat{\mathbf{H}}^p_{\mathrm{OAM}}(k_w)$, and $\widehat{\mathbf{V}}^H_{p,w}$ is the $PU \times PU$ right singular matrix. Denote the rank of $\widehat{\mathbf{H}}^p_{\mathrm{OAM}}(k_w)$ as $\hat{L}_p$; then $\widehat{\mathbf{V}}^H_{p,w}$ can be divided into two parts, where $\widehat{\mathbf{V}}^{(1)}_{p,w}$ holds the first $\hat{L}_p$ right singular vectors and $\widehat{\mathbf{V}}^{(2)}_{p,w}$ holds the last $(PU - \hat{L}_p)$ right singular vectors. Therefore, $\widehat{\mathbf{V}}^{(2)}_{p,w}$ forms an orthogonal basis for the null space of $\widehat{\mathbf{H}}^p_{\mathrm{OAM}}(k_w)$, and its columns are, thus, candidates for $\mathbf{E}_p(k_w)$. As $\hat{L}_p \leq (P-1)U$, $\widehat{\mathbf{V}}^{(2)}_{p,w}$ contains at least U right singular vectors. Thereafter, we choose the last U columns of $\widehat{\mathbf{V}}^{(2)}_{p,w}$ to form $\mathbf{E}_p(k_w)$.

After eliminating the co-mode interference by $\mathbf{E}(k_w)$, the effective OAM channel matrix of downlink MU-OAM wireless backhaul system at the w-th subcarrier becomes

$$\mathbf{H}'_{\text{OAM}}(k_w) = \mathbf{H}_{\text{OAM}}(k_w)\mathbf{E}(k_w) = \begin{bmatrix} \mathbf{H}^1_{\text{OAM}}(k_w)\mathbf{E}_1(k_w) & & \mathbf{0} \\ & \ddots & \\ \mathbf{0} & & \mathbf{H}^P_{\text{OAM}}(k_w)\mathbf{E}_P(k_w) \end{bmatrix}. \tag{5.32}$$

where $\mathbf{H}^p_{\text{OAM}}(k_w)\mathbf{E}_p(k_w)$ is the $U \times U$ effective OAM channel matrix from MBS to the p-th SBS. Comparing (5.27) and (5.32), we can see that the MU-OAM channel is decoupled to P parallel single-user OAM channels by $\mathbf{E}(k_w)$.

5.4.2.2 Inter-mode Interference Elimination

The angles $\{(\theta_p, \varphi_p)|p = 1, 2, \cdots, P\}$ between the transmit and each receive OAM beam cause inter-mode interference, which results in $\mathbf{H}^p_{\text{OAM}}(k_w)\mathbf{E}_p(k_w)$ not being a diagonal matrix [2]. Thus, the purpose of the inter-mode interference elimination is to remove the off-diagonal elements in $\mathbf{H}^p_{\text{OAM}}(k_w)\mathbf{E}_p(k_w)$ so that $\mathbf{H}^p_{\text{OAM}}(k_w)\mathbf{E}_p(k_w)$ can be diagonalized.

We construct the inter-mode interference elimination matrix as

$$\mathbf{G}(k_w) = \text{diag}\{\mathbf{G}_1(k_w), \cdots, \mathbf{G}_p(k_w), \cdots, \mathbf{G}_P(k_w)\}, \tag{5.33}$$

where $\mathbf{G}_p(k_w)$ is the $U \times U$ submatrix for the p-th SBS at the w-th subcarrier. The role of $\mathbf{G}_p(k_w)$ can be viewed as diagonalizing the product $\mathbf{H}^p_{\text{OAM}}(k_w)\mathbf{E}_p(k_w)\mathbf{G}_p(k_w)$. Therefore, $\mathbf{G}_p(k_w)$ can be easily formed by the inverse of the product $\mathbf{H}^p_{\text{OAM}}(k_w)\mathbf{E}_p(k_w)$, i.e., $\mathbf{G}_p(k_w) = \left(\mathbf{H}^p_{\text{OAM}}(k_w)\mathbf{E}_p(k_w)\right)^{-1}$.

Through above analysis, the preprocessing matrix $\mathbf{P}_p(k_w)$ can be designed as the cascade of $\mathbf{E}_p(k_w)$ and $\mathbf{G}_p(k_w)$, i.e., $\mathbf{P}_p(k_w) = \mathbf{E}_p(k_w)\mathbf{G}_p(k_w)$. Correspondingly, the preprocessing matrix $\mathbf{P}(k_w)$ of the downlink MU-OAM system can be expressed as

$$\mathbf{P}(k_w) = \mathbf{E}(k_w)\mathbf{G}(k_w) = [\mathbf{E}_1(k_w)\mathbf{G}_1(k_w), \cdots, \mathbf{E}_P(k_w)\mathbf{G}_P(k_w)]. \tag{5.34}$$

As the effective channel of the downlink MU-OAM system is completely characterized by the position of each SBS, the preprocessing matrix $\mathbf{P}(k_w)$ can be designed only according to the position information of P SBSs instead of all the channel state information (CSI). Therefore, the channel parameters needed by the proposed MU-OAM preprocessing scheme are significantly less than that in the traditional MIMO precoding. If $\{(\hat{r}_p, \hat{\theta}_p, \hat{\varphi}_p)|p = 1, 2, \cdots, P\}$ obtained by the MU-OAM distance and AoA estimation are accurate enough, the effective OAM channel matrix after preprocessing $\mathbf{H}_{\text{OAM}}(k_w)\mathbf{P}(k_w)$ approaches diagonal. Figure 5.5 shows a block diagram of a UCA-based downlink MU-OAM wireless backhaul system with the preprocessing assisted by the MU-OAM distance and AoA estimation at MBS.

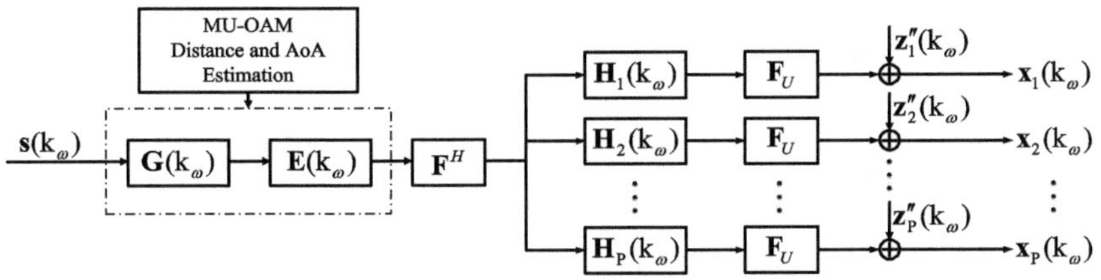

Fig. 5.5 The block diagram of a downlink MU-OAM wireless backhaul system with preprocessing assisted by the MU-OAM distance and AoA estimation at MBS

5.4.3 Spectrum Efficiency of Downlink MU-OAM System

Following the signal model in (5.9), the detected OAM data symbol vector at the p-th SBS can be written as

$$\begin{aligned} \mathbf{x}_p(k_w) =& \mathbf{F}_U\left(\mathbf{H}_p(k_w)\mathbf{F}^H\mathbf{P}(k_w)\mathbf{s}(k_w)+\mathbf{z}_p(k_w)\right) \\ =& \mathbf{F}_U\left(\mathbf{H}_p(k_w)\mathbf{F}^H\sum_{q=1}^{P}\mathbf{P}_q(k_w)\mathbf{s}_q(k_w)+\mathbf{z}_p(k_w)\right) \\ =& \mathbf{H}^p_{\text{OAM}}(k_w)\mathbf{P}_p(k_w)\mathbf{s}_p(k_w)+\mathbf{H}^p_{\text{OAM}}(k_w)\widehat{\mathbf{P}}_p(k_w)\hat{\mathbf{s}}_p(k_w)+\mathbf{z}''_p(k_w) \\ =& \mathbf{s}_p(k_w)+\mathcal{I}^p_{inter}(k_w)+\mathcal{I}^p_{co}(k_w)+\mathbf{z}''_p(k_w). \end{aligned} \tag{5.35}$$

where $\widehat{\mathbf{P}}_p(k_w)$ and $\hat{\mathbf{s}}_p(k_w)$ are, respectively, defined as

$$\begin{aligned} \widehat{\mathbf{P}}_p(k_w) =& [\mathbf{P}_1(k_w),\cdots,\mathbf{P}_{p-1}(k_w),\mathbf{P}_{p+1}(k_w),\cdots,\mathbf{P}_P(k_w)], \\ \hat{\mathbf{s}}^T_p(k_w) =& [\mathbf{s}^T_1(k_w),\cdots,\mathbf{s}^T_{p-1}(k_w),\mathbf{s}^T_{p+1}(k_w),\cdots,\mathbf{s}^T_P(k_w)], \end{aligned}$$

$\mathcal{I}^p_{inter}(k_w)$ and $\mathcal{I}^p_{co}(k_w)$ are, respectively, the U-dimensional inter-mode interference vector and co-mode interference vector induced by the error in preprocessing matrix, and $\mathbf{z}''_p(k_w)$ is the noise vector. Since both the co-mode and inter-mode interferences are eliminated at MBS by the MU-OAM preprocessing, P SBSs can directly obtain detected OAM data symbol vector after separating and despiralizing OAM beams, which can greatly reduce the complexity of each SBS in downlink MU-OAM system. However, considering the inevitable estimation errors of $\{(\hat{r}_p,\hat{\theta}_p,\hat{\varphi}_p)|p=1,2,\cdots,P\}$ in practice, we have to evaluate the effect of these errors on the system SE performance.

Define $\mathbf{R}^p_{\text{inter}}(k_w)$, $\mathbf{R}^p_{\text{co}}(k_w)$, and $\mathbf{R}^p_z(k_w)$ as the $U\times U$ covariance matrices of $\mathcal{I}^p_{inter}(k_w)$, $\mathcal{I}^p_{co}(k_w)$, and $\mathbf{z}''_p(k_w)$, i.e.,

$$\mathbf{R}^p_{\text{inter}}(k_w)=\left(\mathbf{H}^p_{\text{OAM}}(k_w)\mathbf{P}_p(k_w)-\mathbf{I}_U\right)\mathbb{E}\left(\mathbf{s}_p(k_w)\mathbf{s}^H_p(k_w)\right)$$

$$\times \left(\mathbf{H}_{\mathrm{OAM}}^{p}(k_w)\mathbf{P}_p(k_w) - \mathbf{I}_U\right)^H, \tag{5.36}$$

$$\mathbf{R}_{\mathrm{co}}^{p}(k_w) = \sum_{q=1,q\neq p}^{P} (\mathbf{H}_{\mathrm{OAM}}^{p}(k_w)\mathbf{P}_q(k_w))\mathbb{E}\left(\mathbf{s}_q(k_w)\mathbf{s}_q^H(k_w)\right)(\mathbf{H}_{\mathrm{OAM}}^{p}(k_w)\mathbf{P}_q(k_w))^H,$$

$$\mathbf{R}_{z}^{p}(k_w) = \mathbb{E}\left(\mathbf{z}_p''(k_w)(\mathbf{z}_p''(k_w))^H\right), \tag{5.37}$$

where $\mathbf{I}_U$ is the U-dimensional unit matrix. Then, the SINR of the p-th SBS on the u-th OAM mode at the w-th subcarrier can be formulated as

$$\mathrm{SINR}_p(\ell_u, k_w) = \frac{\mathbb{E}\left(|s_p(\ell_u, k_p)|^2\right)}{\left[\mathbf{R}_{\mathrm{inter}}^{p}(k_w)\right]_u + \left[\mathbf{R}_{\mathrm{co}}^{p}(k_w)\right]_u + \left[\mathbf{R}_{z}^{p}(k_w)\right]_u}, \tag{5.38}$$

where $[\cdot]_u$ denotes the u-th diagonal element. Therefore, the SE of the UCA-based downlink MU-OAM wireless backhaul system can be written as

$$C' = \left(1 - \frac{T_p\widetilde{W}}{T_c W}\right)\frac{1}{W}\sum_{p=1}^{P}\sum_{w=1}^{W}\sum_{u=1}^{U}\log_2\left(1 + \mathrm{SINR}_p(\ell_u, k_w)\right). \tag{5.39}$$

5.4.4 *Energy Efficiency of Downlink MU-OAM System*

The EE, defined as the ratio of the system capacity to the total system power consumption, has been recognized as an important performance metric of green communication systems. Due to the high SE of MU-OAM relying on the high transmit power, it is therefore necessary to investigate the EE performance of the MU-OAM wireless backhaul system.

For the proposed MU-OAM system, the consumed total power can be expressed as

$$\mathcal{P}_{\mathrm{total}} = \frac{\mathcal{P}_{\mathrm{T}}}{\rho} + \mathcal{P}_{\mathrm{c}}, \tag{5.40}$$

where $\mathcal{P}_{\mathrm{T}}$ is the total transmit power, ρ is the efficiency of the power amplifier (PA), and $\mathcal{P}_{\mathrm{c}}$ is the total circuit power consumption which is modeled as a linear function of the number of transmission antennas and RF chains. For easier analysis, we consider equal power allocation at MBS and denote the power of the transmit UCA at each subcarrier as $\mathcal{P}_{\mathrm{t}}$; thus, $\mathcal{P}_{\mathrm{T}} = W\mathcal{P}_{\mathrm{t}}$. Besides, for proposed MU-OAM system with the baseband digital structure, the total circuit power consumption includes the powers of the baseband, the RF chains, and the low noise amplifiers (LNAs); thus, $\mathcal{P}_{\mathrm{c}}$ can be written as

$$\begin{aligned}\mathcal{P}_{\mathrm{c}} &= (1+P)\mathcal{P}_{\mathrm{BB}} + (N+MP)\mathcal{P}_{\mathrm{RF}} + MP\mathcal{P}_{\mathrm{LNA}} \\ &= (1+P)\mathcal{P}_{\mathrm{BB}} + 2MP\mathcal{P}_{\mathrm{RF}} + MP\mathcal{P}_{\mathrm{LNA}},\end{aligned} \tag{5.41}$$

where $\mathcal{P}_{\mathrm{BB}}$, $\mathcal{P}_{\mathrm{RF}}$, and $\mathcal{P}_{\mathrm{LNA}}$ are the powers consumed by a baseband processor, an RF chain of the transmitter/receiver, and a LNA, respectively, and $\mathcal{P}_{\mathrm{RF}}$ consists of the powers consumed by a mixer, a local oscillator, a filter, and a DAC/ADC [9]. Then, the EE of the UCA-based MU-OAM wireless backhaul system can be expressed as

$$\eta_{\mathrm{EE}} = \frac{BC'}{\mathcal{P}_{\mathrm{total}}} = \frac{BC'}{W\mathcal{P}_{\mathrm{t}}/\rho + \mathcal{P}_{\mathrm{c}}}, \tag{5.42}$$

where B is the system available bandwidth.

5.5 UCCA-Based Downlink MU-OAM-MIMO Wireless Backhaul System

In this section, we first extend the proposed MU-OAM preprocessing method to the UCCA-based downlink MU-OAM-MIMO wireless backhaul system, which can further increase the JSDCM gain of the MU-OAM communications. After that, we analyze the performance and the computational complexity of the proposed UCCA-based MU-OAM-MIMO wireless backhaul scheme.

5.5.1 Downlink MU-OAM-MIMO Wireless Backhaul Systems With UCCAs

Similar to the single-user OAM wireless communications in Sect. 2.6, we extend the proposed methods to the downlink MU-OAM system combined with MIMO. The transceiver structure of the MU-OAM-MIMO system is similar to that of the MU-OAM system shown in Fig. 5.4 (i.e, replace UCAs with UCCAs).

For the proposed UCCA-based downlink MU-OAM-MIMO system, we assume that the UCCAs at MBS and P SBSs both consist of $\mathfrak{N}$ concentric UCAs with radii $\{R_{t\mathfrak{n}}|\mathfrak{n} = 1, 2, \cdots, \mathfrak{N}\}$ and $\{R_{r\mathfrak{m}}|\mathfrak{m} = 1, 2, \cdots, \mathfrak{N}\}$. Then, the channel matrix of the UCCA-based downlink MU-OAM-MIMO wireless backhaul system can be written as

$$\bar{\mathbf{H}}(k_w) = \left[\bar{\mathbf{H}}_1^T(k_w), \cdots, \bar{\mathbf{H}}_p^T(k_w), \cdots, \bar{\mathbf{H}}_P^T(k_w)\right]^T, \tag{5.43}$$

where $\bar{\mathbf{H}}_p(k_w) = \left[\mathbf{H}^p_{\mathfrak{m},\mathfrak{n}}(k_w)\right]_{\mathfrak{N}\times\mathfrak{N}}$ and $\mathbf{H}^p_{\mathfrak{m},\mathfrak{n}}(k_w)$ denotes the $M \times N$ channel matrix from the $\mathfrak{n}$-th UCA in UCCA of MBS to the $\mathfrak{m}$-th UCA in UCCA of the p-th SBS, $\mathfrak{m}, \mathfrak{n} = 1, 2, \cdots, \mathfrak{N}$, $p = 1, 2, \cdots, P$. Similar to UCA-based downlink MU-OAM systems, each UCA in UCCA of MBS in the downlink MU-OAM-MIMO system is also divided into P groups to simultaneously transmit P OAM-based signals to P SBSs, and then $\mathbf{H}^p_{\mathfrak{m},\mathfrak{n}}(k_w)$ can be expressed as

$$\mathbf{H}^p_{\mathfrak{m},\mathfrak{n}}(k_w) = [\mathbf{H}^{\mathfrak{m},\mathfrak{n}}_{p,1}(k_w), \cdots, \mathbf{H}^{\mathfrak{m},\mathfrak{n}}_{p,q}(k_w), \cdots, \mathbf{H}^{\mathfrak{m},\mathfrak{n}}_{p,P}(k_w)], \tag{5.44}$$

where $\mathbf{H}^{\mathfrak{m},\mathfrak{n}}_{p,q}(k_w)$ is the $M \times M$ channel matrix from the q-th $(1 \leq q \leq P)$ group elements of the $\mathfrak{n}$-th UCA in UCCA of MBS to the $\mathfrak{m}$-th UCA in UCCA of the p-th SBS.

5.5.2 *Preprocessing for Interference Elimination in Downlink MU-OAM-MIMO System*

As the distances and AoAs of $\mathfrak{N}$ concentric UCAs in each SBS are exactly the same, the MU-OAM distance and AoA estimation and MU-OAM preprocessing methods based on a single UCA proposed in Sects. 3.3 and 3.4 can be directly utilized in the UCCA-based system. Meanwhile, the number of training symbols required by the UCCA-based downlink MU-OAM-MIMO wireless backhaul system does not increase with the number of subcarriers, transmit and receive antenna elements, which is opposite in traditional MU-MIMO-OFDM channel estimation.

The effective OAM channel of the MU-OAM-MIMO system can be written as

$$\begin{aligned}\bar{\mathbf{H}}_{\text{OAM}}(k_w) &= \mathbf{I}_P \otimes \left(\mathbf{I}_{\mathfrak{N}} \otimes \mathbf{F}_U\right) \bar{\mathbf{H}}(k_w)\mathbf{I}_{\mathfrak{N}} \otimes \mathbf{F}^H \\ &= \left[\bar{\mathbf{H}}^T_{\text{OAM}_1}(k_w), \cdots, \bar{\mathbf{H}}^T_{\text{OAM}_p}(k_w), \cdots, \bar{\mathbf{H}}^T_{\text{OAM}_P}(k_w)\right]^T,\end{aligned} \tag{5.45}$$

where $\bar{\mathbf{H}}_{\text{OAM}_p}(k_w) = \mathbf{I}_{\mathfrak{N}} \otimes \mathbf{F}_U \bar{\mathbf{H}}_p(k_w)\mathbf{I}_{\mathfrak{N}} \otimes \mathbf{F}^H$ denotes the effective OAM channel matrix from MBS to the p-th SBS. Using the MU-OAM preprocessing method in Sect. 3.4, the preprocessing matrix of MU-OAM-MIMO system can be obtained and is denoted as

$$\bar{\mathbf{P}}(k_w) = \bar{\mathbf{E}}(k_w)\bar{\mathbf{G}}(k_w) = [\bar{\mathbf{E}}_1(k_w)\bar{\mathbf{G}}_1(k_w), \cdots, \bar{\mathbf{E}}_P(k_w)\bar{\mathbf{G}}_P(k_w)]. \tag{5.46}$$

where $\bar{\mathbf{E}}(k_w) = [\bar{\mathbf{E}}_1(k_w), \cdots, \bar{\mathbf{E}}_P(k_w)]$ and $\bar{\mathbf{G}}(k_w) = \text{diag}\{\bar{\mathbf{G}}_1(k_w), \cdots, \bar{\mathbf{G}}_P(k_w)\}$, $\bar{\mathbf{E}}_p(k_w)$ and $\bar{\mathbf{G}}_p(k_w)$ are respectively the $P\mathfrak{N}U \times \mathfrak{N}U$ inter-user interference elimination matrix and $\mathfrak{N}U \times \mathfrak{N}U$ intra-user interference elimination matrix corresponding to the p-th SBS. The block diagram shown in Fig. 5.5 can also be applied to the downlink MU-OAM-MIMO wireless back system as long as $\mathbf{G}(k_w)$, $\mathbf{E}(k_w)$, $\mathbf{F}^H$,

$\mathbf{H}_p(k_w)$, and $\mathbf{F}_U$ are respectively replaced with $\bar{\mathbf{G}}(k_w)$, $\bar{\mathbf{E}}(k_w)$, $\mathbf{I}_{\mathfrak{N}} \otimes \mathbf{F}^H$, $\bar{\mathbf{H}}_p(k_w)$, and $\mathbf{I}_{\mathfrak{N}} \otimes \mathbf{F}_U$.

After that, the detected OAM data symbol vector at the p-th SBS can be written as

$$\begin{aligned}
\bar{\mathbf{x}}_p(k_w) &= \mathbf{I}_{\mathfrak{N}} \otimes \mathbf{F}_U \left(\bar{\mathbf{H}}_p(k_w)\mathbf{I}_{\mathfrak{N}} \otimes \mathbf{F}^H \bar{\mathbf{P}}(k_w)\bar{\mathbf{s}}(k_w) + \bar{\mathbf{z}}_p(k_w) \right) \\
&= \mathbf{I}_{\mathfrak{N}} \otimes \mathbf{F}_U \left(\bar{\mathbf{H}}_p(k_w)\mathbf{I}_{\mathfrak{N}} \otimes \mathbf{F}^H \sum_{q=1}^{P} \bar{\mathbf{P}}_q(k_w)\bar{\mathbf{s}}_q(k_w) + \bar{\mathbf{z}}_p(k_w) \right) \\
&= \bar{\mathbf{H}}_{\mathrm{OAM}_p}(k_w)\bar{\mathbf{P}}_p(k_w)\bar{\mathbf{s}}_p(k_w) + \bar{\mathbf{H}}_{\mathrm{OAM}_p}(k_w)\bar{\mathbf{P}}'_p(k_w)\bar{\mathbf{s}}'_p(k_w) + \bar{\mathbf{z}}''_p(k_w) \\
&= \bar{\mathbf{s}}_p(k_w) + \bar{\mathcal{I}}^1_p(k_w) + \bar{\mathcal{I}}^2_p(k_w) + \bar{\mathbf{z}}''_p(k_w).
\end{aligned} \tag{5.47}$$

where $\bar{\mathbf{s}}(k_w) = [\bar{\mathbf{s}}_1^T(k_w), \bar{\mathbf{s}}_2^T(k_w), \cdots, \bar{\mathbf{s}}_P^T(k_w)]^T$, $\bar{\mathbf{s}}_p(k_w) = [\bar{\mathbf{s}}_{p1}^T(k_w), \bar{\mathbf{s}}_{p2}^T(k_w), \cdots, \bar{\mathbf{s}}_{p\mathfrak{N}}^T(k_w)]^T$ is the data symbol vector transmitted to the p-th SBS and $\bar{\mathbf{s}}_{p\mathfrak{n}}(k_w) = [s_{\mathfrak{n}}^p(\ell_1, k_w), s_{\mathfrak{n}}^p(\ell_2, k_w), \cdots, s_{\mathfrak{n}}^p(\ell_U, k_w)]^T$, $\bar{\mathbf{x}}_p(k_w) = [\bar{\mathbf{x}}_{p1}^T(k_w), \bar{\mathbf{x}}_{p2}^T(k_w), \cdots, \bar{\mathbf{x}}_{p\mathfrak{N}}^T(k_w)]^T$ is the corresponding detected data symbol vector of the p-th SBS and $\bar{\mathbf{x}}_{p\mathfrak{m}}(k_w) = [x_{\mathfrak{m}}^p(\ell_1, k_w), x_{\mathfrak{m}}^p(\ell_2, k_w), \cdots, x_{\mathfrak{m}}^p(\ell_U, k_w)]^T$, and $\bar{\mathbf{P}}'_p(k_w)$ and $\bar{\mathbf{s}}'_p(k_w)$ are defined as

$$\begin{aligned}
\bar{\mathbf{P}}'_p(k_w) &= [\bar{\mathbf{P}}_1(k_w), \cdots, \bar{\mathbf{P}}_{p-1}(k_w), \bar{\mathbf{P}}_{p+1}(k_w), \cdots, \bar{\mathbf{P}}_P(k_w)], \\
\bar{\mathbf{s}}'_p(k_w) &= [\bar{\mathbf{s}}_1^T(k_w), \cdots, \bar{\mathbf{s}}_{p-1}^T(k_w), \bar{\mathbf{s}}_{p+1}^T(k_w), \cdots, \bar{\mathbf{s}}_P^T(k_w)],
\end{aligned}$$

$\bar{\mathbf{P}}_p(k_w) = \bar{\mathbf{E}}_p(k_w)\bar{\mathbf{G}}_p(k_w)$; $\bar{\mathcal{I}}^1_p(k_w)$ and $\bar{\mathcal{I}}^2_p(k_w)$ are, respectively, the $\mathfrak{N}U$-dimensional inter-user interference vector and intra-user interference vector induced by the error in preprocessing matrix; and $\bar{\mathbf{z}}''_p(k_w)$ is the noise vector.

In the presence of interferences and noise, if we define $\bar{\mathbf{R}}_1^p(k_w)$, $\bar{\mathbf{R}}_2^p(k_w)$, and $\bar{\mathbf{R}}_z^p(k_w)$ as the $\mathfrak{N}U \times \mathfrak{N}U$ covariance matrices of $\bar{\mathcal{I}}^1_p(k_w)$, $\bar{\mathcal{I}}^2_p(k_w)$, and $\bar{\mathbf{z}}''_p(k_w)$, i.e.,

$$\begin{aligned}
\bar{\mathbf{R}}_1^p(k_w) &= \sum_{q=1, q\neq p}^{P} (\bar{\mathbf{H}}_{\mathrm{OAM}_p}(k_w)\bar{\mathbf{P}}_q(k_w))\mathbb{E}\left(\bar{\mathbf{s}}_q(k_w)\bar{\mathbf{s}}_q^H(k_w)\right)(\bar{\mathbf{H}}_{\mathrm{OAM}_p}(k_w)\bar{\mathbf{P}}_q(k_w))^H, \\
\bar{\mathbf{R}}_2^p(k_w) &= \left(\bar{\mathbf{H}}_{\mathrm{OAM}_p}(k_w)\bar{\mathbf{P}}_p(k_w) - \mathbf{I}_{\mathfrak{N}U}\right)\mathbb{E}\left(\bar{\mathbf{s}}_p(k_w)\bar{\mathbf{s}}_p^H(k_w)\right) \\
&\quad \times \left(\bar{\mathbf{H}}_{\mathrm{OAM}_p}(k_w)\bar{\mathbf{P}}_p(k_w) - \mathbf{I}_{\mathfrak{N}U}\right)^H, \\
\bar{\mathbf{R}}_z^p(k_w) &= \mathbb{E}\left(\bar{\mathbf{z}}''_p(k_w)(\bar{\mathbf{z}}''_p(k_w))^H\right).
\end{aligned} \tag{5.48}$$

Then, the SINR of the p-th SBS on the u-th mode OAM generated by the $\mathfrak{n}$-th UCA of MBS at the w-th subcarrier can be formulated as

$$\mathrm{SINR}_{\mathfrak{n}}^{p}(\ell_u, k_w) = \frac{\mathbb{E}\left(\left|s_{\mathfrak{n}}^{p}(\ell_u, k_w)\right|^2\right)}{\left[\bar{\mathbf{R}}_1^{p}(k_w)\right]_{\kappa} + \left[\bar{\mathbf{R}}_2^{p}(k_w)\right]_{\kappa} + \left[\bar{\mathbf{R}}_z^{p}(k_w)\right]_{\kappa}}, \tag{5.49}$$

where $\kappa = (\mathfrak{n} - 1)U + u, \mathfrak{n} = 1, 2, \cdots, \mathfrak{N}$. Therefore, the SE of the UCCA-based downlink MU-OAM-MIMO wireless backhaul system can be written as

$$\begin{aligned}\bar{C}' =&\left(1 - \frac{T_P \widetilde{W}}{T_C W}\right)\frac{1}{W}\sum_{p=1}^{P}\sum_{w=1}^{W}\sum_{u=1}^{U}\log_2\left(1 + \mathrm{SINR}_{\mathfrak{n}}^{p}(\ell_u, k_w)\right)\\ &+ \frac{1}{W}\sum_{p=1}^{P}\sum_{\mathfrak{n}=2}^{\mathfrak{N}}\sum_{w=1}^{W}\sum_{u=1}^{U}\log_2\left(1 + \mathrm{SINR}_{\mathfrak{n}}^{p}(\ell_u, k_w)\right).\end{aligned} \tag{5.50}$$

The first term in (5.50) accounts for the SE of one UCA with training overhead for multi-user distance and AoA estimation, and the second term is the total SE of the other UCAs in UCCA at MBS.

Corresponding, the EE of the UCCA-based MU-OAM-MIMO wireless backhaul system can be written as

$$\bar{\eta}_{\mathrm{EE}} = \frac{B\bar{C}'}{\bar{\mathcal{P}}_{\mathrm{total}}} = \frac{B\bar{C}'}{W\mathfrak{N}\bar{\mathcal{P}}_{\mathrm{t}}/\rho + \bar{\mathcal{P}}_{\mathrm{c}}}, \tag{5.51}$$

where $\bar{\mathcal{P}}_{\mathrm{t}}$ is the transmit power of each transmit UCA in MBS at each subcarrier, $\bar{\mathcal{P}}_{\mathrm{c}} = (1 + P)\mathcal{P}_{\mathrm{BB}} + 2\mathfrak{N}MP\mathcal{P}_{\mathrm{RF}} + \mathfrak{N}MP\mathcal{P}_{\mathrm{LNA}}$.

5.5.3 *Performance and Complexity Discussions*

In this part, we mainly analyze the training overhead and computational complexity of the proposed downlink MU-OAM-MIMO wireless backhaul system with P SBSs working on W subcarriers and U OAM modes, which is similar in structure to the traditional downlink MU-MIMO wireless backhaul system using W subcarriers and $P\mathfrak{N}M \times \mathfrak{N}N$ antenna elements, where $\mathfrak{N}N = P\mathfrak{N}M$. In the traditional MU-MIMO system, to correctly recover the transmitted data symbols, channel estimation has to determine the $P\mathfrak{N}M \times \mathfrak{N}N$ channel coefficients for each subcarrier, i.e., $W \times (P\mathfrak{N}M)^2$ unknown variables in total. Thus, the required number of training symbols increases proportional to W, P, $\mathfrak{N}$, and M in large-scale antenna systems. However, by taking advantage of the geometrical relationship of the UCCA-based downlink MU-OAM-MIMO wireless backhaul system, given that the number of elements and the radii of $\mathfrak{N}$ UCAs in UCCAs of SBSs are known to MBS, the proposed MU-OAM-MIMO system only needs determining $3P$ unknown parameters, i.e., $\{(r_p, \theta_p, \varphi_p)|p = 1, 2, \cdots, P\}$ to recover the transmitted data symbols. Since $\widetilde{W}$ and

$\widetilde{U}$ have only effect on the accuracy of the MU-OAM distance and AoA estimation, when $\{(\hat{r}_p, \hat{\theta}_p, \hat{\varphi}_p) | p = 1, 2, \cdots, P\}$ are accurate enough, the number of training symbols required by the downlink MU-OAM-MIMO system does not need to increase with W, P, $\mathfrak{R}$, and M even in large-scale antenna systems. Therefore, compared with the traditional downlink MU-MIMO wireless backhaul system, the SE of the proposed downlink MU-OAM-MIMO system can be improved due to much less training overhead.

For the downlink MU-OAM-MIMO wireless backhaul system, the complexity of the proposed MU-OAM distance and AoA estimation is determined by the complexity of (5.24) and (5.20) corresponding to the values of P, $\widetilde{W}$, and $\widetilde{U}$, and the complexity of the applied MU-OAM preprocessing scheme is determined by the complexity of the SVD for $\hat{\bar{\mathbf{H}}}^{\mathrm{T}}_{\mathrm{OAM}_1}(k_w) = [\bar{\mathbf{H}}^{\mathrm{T}}_{\mathrm{OAM}_1}(k_w), \cdots, \bar{\mathbf{H}}^{\mathrm{T}}_{\mathrm{OAM}_{p-1}}(k_w), \bar{\mathbf{H}}^{\mathrm{T}}_{\mathrm{OAM}_{p+1}}(k_w), \cdots \bar{\mathbf{H}}^{\mathrm{T}}_{\mathrm{OAM}_P}(k_w)]^{\mathrm{T}}$ corresponding to the values of W, P, $\mathfrak{R}$, and U. The specific computational complexity of the proposed distance and AoA estimation and preprocessing for the downlink MU-OAM-MIMO wireless backhaul system is compared with that of traditional downlink MU-MIMO wireless backhaul system in Table 5.1, and the total computational complexity comparison between the downlink MU-OAM-MIMO system and the downlink MU-MIMO system is shown in Fig. 5.6. It can be seen from the figure that the complexity of the downlink MU-OAM-MIMO wireless backhaul system is considerably lower than the traditional downlink MU-MIMO wireless backhaul systems when $\mathfrak{R}$ and M are large. For example, when $W = 128$, $\widetilde{W} = 64$, $U = \widetilde{U} = 30$, $P = 3$, $\mathfrak{R} = 4$ and

Table 5.1 The complexity comparison between downlink MU-OAM-MIMO and downlink MU-MIMO

Scheme	Complexity	
MU-OAM-MIMO	Position estimation	$O\left(\widetilde{W}\widetilde{U}^3 + \widetilde{W}P^3 + \widetilde{W}\widetilde{U}^2 P\right)$
	Preprocessing	$O\left(WP^4\mathfrak{R}^3U^3\right)$
MU-MIMO	Channel Estimation	$O\left(WP^3\mathfrak{R}^3M^3\right)$
	Preprocessing	$O\left(WP^4\mathfrak{R}^3M^3\right)$

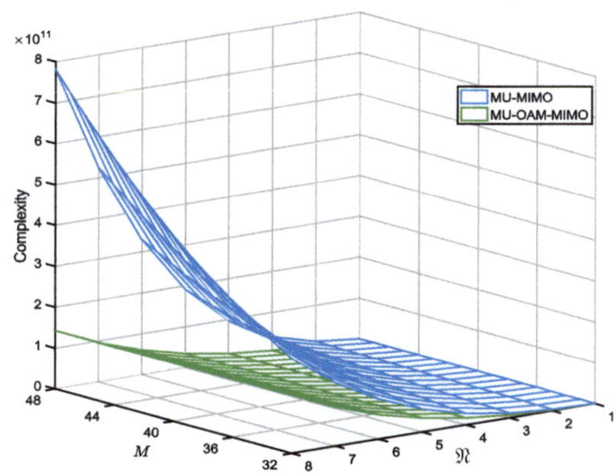

Fig. 5.6 The complexities of the downlink MU-OAM-MIMO wireless backhaul system and the traditional downlink MU-MIMO wireless backhaul system vs. M and $\mathfrak{R}$ at $W = 128$, $\widetilde{W} = 64$, $U = \widetilde{U} = 30$, $P = 3$

$M = 32$, the MU-OAM-MIMO system has lower computational complexity than the traditional MU-MIMO system.

5.6 Summary

In this section, we show the performances of the proposed methods by numerical simulations. We first verify the proposed recursive ESPRIT-based distance and AoA estimation method and then compare the BER performances of the applied MU-OAM preprocessing scheme with those in the ideal downlink MU-OAM channel. Besides, the SEs of the UCA-based downlink MU-OAM wireless backhaul system, the UCCA-based downlink MU-OAM-MIMO wireless backhaul system, and the traditional downlink MU-MIMO wireless backhaul system are compared under different multiplexed OAM modes. At last, the EEs of the UCA-based MU-OAM system and the UCCA-based MU-OAM-MIMO system are illustrated and compared with that of the traditional MU-MIMO system. Unless otherwise stated, the SNRs in all the figures are defined as the ratio of the received signal power versus the noise power.

In Fig. 5.7, we choose $\widetilde{W} = 64$ subcarriers at $f = 9$GHz and $\widetilde{U} = 20$ OAM modes with $\ell_1, \ell_2, \cdots, \ell_{20} = -10, -9, \cdots, +9$, $M = 21$, $P = 3$, $R_t = 30\lambda$, $R_{r_1} = R_{r_2} = R_{r_3} = 15\lambda$, $\lambda = c/f$, $(r_1, \theta_1, \varphi_1) = (12\text{m}, 18^\circ, 2^\circ)$, $(r_2, \theta_2, \varphi_2) = (24\text{m}, 10^\circ, 10^\circ)$ and $(r_3, \theta_3, \varphi_3) = (36\text{m}, 2^\circ, 18^\circ)$. Then, by using the proposed recursive ESPRIT-based distance and AoA estimation method, the estimated locations of P SBSs in the downlink MU-OAM wireless backhaul system are shown in Fig. 5.7. It is obvious to

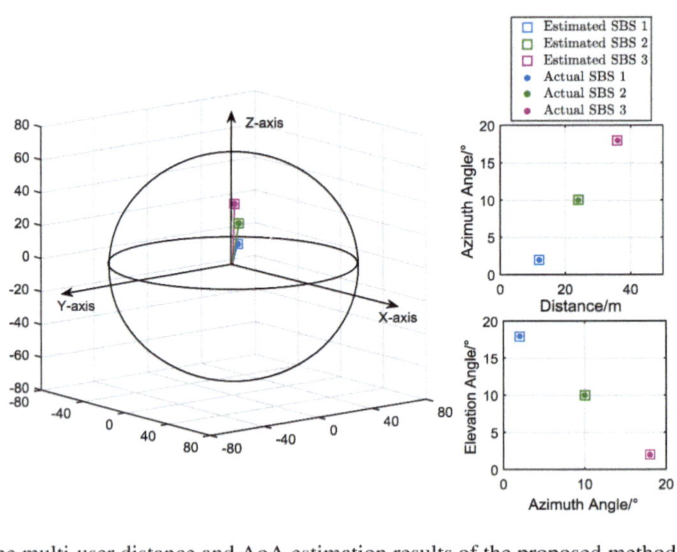

Fig. 5.7 The multi-user distance and AoA estimation results of the proposed method at 20 dB

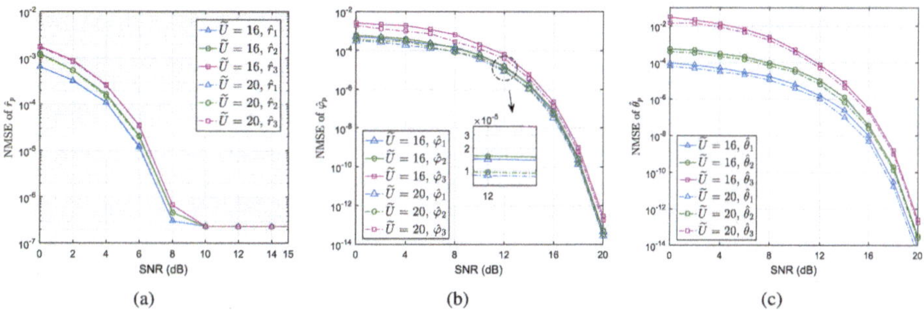

Fig. 5.8 The NMSEs of $\{(\hat{r}_p, \hat{\theta}_p, \hat{\varphi}_p)|p = 1, 2, 3\}$ vs. SNR under different $\widetilde{U}$ at $\widetilde{W} = 64$

Fig. 5.9 The BER comparison of MU-OAM pre-processing

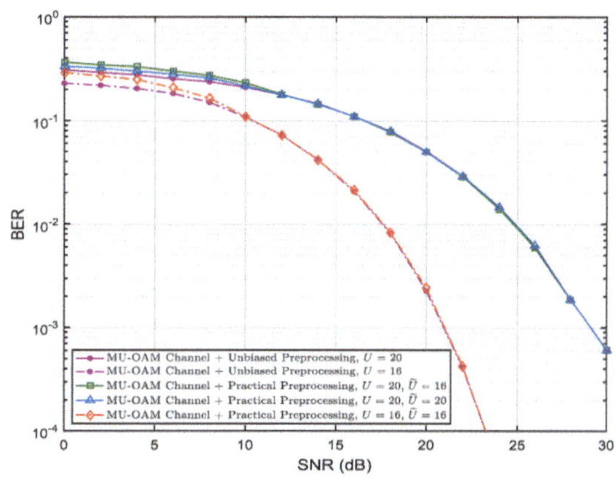

see from the figure that the estimated locations of P SBSs are very close to the actual locations at 20 dB. Figure 5.8 illustrates the NMSEs of $\{(\hat{r}_p, \hat{\theta}_p, \hat{\varphi}_p)|p = 1, 2, 3\}$ versus SNR under different $\widetilde{U}$. It can be seen from Fig. 5.8 that the estimation errors of $\{(\hat{r}_p, \hat{\theta}_p, \hat{\varphi}_p)|p = 1, 2, 3\}$ decrease with the increase of SNR, which proves the validity of the proposed method. Meanwhile, as $\widetilde{U}$ increase, the NMSEs of $\hat{\varphi}_p$ and $\hat{\theta}_p$ slightly decrease.

After obtaining $\{(\hat{r}_p, \hat{\theta}_p, \hat{\varphi}_p)|p = 1, 2, 3\}$, Fig. 5.9 compares the BERs of the applied MU-OAM preprocessing scheme with those in the ideal downlink MU-OAM channel under QPSK modulation. The ideal downlink MU-OAM channel is the downlink MU-OAM channel that is unbiased preprocessed by the actual distances and AoAs of P SBSs. When $U = 20$, the BERs of the applied MU-OAM preprocessing scheme are very close to those in the ideal downlink MU-OAM channel at high SNR. Meanwhile, when using fewer OAM modes (i.e., $\widetilde{U} = 16$ with $\ell_1, \ell_2, \cdots, \ell_{16} = -8, -7, \cdots, +7$) in the MU-OAM distance and AoA estimation, the BER of the applied MU-OAM preprocessing scheme slightly increases at low SNR due to slightly higher estimation errors. Besides, comparing $U = 16$ with $U = 20$, the BERs of both the practical MU-OAM channel and the ideal MU-OAM channel all become much better because of abandoning the high-order OAM modes with smaller gains.

In Fig. 5.10, T_c is assumed to be 512 OFDM symbols, $W = 128$. Then, the SEs of the UCA-based downlink MU-OAM wireless backhaul system under different values of U and $\widetilde{U}$ are shown in Fig. 5.10. It can be seen from the figure that in contrast to the MU-OAM system without preprocessing, the SEs of the MU-OAM system with preprocessing by the MU-OAM distance and AoA estimation are greatly improved, which get close to the SEs of the ideal downlink MU-OAM system at high SNR. Meanwhile, when 20 OAM modes rather than 16 OAM modes are being used in the MU-OAM distance and AoA estimation, the SE of the downlink MU-OAM system increases slightly at low SNR due to lower estimation error. Furthermore, comparing $U = 20$ with $U = 16$, the SEs of both the practical downlink MU-OAM system and the ideal downlink MU-OAM system all become higher due to that more OAM modes are multiplexed to transmit data.

In Fig. 5.11, T_c is also assumed to be 512 OFDM symbols, $M = 21$, $P = 3$, $\mathfrak{R} = 4$, $W = 128$ and $\widetilde{W} = 64$. The SEs of the UCCA-based downlink MU-OAM-MIMO wireless backhaul system and the traditional downlink MU-MIMO wireless

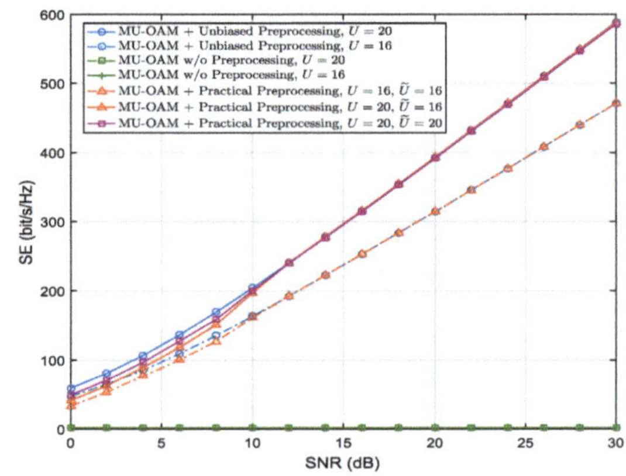

Fig. 5.10 The SEs of the UCA-based downlink MU-OAM wireless backhaul system

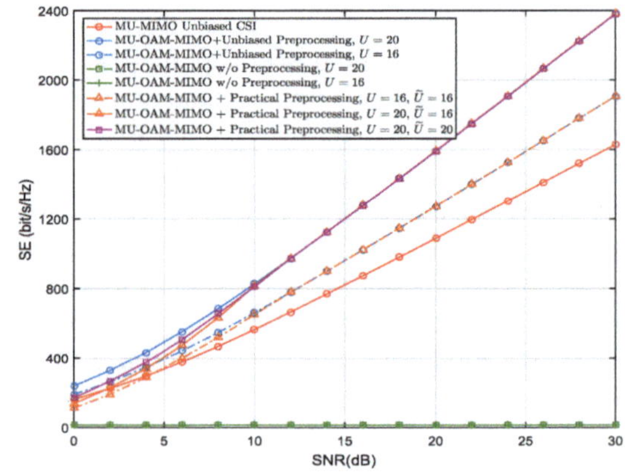

Fig. 5.11 The SEs of the UCCA-based downlink MU-OAM-MIMO wireless backhaul system and the traditional downlink MU-MIMO wireless backhaul system

backhaul system are compared in Fig. 5.11. We assume that the traditional channel estimation of the MU-MIMO system requires $P\mathfrak{N}M = 252$ training symbols per subcarrier or subchannel. From the figure, one can get the same conclusion as in Fig. 5.10, which verifies that the proposed multi-user distance and AoA estimation method and the MU-OAM preprocessing scheme can be applied to the UCCA-based downlink MU-OAM-MIMO wireless backhaul system. More importantly, it is obvious that the SEs of the UCCA-based downlink MU-OAM-MIMO system are about 30% higher than the SE of the downlink MU-MIMO system with traditional channel estimation due to the greatly reduced number of training symbols.

In Fig. 5.12, we choose the system available bandwidth $B = 190$MHz with $W = 128$ subcarriers, subcarrier interval $\Delta f = 1.48$MHz, $\mathcal{P}_{\mathrm{BB}} = 200$mW, $\mathcal{P}_{\mathrm{RF}} = 250$mW, $\mathcal{P}_{\mathrm{LNA}} = 20$mW and $\rho = 0.35$ [3, 9, 12]. Besides, we assume that all systems in Fig. 5.12 have the same total transmit power. Then, the EE-SE relationship of the UCA-based downlink MU-OAM wireless backhaul system under different values of U and $\widetilde{U}$ is shown in Fig. 5.12a and the EE of the UCCA-based MU-OAM-MIMO system is illustrated and compared with that of the traditional MU-MIMO system in Fig. 5.12b. It can be seen from Fig. 5.12a that the EE of the proposed MU-OAM system does not always increase with the SE, and there are trade-offs between the EE and the SE. Meanwhile, under the same transmit power, the EE of the MU-OAM system with $U = 20$ is slightly higher than that with $U = 16$ owing to slightly higher capacity. Besides, when using fewer OAM modes (i.e., $\widetilde{U} = 16$) in the training stage,

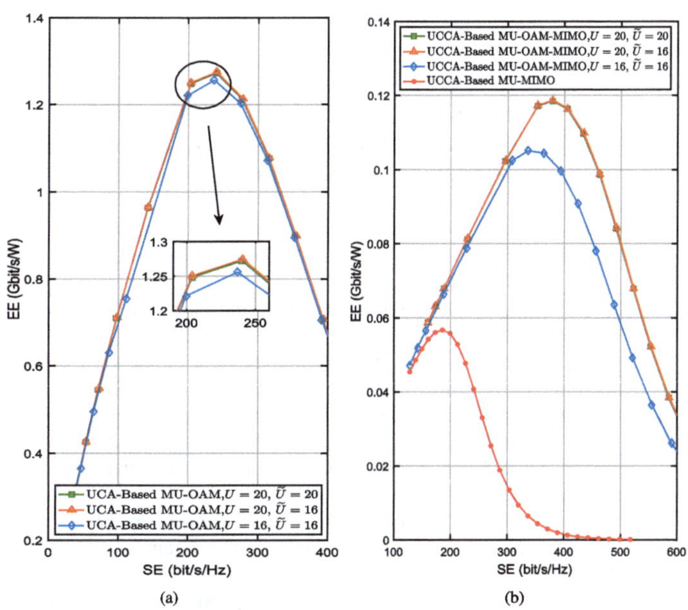

Fig. 5.12 The EE-SE relationships. (**a**) UCA-based downlink MU-OAM wireless backhaul system. (**b**) UCCA-based downlink MU-OAM-MIMO wireless backhaul system and the traditional UCCA-based downlink MU-MIMO wireless backhaul system with $\mathfrak{N} = 4$

the EE of the MU-OAM system slightly increases due to less training overhead. From Fig. 5.12b, one can get the same conclusion as in Fig. 5.12a. Moreover, under the same transmit power, the EE of the MU-OAM-MIMO system is lower than that of the MU-OAM system due to larger circuit power consumption. More importantly, it is obvious that the EE of the MU-OAM-MIMO system is almost twice that of the traditional MU-MIMO system due to the greatly reduced number of training symbols.

In this chapter, we propose a UCA-based downlink MU-OAM wireless backhaul system, which enables joint spatial division and coaxial multiplexing. Then, we design an overall communication scheme for the downlink MU-OAM wireless backhaul system including the MU-OAM distance and AoA estimation and the preprocessing of MU-OAM signals. The proposed distance and AoA estimation method based on the recursive ESPRIT algorithm is shown being able to simultaneously estimate the position of each SBS with a flexible number of training symbols. Thereafter, we apply a MU-OAM preprocessing scheme for the downlink MU-OAM wireless backhaul system, which can effectively eliminate both the co-mode and inter-mode interferences in the downlink MU-OAM channel. Since both interferences are eliminated through preprocessing at MBS, the complexity of SBSs in proposed system can be reduced. At last, the proposed methods are extended to the UCCA-based downlink MU-OAM-MIMO wireless backhaul system, which exhibits higher SE and EE than the large-scale MU-MIMO wireless backhaul system with traditional channel estimation.

References

1. M. Baek, S. Kwak, J. Jung, H.M. Kim, D. Choi, Implementation methodologies of deep learning-based signal detection for conventional MIMO transmitters. IEEE Trans. Broadcast. **65**(3), 636–642 (2019)
2. R. Chen, W.X. Long, X. Wang, L. Jiandong, Multi-mode OAM radio waves: Generation, angle of arrival estimation and reception with UCAs. IEEE Trans. Wireless Commun. **19**(10), 6932–6947 (2020)
3. R. Chen, H. Zhou, W.X. Long, M. Moretti, Spectral and energy efficiency of line-of-sight OAM-MIMO communication systems. China Commun. **17**(9), 119–127 (2020)
4. R. Chen, H. Zhou, M. Moretti, X. Wang, J. Li, Orbital angular momentum waves: Generation, detection, and emerging applications. IEEE Commun. Surv. Tuts. **22**(2), 840–868 (2020)
5. R. Chen, H. Xu, M. Moretti, J. Li, Beam steering for the misalignment in UCA-based OAM communication systems. IEEE Wirel. Commun. Lett. **7**(4), 582–585 (2018)
6. R. Chen, W. Yang, H. Xu, J. Li, A 2-D FFT-based transceiver architecture for OAM-OFDM systems with UCA antennas. **67**(6), 5481–5485 (2018)
7. E. Garro, C. Barjau, D. Gomez-Barquero, J. Kim, S.I. Park, N. Hur, Layered division multiplexing with co-located multiple-input multiple-output schemes. IEEE Trans. Broadcast. **66**(1), 9–20 (2020)
8. E. Hossain, M. Rasti, H. Tabassum, A. Abdelnasser, Evolution toward 5G multi-tier cellular wireless networks: an interference management perspective. IEEE Wirel. Commun. **21**(3), 118–127 (2014)

9. C. Lin, G.Y. Li, Energy-efficient design of indoor mmwave and sub-THz systems with antenna arrays. IEEE Trans. Wirel. Commun. **15**(7), 4660–4672 (2016)
10. K. Liu, Y. Cheng, Z. Yang, H. Wang, Y. Qin, X. Li, Orbital-angular-momentum-based electromagnetic vortex imaging. IEEE Antennas Wirel. Propag. Lett. **14**, 711–714 (2015)
11. F. Eslampanahi Mahmouli, S.D. Walker, 4-Gbps uncompressed video transmission over a 60-GHz orbital angular momentum wireless channel. IEEE Wirel. Commun. Lett. **2**(2), 223–226 (2013)
12. R. Méndez-Rial, C. Rusu, N. González-Prelcic, A. Alkhateeb, R.W. Heath, Hybrid MIMO architectures for millimeter wave communications: phase shifters or switches? IEEE Access **4**, 247–267 (2016)
13. S.M. Mohammadi, L.K.S. Daldorff, et al., Orbital angular momentum in radio-a system study. **58**(2), 565–572 (2010)
14. Y. Ren, L. Li, G. Xie, Y. Yan, Y. Cao, H. Huang, N. Ahmed, Z. Zhao, P. Liao, C. Zhang, G. Caire, A.F. Molisch, M. Tur, A.E. Willner, Line-of-sight millimeter-wave communications using orbital angular momentum multiplexing combined with conventional spatial multiplexing. **16**(5), 3151–3161 (2017)
15. T. Shitomi, E. Garro, K. Murayama, D. Gomez-Barquero, MIMO scattered pilot performance and optimization for ATSC 3.0. IEEE Trans. Broadcast. **64**(2), 188–200 (2018)
16. U. Siddique, H. Tabassum, E. Hossain, D.I. Kim. Wireless backhauling of 5G small cells: challenges and solution approaches. IEEE Wirel. Commun. **22**(5), 22–31 (2015)
17. F. Tamburini, E. Mari, A. Sponselli, B. Thidé, A. Bianchini, F. Romanato, Encoding many channels in the same frequency through radio vorticity: first experimental test. New J. Phys. **14**(3), 033001 (2012)
18. N. Wang, E. Hossain, V.K. Bhargava, Backhauling 5G small cells: a radio resource management perspective. IEEE Wirel. Commun. **22**(5), 41–49 (2015)
19. Z. Wu, X. Gao, C. Jiang, Nonbinary LDPC-coded spatial multiplexing for rate-2 MIMO of DVB-NGH system. IEEE Trans. Broadcast. **64**(2), 201–210 (2018)
20. J. Xiong, Y. Fang, P. Cheng, Z. Shi, W. Zhang, Distributed caching in converged networks: a deep reinforcement learning approach. IEEE Trans. Broadcast. **67**(1), 201–211 (2020)
21. Y. Yagi, H. Sasaki, T. Yamada, D. Lee, 200 Gb/s wireless transmission using dual-polarized OAM-MIMO multiplexing with uniform circular array on 28 GHz band. IEEE Antennas Wirel. Propag. Lett. **20**(5), 833–837 (2021)
22. Y. Yan, G. Xie, M.P.J. Lavery, H. Huang, N Ahmed, C. Bao, Y. Ren, Y. Cao, L. Li, Z. Zhao, A.F. Molisch, M. Tur, M.J. Padgett, A.E. Willner, High-capacity millimetre-wave communications with orbital angular momentum multiplexing. Nat. Commun. **5**, 4876 (2014)
23. Y. Yan, L. Li, Z. Zhao, G. Xie, Z. Wang, Y. Ren, N. Ahmed, S. Sajuyigbe, S. Talwar, M. Tur, N. Ashrafi, S. Ashrafi, A.F. Molisch, A.E. Willner, 32-Gbit/s 60-GHz millimeter-wave wireless communication using orbital angular momentum and polarization multiplexing, in *Proc. IEEE Int. Conf. Commun.* (2016)
24. C. Zhang, Y. Zhao, Orbital angular momentum nondegenerate index mapping for long distance transmission. **18**(11), 5027–5036 (2019)
25. W. Zhang, S. Zheng, X. Hui, R. Dong, X. Jin, H. Chi, X. Zhang, Mode division multiplexing communication using microwave orbital angular momentum: an experimental study. **16**(2), 1308–1318 (2017)
26. Y. Zhao, C. Zhang, Compound angular lens for radio orbital angular momentum coaxial separation and convergence. IEEE Antennas Wirel. Propag. Lett. **18**(10), 2160–2164 (2019)

[illegible] (2021)

[illegible] (2014)

[illegible]

[illegible] (2019)

[illegible]

[illegible]

Chapter 6
Circular Array-Based Terahertz Communications for Point-to-Multipoint OAM Wireless Backhaul

6.1 Introduction

The development trend of wireless communication systems is toward higher carrier frequency, larger bandwidth, and faster data rate. Millimeter-wave (mmWave) communications have been used in 5G, which provide a large bandwidth ranging from hundreds of megahertz (MHz) to several gigahertz (GHz) [12, 24]. Although the data rate of the mmWave frequency band can reach several Gbps [1], it still cannot meet the requirement of explosive wireless services in the future, e.g., sensory transmission, extended reality, digital twin, etc. [7, 20]. Therefore, 6G is likely to utilize terahertz (THz) frequency band (0.1–10 THz) [9, 19], which can satisfy the ultrahigh data rate requirement of exceeding 100 Gbps. Although the propagation of THz signal encounters the high path loss caused by high carrier frequency and the inter-symbol interference caused by wide bandwidth, the utilization of massive multiple-input multiple-output (MIMO) and orthogonal frequency division multiplexing (OFDM) can alleviate the negative effects of these problems [10, 17, 22]. A number of applications of THz communication in cellular, vehicle, and medical communication scenarios have been prospected [2, 4, 8, 21].

Only combining with THz can OAM unleash its potential in achieving high spectrum efficiency because of high carrier frequency reducing the divergence angle of the OAM beam to a large extent. Since both THz and OAM are potential 6G techniques in achieving extremely high data rate, their combination is natural. In this chapter, we propose a UCA-based THz multi-user OAM (MU-OAM) communication system including simple downlink (DL) and uplink (UL) transmission schemes based on our proved decomposition of THz MU-OAM channel matrix that simplifies the signal detection to the despiralization and amplitude detection (AD). Then, we derive the achievable sum rate of the THz MU-OAM system and analyze the effect of carrier frequency on inter-mode interference (IMI) and inter-user interference (IUI), respectively. Mathematical analysis reveals that the IMI and IUI decrease with the increase of frequency. The proposed MU-OAM

R. Chen et al., *Circular Array-Based Radio Frequency OAM Communications*, Wireless Networks, https://doi.org/10.1007/978-3-031-37589-7_6

communication scheme has lower computational complexity than traditional MU-MIMO scheme but retains good bit error rate (BER) and achievable sum rate performances, which has, thanks to the narrower OAM beamwidth in THz frequency band, less interference than lower-frequency bands.

The remainder of this chapter is organized as follows. In Sect. 6.2, we present the THz channel model and formulate the UCA-based downlink and uplink THz MU-OAM channel. In Sect. 6.3, we propose the THz MU-OAM DL and UL transmission schemes with our proved decomposition of THz MU-OAM channel matrix. Then, we investigate the effect of carrier frequency on IMI and IUI of the MU-OAM system and compare the complexities of the proposed MU-OAM communication scheme with the traditional MU-MIMO scheme in Sect. 6.4. Simulation results are shown in Sect. 6.5, and conclusions are summarized in Sect. 6.6.

6.2 THz MU-OAM Channel Model

The present researches hold that OAM wireless communications are most applicable to LoS scenarios [15, 18, 23, 25]. Therefore, we only focus on the LoS propagation channel model in the proposed THz MU-OAM system. Since the free-space loss is proportional to the square of the frequency, the THz wave experiences larger path loss during its propagation in contrast to microwave and mmWave. Moreover, due to the shorter wavelength, the molecular absorption effect caused by the tiny particles and water vapor in the atmosphere is more obvious. The very high molecular absorption and free-space loss properties of THz radiation result in a very high frequency-selective path loss for LoS links [16]. Therefore, we model the LoS path loss deterministically as the sum of free-space loss and molecular absorption loss in the proposed THz MU-OAM system. According to Kirchhoff scattering theory and ray tracing technology, the LoS path loss can be expressed as [16]:

$$A_{\mathrm{LoS}}^{dB}(k,d) = A_{\mathrm{abs}}^{dB}(k,d) + A_{\mathrm{spread}}^{dB}(k,d), \tag{6.1}$$

$$A_{\mathrm{abs}}^{dB}(k,d) = \alpha_{\mathrm{molec}}\left(k, T_{\mathrm{K}}, p\right) d\, 10 \lg e, \tag{6.2}$$

$$A_{\mathrm{spread}}^{dB}(k,d) = 20 \lg\left(2kd\right), \tag{6.3}$$

where $k = 2\pi/\lambda$ is the wave number, $\lambda = c/f$ is the wavelength, d is the distance between the transmit and receive antennas, A_{abs}^{dB} is the molecular absorption loss, and A_{spread}^{dB} represents the free-space loss in the THz band. The molecular absorption coefficient α_{molec} is influenced by frequency f, temperature T_{K}, and pressure p on molecular absorption.

Thus, the equivalent THz LoS channel can be expressed as

$$h(k,d) = h_{\mathrm{LoS}}(k,d) \exp\left(-jkd\right), \tag{6.4}$$

and the magnitude of the LoS path $h_{\mathrm{LoS}}(k,d)$ following from (6.1), (6.2) and (6.3) is

$$h_{\mathrm{LoS}}(k,d) = h_{\mathrm{abs}}(k,d) \cdot h_{\mathrm{spread}}(k,d), \tag{6.5}$$

where $h_{\mathrm{abs}}(k,d) = \mathrm{e}^{-\frac{1}{2}\alpha_{\mathrm{molec}}(k,T_{\mathrm{K}},p)d}$ and $h_{\mathrm{spread}}(k,d) = \frac{1}{2kd}$ are the amplitude attenuations caused by molecular absorption and free-space propagation, respectively. In order to simplify the expression, denote $\mathrm{e}^{-\frac{1}{2}\alpha_{\mathrm{molec}}(k,T_{\mathrm{K}},p)d}$ as $\beta(k)$. Then, (6.4) can be rewritten as

$$h(k,d) = \beta(k)\frac{1}{2kd}\exp(-jkd). \tag{6.6}$$

Based on the channel model, we consider a THz MU-OAM communication system, which consists of a BS equipped with a PN-element UCA and P users each with a N-element UCA as shown in Fig. 6.1. We assume that the plane of each user UCA is perpendicular to the line between its own center and the center of the BS UCA through mechanical beam steering [6] and apply the electronic beam steering [15] to make the multiple beams of the BS toward different users. The geometric relationship between the BS UCA and the pth user UCA is shown in Fig. 6.2. The coordinate of the center of the pth user UCA is denoted as $(r_p, \varphi_p, \theta_p)$ in $\bar{\mathrm{Z}}-\bar{\mathrm{X}}\bar{\mathrm{O}}\bar{\mathrm{Y}}$, and the distance between the mth element of the pth user UCA and the nth element of the BS UCA can be written as

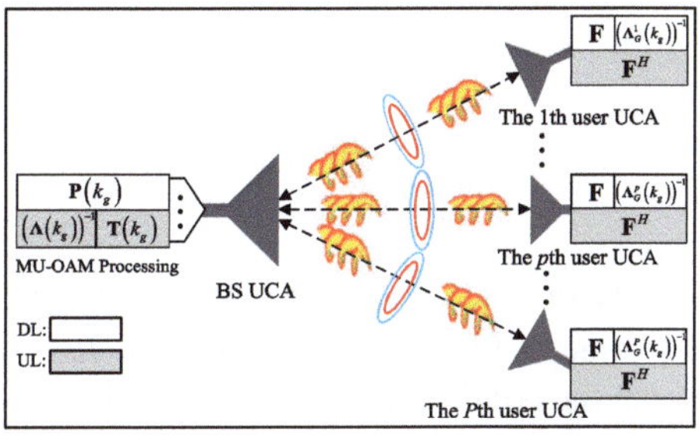

Fig. 6.1 The proposed THz MU-OAM communication system

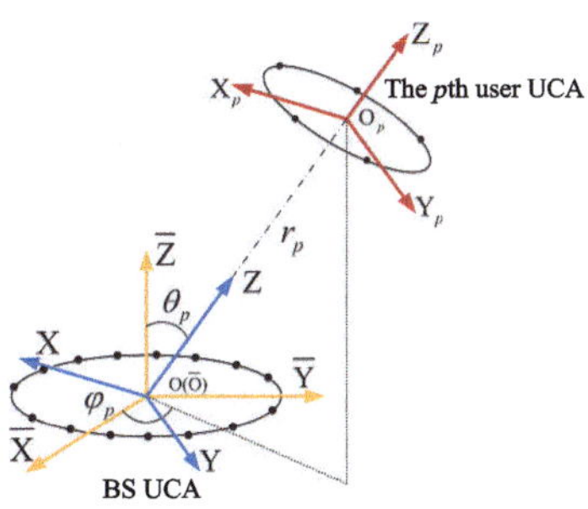

Fig. 6.2 Relative position of BS UCA and the pth user UCA

$$
\begin{aligned}
d_{m,n}^{p} = \Big(& R_t^2 + R_p^2 + r_p^2 - 2r_p R_t \sin\theta_p \cos(\varphi_p - \omega_n) \\
& - 2R_t R_p \big(\cos\theta_p \cos(\upsilon_m^p - \omega_n) + (1 - \cos\theta_p) \\
& \times \sin(\varphi_p - \upsilon_m^p) \sin(\varphi_p - \omega_n) \big) \Big)^{\frac{1}{2}},
\end{aligned} \tag{6.7}
$$

where R_p is the radius of the pth user UCA, R_t is the radius of BS UCA, $\omega_n = \frac{2\pi(n-1)}{NP} + \omega_0, n = 1, \cdots, NP, \upsilon_m^p = \frac{2\pi(m-1)}{N} + \upsilon_0^p, m = 1, \cdots, N, p = 1, \cdots, P$, and ω_0 and υ_0^p are, respectively, the initial angles of the first reference antenna elements in BS UCA and pth user UCA. To simplify the analysis, we assume $\omega_0 = 0$ and $\upsilon_0^p = 0$ here. The detailed calculation process of $d_{m,n}^p$ is given in Appendix A. By assuming that all the P user UCAs are located in the far-field distance region of the BS UCA, i.e., $r_p \gg R_t$, and $r_p \gg R_p$. We can approximate $d_{m,n}^p$ as

$$
\begin{aligned}
d_{m,n}^{p} & \overset{(a)}{\approx} \sqrt{R_t^2 + R_p^2 + r_p^2} - \frac{r_p R_t}{\sqrt{R_t^2 + R_p^2 + r_p^2}} \sin\theta_p \cos(\varphi_p - \omega_n) \\
& - \frac{R_t R_p}{\sqrt{R_t^2 + R_p^2 + r_p^2}} \cos\theta_p \cos(\upsilon_m^p - \omega_n) \\
& - \frac{R_t R_p}{\sqrt{R_t^2 + R_p^2 + r_p^2}} (1 - \cos\theta_p) \sin(\varphi_p - \upsilon_m^p) \sin(\varphi_p - \omega_n) \\
& \overset{(b)}{\approx} r_p - R_t \sin\theta_p \cos(\varphi_p - \omega_n) - \frac{R_t R_p}{r_p} \cos\theta_p \cos(\upsilon_m^p - \omega_n) \\
& - \frac{R_t R_p}{r_p} (1 - \cos\theta_p) \sin(\varphi_p - \upsilon_m^p) \sin(\varphi_p - \omega_n),
\end{aligned} \tag{6.8}
$$

where (a) uses the method of completing a square and the condition $r_p \gg R_t, R_p$ as same as the simple case $\sqrt{a^2 - 2b} \approx a - \frac{b}{a}, a \gg b$ and (b) is directly obtained from the condition $r_p \gg R_t, R_p$. Then, substituting (6.8) into (6.6) and abbreviating $h(k, d_{m,n}^p)$ to $h_{m,n}^p(k)$, we thus have

$$
\begin{aligned}
h_{m,n}^{p}(k) \overset{(a)}{\approx} & \beta(k) \frac{1}{2kr_p} \exp\Big(-jkr_p + jkR_t \sin\theta_p \cos(\varphi_p - \omega_n) \\
& + jk \frac{R_t R_p}{r_p} \Big(\cos\theta_p \cos(\upsilon_m^p - \omega_n) + (1 - \cos\theta_p) \\
& \times \sin(\varphi_p - \upsilon_m^p) \sin(\varphi_p - \omega_n) \Big) \Big),
\end{aligned} \tag{6.9}
$$

where the approximation (a) neglects a few small terms in the denominator, and thus only $4\pi r_p$ is left.

6.2.1 Downlink THz MU-OAM Channel Model

In the considered THz MU-OAM communication system, we apply OFDM for higher data rate and easier equalization in the ultra-large bandwidth and assume that there are G subcarriers. Then, the DL THz OAM channel matrix connecting the NP transmit antenna elements of BS UCA with the N receive antenna elements of the pth user UCA at the gth $(1 \leq g \leq G)$ subcarrier can be expressed as

$$\mathbf{H}^p(k_g) = [h^p_{m,n}(k_g)]_{N\times NP}, \quad p = 1, 2, \ldots, P. \tag{6.10}$$

In the MU-OAM system, the BS simultaneously transmits P groups of multi-mode OAM signals to P users in the multi-user DL channel. Thus, the NP elements of the BS UCA could be divided into P groups, and each group contains N elements. Therefore, $\mathbf{H}^p(k_g)$ can be rewritten as

$$\mathbf{H}^p(k_g) = \left[\mathbf{H}^p_1(k_g), \mathbf{H}^p_2(k_g), \cdots, \mathbf{H}^p_P(k_g)\right], \tag{6.11}$$

where

$$\mathbf{H}^p_i(k_g) = \begin{bmatrix} h^p_{1,i} & h^p_{1,i+P} & \cdots & h^p_{1,i+P\times(N-1)} \\ h^p_{2,i} & h^p_{2,i+P} & \cdots & h^p_{2,i+P\times(N-1)} \\ \vdots & \vdots & \ddots & \vdots \\ h^p_{N,i} & h^p_{N,i+P} & \cdots & h^p_{N,i+P\times(N-1)} \end{bmatrix} \tag{6.12}$$

is the $N \times N$ submatrix that denotes the DL channel from the ith $(1 \leq i \leq P)$ group of elements of BS UCA to the pth user UCA at the gth subcarrier.

6.2.2 Uplink THz MU-OAM Channel Model

The UL THz OAM channel matrix connecting the N transmit antennas of the pth user UCA with the NP receive antennas of the BS UCA at the gth subcarrier can be expressed as

$$\bar{\mathbf{H}}^p(k_g) = \left(\mathbf{H}^p(k_g)\right)^T = [h^p_{n,m}]_{NP\times N}, \quad p = 1, 2, \ldots, P. \tag{6.13}$$

P users transmit P groups of multi-mode OAM signals to the BS simultaneously in the multi-user UL channel. The same operation as DL transmission divides the

NP elements of the BS UCA into P groups, and each group contains N elements. Therefore, $\bar{\mathbf{H}}^p(k_g)$ can be rewritten as

$$\bar{\mathbf{H}}^p(k_g)=\left[(\bar{\mathbf{H}}_1^p(k_g))^T, (\bar{\mathbf{H}}_2^p(k_g))^T, \cdots, (\bar{\mathbf{H}}_P^p(k_g))^T\right]^T, \tag{6.14}$$

where $\bar{\mathbf{H}}_i^p(k_g) = (\mathbf{H}_i^p(k_g))^T$ is $N \times N$ submatrix that denotes the channel from the pth user UCA to the ith group elements of BS UCA.

6.3 THz MU-OAM Communication Scheme

It is assumed that the position of each user can be obtained by the OAM-based distance and angle of arrival (AoA) estimation method [5, 14, 15]. As we can see from (6.10) and (6.13), the channel between the BS and the user is completely determined by the user's position. Therefore, the channel state information (CSI) can be completely obtained by OAM-based distance and AoA estimation. Then, the DL and UL THz MU-OAM communication schemes based on the relative estimated positions of the transmitter and receiver are proposed.

6.3.1 Downlink THz MU-OAM Transmission and Reception

Due to the inevitable errors introduced by the position estimation, we assume that the BS has only an estimate of the practical channel matrix $\mathbf{H}^p(k_g)$ that we denote by $\hat{\mathbf{H}}^p(k_g)$. Therefore, $\mathbf{H}^p(k_g)$ can be expressed as

$$\mathbf{H}^p\left(k_g\right) = \hat{\mathbf{H}}^p\left(k_g\right) + \Delta\mathbf{H}^p\left(k_g\right), \tag{6.15}$$

where $\Delta\mathbf{H}^p\left(k_g\right)$ is the $N \times NP$ dimensional channel estimation error matrix, which is generated from the position parameter estimation errors of the pth user. Let $\hat{\theta}_p$, $\hat{\varphi}_p$, and $\hat{r}_p$ denote the estimators of the position parameters θ_p, φ_p, and r_p of the pth user, respectively, and $\Delta\theta_p = \theta_p - \hat{\theta}_p$, $\Delta\varphi_p = \varphi_p - \hat{\varphi}_p$, $\Delta r_p = r_p - \hat{r}_p$. Denote by σ_θ^2, σ_φ^2, and σ_r^2 the variances of $\Delta\theta_p$, $\Delta\varphi_p$, and Δr_p. The element of $\Delta\mathbf{H}^p\left(k_g\right)$ can be expressed as

$$\begin{aligned}
&\Delta h_{m,n}^p(k_g)\\
&= h_{m,n}^{p,k_g}(\theta_p, \varphi_p, r_p) - h_{m,n}^{p,k_g}(\hat{\theta}_p, \hat{\varphi}_p, \hat{r}_p)\\
&= \beta(k_g)\frac{1}{2kr_p}\exp\Big(-jkr_p+jkR_t\sin\theta_p\cos(\varphi_p-\omega_n)
\end{aligned}$$

$$+ jk\frac{R_t R_p}{r_p}\Big(\cos\theta_p \cos(\upsilon_m^p - \omega_n) + (1-\cos\theta_p)$$

$$\times \sin(\varphi_p - \upsilon_m^p)\sin(\varphi_p - \omega_n)\Big)\Big) - \beta(k_g)\frac{1}{2k\hat{r}_p}$$

$$\times \exp\Big(-jk\hat{r}_p + jkR_t \sin\hat{\theta}_p \cos(\hat{\varphi}_p - \omega_n)$$

$$+ jk\frac{R_t R_p}{\hat{r}_p}\Big(\cos\hat{\theta}_p \cos(\upsilon_m^p - \omega_n) + (1-\cos\hat{\theta}_p) \times \sin(\hat{\varphi}_p - \upsilon_m^p)\sin(\hat{\varphi}_p - \omega_n)\Big)\Big). \tag{6.16}$$

We assume that $\Delta \mathbf{H}^p(k_g)$ is independent of the transmit signal and the Gaussian noise. In our DL channel model, the transmitter preprocessing matrix is designed based on the knowledge of the estimated channel matrix $\hat{\mathbf{H}}^p(k_g)$.

The OAM beam is generated by the BS UCA through feeding its antenna elements with the same input signal, but with a successive phase shift from element to element. Thus, the phase has the increment of $2\pi\ell$ after a full turn, where ℓ is an unbounded integer termed as OAM mode number [3]. At the user UCA, the opposite operations will be performed to recover the signal. Considering the multiplexing of U $(1 \le U \le N)$ orthogonal OAM modes, the spiralization at BS UCA is equivalent to the transmitting signal multiplied by a U-dimensional inverse discrete Fourier transform (IDFT) matrix $\mathbf{F}^H$, and the despiralization at user UCA is equivalent to the receiving signal multiplied by a U-dimensional discrete Fourier transform (DFT) matrix $\mathbf{F}$ [11]. When the beam at the BS is aligned with the pth user UCA, the estimated channel matrix $\{\hat{\mathbf{H}}_i^p(k_g)|i = 1, 2, \cdots, P\}$ can be decomposed into

$$\hat{\mathbf{H}}_i^p(k_g) = \mathbf{F}^H \hat{\mathbf{\Lambda}}_i^p(k_g)\mathbf{F}, \tag{6.17}$$

where $\hat{\mathbf{\Lambda}}_i^p(k_g)$ is a $U \times U$ dimensional diagonal matrix, $\mathbf{F} = [\mathbf{f}^H(\ell_1), \mathbf{f}^H(\ell_2), \cdots, \mathbf{f}^H(\ell_U)]^H$, and $\mathbf{f}(\ell_u) = \frac{1}{\sqrt{N}}[1, e^{-j\frac{2\pi\ell_u}{N}}, \cdots, e^{-j\frac{2\pi\ell_u(N-1)}{N}}]$. However, when the beam at the BS is misaligned with the pth user UCA, the transmitter needs a beam steering matrix to implement the diagonalization decomposition mentioned above [6, 15]

$$\hat{\mathbf{H}}_i^p(k_g) = \mathbf{F}^H \hat{\mathbf{\Lambda}}_i^p(k_g)(\hat{\mathbf{B}}_i^p(k_g) \odot \mathbf{F}), \tag{6.18}$$

where $\hat{\mathbf{B}}_i^p(k_g) = \mathbf{1} \otimes \hat{\mathbf{b}}_i^p(k_g)$ is the transmit beam steering matrix of the ith group elements of BS UCA, and $\hat{\mathbf{b}}_i^p(k_g) = [e^{j\hat{W}_i^p(k_g)}, e^{j\hat{W}_{i+P}^p(k_g)}, \cdots, e^{j\hat{W}_{i+P\times(N-1)}^p(k_g)}]$. In order to obtain the expression of $\hat{W}_{i+P\times(n-1)}^p(k_g)$, $n = 1, 2, \ldots, N$, we denote the second item in (6.9) as $\eta(\omega_n, \upsilon_m^p)$, where ω_n and υ_m^p are respectively the azimuth angle of the nth antenna element in BS UCA and that of the mth antenna element in the pth user UCA. Since both addend items of $\eta(\omega_n, \upsilon_m^p)$ contain ω_n and υ_m^p, the phase shift $\eta(\omega_n, \upsilon_m^p)$ caused by ω_n and υ_m^p cannot be compensated either at the transmitter or at the receiver. But the value of $\eta(\omega_n, \upsilon_m^p)$ is too small in the case of $r_p \gg R_t, R_p$ to be ignored when beam steering is performed at the transmitter. Therefore, we obtain

$$\hat{W}^p_{i+P\times(n-1)}(k_g) = -k_g R_t \sin\hat{\theta}_p \cos\left(\hat{\varphi}_p - \omega_{i+P\times(n-1)}\right),$$
$$i = 1, 2, \cdots, P, n = 1, 2, \ldots, N. \tag{6.19}$$

After that, $\hat{\mathbf{H}}^p(k_g)$ can be expressed as

$$\hat{\mathbf{H}}^p(k_g) = \mathbf{F}^H\Big[\hat{\boldsymbol{\Lambda}}_1^p(k_g)(\hat{\mathbf{B}}_1^p(k_g) \odot \mathbf{F}), \hat{\boldsymbol{\Lambda}}_2^p(k_g)(\hat{\mathbf{B}}_2^p(k_g) \odot \mathbf{F}), \cdots,$$
$$\hat{\boldsymbol{\Lambda}}_P^p(k_g)(\hat{\mathbf{B}}_P^p(k_g) \odot \mathbf{F})\Big]. \tag{6.20}$$

In the following, we first introduce Theorem 3 on the decomposition of $\hat{\mathbf{H}}^p(k_g)$, which lay the foundation for subsequent MU-OAM preprocessing, despiralization, and AD.

Theorem 3 *When $\eta(\omega_n, \upsilon_m^p)$ is equal to zero, the THz OAM estimated channel matrix of the pth user at the gth subcarrier can be decomposed as*

$$\hat{\mathbf{H}}^p(k_g) = \mathbf{F}^H \hat{\boldsymbol{\Lambda}}_G^p(k_g)(\hat{\mathbf{V}}^p(k_g))^H, \tag{6.21}$$

where

$$\hat{\mathbf{V}}^p(k_g) = \big[(\hat{\boldsymbol{\Lambda}}_G^p(k_g))^{-1}\hat{\boldsymbol{\Lambda}}_1^p(k_g)(\hat{\mathbf{B}}_1^p(k_g) \odot \mathbf{F}), (\hat{\boldsymbol{\Lambda}}_G^p(k_g))^{-1}\hat{\boldsymbol{\Lambda}}_2^p(k_g)(\hat{\mathbf{B}}_2^p(k_g) \odot \mathbf{F}),$$
$$\cdots, (\hat{\boldsymbol{\Lambda}}_G^p(k_g))^{-1}\hat{\boldsymbol{\Lambda}}_P^p(k_g)(\hat{\mathbf{B}}_P^p(k_g) \odot \mathbf{F})\big]^H \tag{6.22}$$

is a semi-unitary matrix and $\hat{\boldsymbol{\Lambda}}_G^p(k_g)$ is a $U \times U$ dimensional real matrix.

Proof Construct $\hat{\boldsymbol{\Lambda}}_G^p(k_g)$ as

$$\hat{\boldsymbol{\Lambda}}_G^p(k_g) = [\xi_{k_g}^p(u, v)]_{U\times U}, \quad u, v = 1, 2, \ldots, U, \tag{6.23}$$

where $\xi_{k_g}^p(u, v) = \sqrt{\sum_{i=1}^P |\hat{A}_{i,k_g}^p(u, v)|^2}$, $p = 1, 2, \cdots, P$, are positive real numbers. Then, we prove that $\hat{\mathbf{V}}^p(k_g)$ is a semi-unitary matrix.

$$(\hat{\mathbf{V}}^p(k_g))^H\hat{\mathbf{V}}^p(k_g)$$
$$= \sum_{i=1}^P(\hat{\boldsymbol{\Lambda}}_G^p(k_g))^{-1}(\hat{\boldsymbol{\Lambda}}_i^p(k_g))^H(\mathbf{F}_U \odot \hat{\mathbf{B}}_i^p(k_g))(\mathbf{F}_U \odot \hat{\mathbf{B}}_i^p(k_g))^H\hat{\boldsymbol{\Lambda}}_i^p(k_g)(\hat{\boldsymbol{\Lambda}}_G^p(k_g))^{-1}$$

$$
\begin{aligned}
&= (\hat{\mathbf{\Lambda}}_G^p(k_g))^{-1} \left(\sum_{i=1}^{P} (\hat{\mathbf{\Lambda}}_i^p(k_g))^H \hat{\mathbf{\Lambda}}_i^p(k_g) \right) (\hat{\mathbf{\Lambda}}_G^p(k_g))^{-1} \\
&= (\hat{\mathbf{\Lambda}}_G^p(k_g))^{-1} (\hat{\mathbf{\Lambda}}_G^p(k_g))^2 (\hat{\mathbf{\Lambda}}_G^p(k_g))^{-1} \\
&= \mathbf{I}_U,
\end{aligned}
\tag{6.24}
$$

where $\mathbf{I}_U$ is U-dimensional identity matrix. Thus $\hat{\mathbf{V}}^p(k_g)$ is a semi-unitary matrix.

Then, we formulate the DL signal model. Denote $\mathbf{x}^p(k_g) = [x_{\ell_1}^p(k_g), x_{\ell_2}^p(k_g), \cdots, x_{\ell_U}^p(k_g)]^T$ as the U-dimensional signal vector transmitted to the pth user UCA from BS at the gth subcarrier, and all the signals transmitted to P users can be written as the PU-dimensional vector $\mathbf{x}(k_g) = [(\mathbf{x}^1(k_g))^T, (\mathbf{x}^2(k_g))^T, \cdots, (\mathbf{x}^P(k_g))^T]^T$. Then, the received signal vector at the pth user can be expressed as

$$
\begin{aligned}
\mathbf{y}^p(k_g) = \mathbf{H}^p(k_g)\hat{\mathbf{P}}(k_g)\mathbf{x}(k_g) + \mathbf{n}^p(k_g) &= \left(\hat{\mathbf{H}}^p\left(k_g\right) + \Delta\mathbf{H}^p\left(k_g\right)\right) \\
&\times \hat{\mathbf{P}}(k_g)\mathbf{x}(k_g) + \mathbf{n}^p(k_g),
\end{aligned}
\tag{6.25}
$$

where $\hat{\mathbf{P}}(k_g) = [\hat{\mathbf{P}}_1(k_g), \hat{\mathbf{P}}_2(k_g), \cdots, \hat{\mathbf{P}}_P(k_g)]$ is the DL transmission preprocessing matrix at BS and $\mathbf{n}^p(k_g)$ is the additive white Gaussian noise (AWGN) of the pth user UCA. Based on Theorem 3, $\mathbf{y}^p(k_g)$ can be rewritten as

$$
\mathbf{y}^p(k_g) = \left(\mathbf{F}^H \hat{\mathbf{\Lambda}}_G^p(k_g)(\hat{\mathbf{V}}^p(k_g))^H + \Delta\mathbf{H}^p\left(k_g\right)\right) \hat{\mathbf{P}}(k_g)\mathbf{x}(k_g) + \mathbf{n}^p(k_g). \tag{6.26}
$$

Then, the total received signals of the DL THz MU-OAM system can be written as

$$
\begin{aligned}
\mathbf{y}(k_g) &= \left[(\mathbf{y}^1(k_g))^T, (\mathbf{y}^2(k_g))^T, \cdots, (\mathbf{y}^P(k_g))^T\right]^T \\
&= \left(\mathbf{U}\hat{\mathbf{\Lambda}}(k_g)\hat{\mathbf{V}}(k_g)^H + \Delta\mathbf{H}\left(k_g\right)\right) \hat{\mathbf{P}}(k_g)\mathbf{x}(k_g) + \mathbf{n}(k_g),
\end{aligned}
\tag{6.27}
$$

where

$$
\begin{aligned}
\mathbf{U} &= \mathrm{diag}\left\{\mathbf{F}^H, \mathbf{F}^H, \cdots, \mathbf{F}^H\right\}, \\
\hat{\mathbf{\Lambda}}(k_g) &= \mathrm{diag}\left\{\hat{\mathbf{\Lambda}}_G^1(k_g), \hat{\mathbf{\Lambda}}_G^2(k_g), \cdots, \hat{\mathbf{\Lambda}}_G^P(k_g)\right\}, \\
\hat{\mathbf{V}}(k_g) &= \left[\hat{\mathbf{V}}^1(k_g), \hat{\mathbf{V}}^2(k_g), \cdots, \hat{\mathbf{V}}^P(k_g)\right], \\
\Delta\mathbf{H}\left(k_g\right) &= \left[(\Delta\mathbf{H}^1(k_g))^T, (\Delta\mathbf{H}^2(k_g))^T, \cdots, (\Delta\mathbf{H}^P(k_g))^T\right]^T,
\end{aligned}
$$

Algorithm 1 Downlink THz MU-OAM transmission and reception

Input: $\mathbf{x}^p, N, P, \hat{\varphi}_p, \hat{\theta}_p, R_p, R_t, \hat{r}_p, \upsilon_m^p, \omega_n$
Output: $\hat{\mathbf{x}}^p$
procedure
$\mathbf{x} \leftarrow [(\mathbf{x}^1)^T, (\mathbf{x}^2)^T, \cdots, (\mathbf{x}^P)^T]^T$
$\hat{\mathbf{H}}^p, \hat{\mathbf{B}}^p \leftarrow N, P, \hat{\varphi}_p, \hat{\theta}_p, R_p, R_t, \hat{r}_p, \upsilon_m^p, \omega_n$
$\hat{\mathbf{\Lambda}}^p \leftarrow \mathbf{F}\hat{\mathbf{H}}^p(\hat{\mathbf{B}}^p \odot \mathbf{F})^H$
$\xi_{k_g}^p(u, v) = \sqrt{\sum_{i=1}^P |\hat{\Lambda}_{i,k_g}^p(u, v)|^2}, p = 1, 2, \cdots, P$
$\hat{\mathbf{\Lambda}}_G^p(k_g) = [\xi_{k_g}^p(u, v)]_{U\times U}, u, v = 1, 2, \ldots, U$
$\hat{\mathbf{V}}^p \leftarrow ((\hat{\mathbf{\Lambda}}_G^p)^{-1}\hat{\mathbf{\Lambda}}^p(\hat{\mathbf{B}}^p \odot \mathbf{F}))^H$
$\hat{\mathbf{P}} \leftarrow \left([\hat{\mathbf{V}}^1, \hat{\mathbf{V}}^2, \cdots, \hat{\mathbf{V}}^P]^H\right)^\dagger$
$\mathbf{y}^p \leftarrow \mathbf{H}^p\hat{\mathbf{P}}\mathbf{x} + \mathbf{n}^p$
$\hat{\mathbf{x}}^p \leftarrow (\hat{\mathbf{\Lambda}}_G^p)^{-1}\mathbf{F}\mathbf{y}^p$
return $\hat{\mathbf{x}}^p$
end procedure

$$\mathbf{n}(k_g) = \left[(\mathbf{n}^1(k_g))^T, (\mathbf{n}^2(k_g))^T, \cdots, (\mathbf{n}^P(k_g))^T\right]^T. \tag{6.28}$$

In a simple way, the preprocessing matrix at the transmitter can be designed as

$$\hat{\mathbf{P}}(k_g) = \left([\hat{\mathbf{V}}^1(k_g), \hat{\mathbf{V}}^2(k_g), \cdots, \hat{\mathbf{V}}^P(k_g)]^H\right)^\dagger. \tag{6.29}$$

Upon substituting (6.29) into (6.27), we arrive at

$$\mathbf{y}(k_g) = \mathbf{U}\hat{\mathbf{\Lambda}}(k_g)\mathbf{x}(k_g) + \Delta\mathbf{H}\left(k_g\right)\hat{\mathbf{P}}(k_g)\mathbf{x}(k_g) + \mathbf{n}(k_g). \tag{6.30}$$

To be more specific, the received signal vector at the pth user is

$$\mathbf{y}^p(k_g) = \mathbf{F}^H\hat{\mathbf{\Lambda}}_G^p(k_g)\mathbf{x}^p(k_g) + \Delta\mathbf{H}^p\left(k_g\right)\hat{\mathbf{P}}(k_g)\mathbf{x}(k_g) + \mathbf{n}^p(k_g). \tag{6.31}$$

Therefore, the signal detection at the pth user can be simplified as an AD matrix $(\hat{\mathbf{\Lambda}}_G^p(k_g))^{-1}$ after despiralization $\mathbf{F}$. Then, the signal vector at the pth user can be recovered as

$$\hat{\mathbf{x}}^p(k_g) = \mathbf{x}^p(k_g) + \mathcal{L}^p(k_g) + (\hat{\mathbf{\Lambda}}_G^p(k_g))^{-1}\mathbf{F}\mathbf{n}^p(k_g), \tag{6.32}$$

where $\mathcal{L}^p(k_g) = (\hat{\mathbf{\Lambda}}_G^p(k_g))^{-1}\mathbf{F}\Delta\mathbf{H}^p\left(k_g\right)\hat{\mathbf{P}}(k_g)\mathbf{x}(k_g)$ is the interference due to the estimation errors. Thus, the DL MU-OAM signal transmission and reception are completed, whose detailed procedure is summarized in Algorithm 1.

6.3.2 Uplink THz MU-OAM Transmission and Reception

Similarly, we assume that the BS has only an estimate of the practical channel matrix $\bar{\mathbf{H}}^p(k_g)$ that we denote by $\hat{\bar{\mathbf{H}}}^p(k_g)$ in the UL THz MU-OAM system. Therefore, $\bar{\mathbf{H}}^p(k_g)$ can be expressed as

$$\bar{\mathbf{H}}^p(k_g) = \hat{\bar{\mathbf{H}}}^p(k_g) + \left(\Delta \mathbf{H}^p\left(k_g\right)\right)^T. \tag{6.33}$$

In our UL channel model, the receiver post-processing matrix is designed based on the knowledge of the estimated channel matrix $\hat{\bar{\mathbf{H}}}^p(k_g)$. When the beam at the pth user is aligned with the BS UCA, the estimated channel matrix $\{\hat{\bar{\mathbf{H}}}_i^p(k_g)|i = 1, \cdots, P\}$ can be decomposed into

$$\hat{\bar{\mathbf{H}}}_i^p(k_g) = \mathbf{F}^H \hat{\mathbf{\Lambda}}_i^p(k_g)\mathbf{F}. \tag{6.34}$$

Otherwise, when the beam at the pth user is misaligned with the BS UCA, the beam steering matrix at receiver is

$$\hat{\bar{\mathbf{H}}}_i^p(k_g) = (\hat{\mathbf{B}}_i^p(k_g) \odot \mathbf{F}^H)\hat{\mathbf{\Lambda}}_i^p(k_g)\mathbf{F}. \tag{6.35}$$

After that, $\hat{\bar{\mathbf{H}}}^p(k_g)$ can be expressed as

$$\begin{aligned}\hat{\bar{\mathbf{H}}}^p(k_g) = \Big[&(\hat{\mathbf{B}}_1^p(k_g) \odot \mathbf{F}^H)\hat{\mathbf{\Lambda}}_1^p(k_g);\ \ (\hat{\mathbf{B}}_2^p(k_g) \odot \mathbf{F}^H)\hat{\mathbf{\Lambda}}_2^p(k_g);\ \ \cdots;\\ &(\hat{\mathbf{B}}_P^p(k_g) \odot \mathbf{F}^H)\hat{\mathbf{\Lambda}}_P^p(k_g)\Big]\mathbf{F},\end{aligned} \tag{6.36}$$

where the semicolon in (6.36) represents that the sub-block matrix is placed in columns. Based on Theorem 3, (6.36) can be decomposed as

$$\hat{\bar{\mathbf{H}}}^p(k_g) = \left(\hat{\bar{\mathbf{V}}}^p(k_g)\right)^H \hat{\mathbf{\Lambda}}_G^p(k_g)\mathbf{F}, \tag{6.37}$$

where

$$\begin{aligned}\hat{\bar{\mathbf{V}}}^p(k_g) = \big[&(\hat{\mathbf{B}}_1^p(k_g) \odot \mathbf{F}^H)\hat{\mathbf{\Lambda}}_1^p(k_g)(\hat{\mathbf{\Lambda}}_G^p(k_g))^{-1}; (\hat{\mathbf{B}}_2^p(k_g) \odot \mathbf{F}^H)\hat{\mathbf{\Lambda}}_2^p(k_g)(\hat{\mathbf{\Lambda}}_G^p(k_g))^{-1};\\ &\cdots;\ \ (\hat{\mathbf{B}}_P^p(k_g) \odot \mathbf{F}^H)\hat{\mathbf{\Lambda}}_P^p(k_g)(\hat{\mathbf{\Lambda}}_G^p(k_g))^{-1}\big]^H.\end{aligned} \tag{6.38}$$

Denote $\bar{\mathbf{x}}^p(k_g) = [\bar{x}_{\ell_1}^p(k_g), \bar{x}_{\ell_2}^p(k_g), \cdots, \bar{x}_{\ell_U}^p(k_g)]^T$ as the U-dimensional signal vector transmitted to BS UCA from the pth user UCA at the gth subcarrier, and all the signals transmitted to BS from all P users can be written as the PU-dimensional vector $\bar{\mathbf{x}}(k_g) = [(\bar{\mathbf{x}}^1(k_g))^T, (\bar{\mathbf{x}}^2(k_g))^T, \cdots, (\bar{\mathbf{x}}^P(k_g))^T]^T$. Then, the received signal vector at BS can be expressed as

$$\begin{aligned}\bar{\mathbf{y}}(k_g) &= \sum_{p=1}^{P} \bar{\mathbf{H}}^p(k_g)\mathbf{F}^H\bar{\mathbf{x}}^p(k_g) + \bar{\mathbf{n}}(k_g) \\ &= \sum_{p=1}^{P} \left(\hat{\bar{\mathbf{H}}}^p(k_g) + \left(\Delta\mathbf{H}^p\left(k_g\right)\right)^T\right)\mathbf{F}^H\bar{\mathbf{x}}^p(k_g) + \bar{\mathbf{n}}(k_g),\end{aligned} \tag{6.39}$$

where $\mathbf{F}^H$ is U-dimensional IDFT matrix that spiralizes $\bar{\mathbf{x}}^p(k_g)$ for the pth user and $\bar{\mathbf{n}}(k_g)$ is the AWGN. Based on (6.37), the vector of the UL received signal can be rewritten as

$$\bar{\mathbf{y}}(k_g) = \sum_{p=1}^{P} \left(\left(\hat{\bar{\mathbf{V}}}^p(k_g)\right)^H \hat{\mathbf{\Lambda}}_G^p(k_g)\mathbf{F} + \left(\Delta\mathbf{H}^p\left(k_g\right)\right)^T\right)\mathbf{F}^H\bar{\mathbf{x}}^p(k_g) + \bar{\mathbf{n}}(k_g). \tag{6.40}$$

Let us define

$$\bar{\mathbf{U}} = \mathrm{diag}\{\mathbf{F}, \mathbf{F}, \cdots, \mathbf{F}\},$$

$$\hat{\bar{\mathbf{V}}}(k_g) = \left[\left(\hat{\bar{\mathbf{V}}}^1(k_g)\right)^H, \left(\hat{\bar{\mathbf{V}}}^2(k_g)\right)^H, \cdots, \left(\hat{\bar{\mathbf{V}}}^P(k_g)\right)^H\right]. \tag{6.41}$$

Then, the received signal of the UL THz MU-OAM system can be expressed as

$$\bar{\mathbf{y}}(k_g) = \left(\hat{\bar{\mathbf{V}}}(k_g)\hat{\mathbf{\Lambda}}(k_g) + \left(\Delta\mathbf{H}\left(k_g\right)\right)^T\bar{\mathbf{U}}\right)\bar{\mathbf{x}}(k_g) + \bar{\mathbf{n}}(k_g). \tag{6.42}$$

At the UL receiver of the BS, the received vector $\bar{\mathbf{y}}(k_g)$ is processed by using the post-processing matrix $\hat{\mathbf{T}}(k_g)$ to restore the transmitted UL data from the pth user. The data estimates of all P UL users are presented by a single vector as

$$\hat{\bar{\mathbf{x}}}(k_g) = \left[\left(\hat{\bar{\mathbf{x}}}^1(k_g)\right)^T, \left(\hat{\bar{\mathbf{x}}}^2(k_g)\right)^T, \cdots, \left(\hat{\bar{\mathbf{x}}}^P(k_g)\right)^T\right]^T = \hat{\mathbf{T}}(k_g)\bar{\mathbf{y}}(k_g), \tag{6.43}$$

where $\hat{\mathbf{T}}(k_g) = [(\hat{\bar{\mathbf{V}}}^1(k_g))^H, (\hat{\bar{\mathbf{V}}}^2(k_g))^H, \cdots, (\hat{\bar{\mathbf{V}}}^P(k_g))^H]^\dagger$. Substituting (6.42) into (6.43), we can get

$$\hat{\bar{\mathbf{x}}}(k_g) = \hat{\mathbf{\Lambda}}(k_g)\bar{\mathbf{x}}(k_g) + \hat{\mathbf{T}}(k_g)\left(\Delta\mathbf{H}\left(k_g\right)\right)^T\bar{\mathbf{U}}\bar{\mathbf{x}}(k_g) + \hat{\mathbf{T}}(k_g)\bar{\mathbf{n}}(k_g). \tag{6.44}$$

Algorithm 2 Uplink THz MU-OAM transmission and reception

Input: $\bar{\mathbf{x}}^p, N, P, \hat{\varphi}_p, \hat{\theta}_p, R_p, R_t, \hat{r}_p, \upsilon_m^p, \omega_n$
Output: $\hat{\hat{\mathbf{x}}}$

procedure
$\bar{\mathbf{x}} \leftarrow [(\bar{\mathbf{x}}^1)^T, (\bar{\mathbf{x}}^2)^T, \cdots, (\bar{\mathbf{x}}^P)^T]^T$
$\hat{\hat{\mathbf{H}}}^p, \hat{\mathbf{B}}^p \leftarrow N, P, \hat{\varphi}_p, \hat{\theta}_p, R_p, R_t, \hat{r}_p, \upsilon_m^p, \omega_n$
$\hat{\mathbf{\Lambda}}^p \leftarrow (\hat{\mathbf{B}}^p \odot \mathbf{F}^H)^H \hat{\hat{\mathbf{H}}}^p \mathbf{F}^H$
$\xi_{k_g}^p(u, v) = \sqrt{\sum_{i=1}^P |\hat{\Lambda}_{i,k_g}^p(u, v)|^2}, p = 1, 2, \cdots, P$
$\hat{\mathbf{\Lambda}}_G^p(k_g) = [\xi_{k_g}^p(u, v)]_{U\times U}, u, v = 1, 2, \ldots, U$
$\hat{\mathbf{\Lambda}} \leftarrow \text{diag}\{\hat{\mathbf{\Lambda}}_G^1, \hat{\mathbf{\Lambda}}_G^2, \cdots, \hat{\mathbf{\Lambda}}_G^P\}$
$\hat{\hat{\mathbf{V}}}^p \leftarrow \left((\hat{\mathbf{B}}^p \odot \mathbf{F}^H)\hat{\mathbf{\Lambda}}^p(\hat{\mathbf{\Lambda}}_G^p)^{-1}\right)^H$
$\bar{\mathbf{y}} \leftarrow \sum_{p=1}^P \bar{\mathbf{H}}^p \mathbf{F}^H \bar{\mathbf{x}}^p + \bar{\mathbf{n}}$
$\hat{\mathbf{T}} \leftarrow [(\hat{\hat{\mathbf{V}}}^1)^H, (\hat{\hat{\mathbf{V}}}^2)^H, \cdots, (\hat{\hat{\mathbf{V}}}^P)^H]^\dagger$
$\hat{\hat{\mathbf{x}}} \leftarrow (\hat{\mathbf{\Lambda}})^{-1}\hat{\mathbf{T}}\bar{\mathbf{y}}$
return $\hat{\hat{\mathbf{x}}}$
end procedure

Then, the signal vector after the AD matrix $(\hat{\mathbf{\Lambda}}(k_g))^{-1}$ can be recovered as

$$\hat{\hat{\mathbf{x}}}(k_g) = \bar{\mathbf{x}}(k_g) + \bar{\mathcal{L}}(k_g) + (\hat{\mathbf{\Lambda}}(k_g))^{-1}\hat{\mathbf{T}}(k_g)\bar{\mathbf{n}}(k_g), \tag{6.45}$$

where $\bar{\mathcal{L}}(k_g) = (\hat{\mathbf{\Lambda}}(k_g))^{-1}\hat{\mathbf{T}}(k_g)\left(\Delta\mathbf{H}\left(k_g\right)\right)^T \bar{\mathbf{U}}\bar{\mathbf{x}}(k_g)$ is the interference due to the estimation error. Now, the UL MU-OAM signal transmission and reception are completed. The detailed procedure is summarized in Algorithm 2.

6.3.3 Low-Complexity Implementation

A salient feature of the THz MU-OAM system is that the channel matrices are completely characterized by the three parameters, i.e., the azimuth angle, the elevation angle, and the distance, independent of the numbers of subcarriers and antennas. Compared with the traditional MIMO-OFDM system, the proposed system significantly reduces the burden by avoiding estimating large channel matrices and performing singular value decomposition (SVD). In order to obtain the preprocessing matrix $\hat{\mathbf{P}}(k_g)$ at the transmitter and the AD matrix $(\hat{\mathbf{\Lambda}}_G^p(k_g))^{-1}$ at the receiver in DL, we need to calculate the values of elements of the estimated effective MU-OAM channel matrix. Based on (6.21), the estimated effective MU-OAM channel matrix of the pth user at the gth subcarrier is

$$\hat{\mathbf{\Lambda}}_G^p(k_g)=\mathbf{F}\hat{\mathbf{H}}^p(k_g)\hat{\mathbf{V}}^p(k_g)=\left[\left(\sum_{i=1}^{P}\left|\hat{\Lambda}_{i,k_g}^p(u,v)\right|^2\right)^{\frac{1}{2}}\right]_{U\times U}, \tag{6.46}$$

where $\hat{\Lambda}_{i,k_g}^p(u,v)$ is the uth-row and vth-column element of $\hat{\mathbf{\Lambda}}_i^p(k_g)$, and it can be obtained as

$$\begin{aligned}
&\hat{\Lambda}_{i,k_g}^p(u,v)\\
&=\frac{1}{N}\sum_{m=1}^{N}\sum_{n=1}^{N}\hat{h}_{m,n}^p(k_g)\exp\left(-j\ell_u\upsilon_m^p+j\ell_v\omega_n+j\hat{W}_n^p(k_g)\right)\\
&=\beta(k_g)\frac{1}{2k_g\hat{r}_pN}\sum_{m=1}^{N}\sum_{n=1}^{N}\exp\Big(-jk_g\hat{r}_p+jk_gR_t\sin\hat{\theta}_p\cos(\hat{\varphi}_p-\omega_n)\\
&\quad+jk_g\frac{R_tR_p}{\hat{r}_p}\big(\cos\hat{\theta}_p\cos(\upsilon_m^p-\omega_n)\\
&\quad+(1-\cos\hat{\theta}_p)\sin(\hat{\varphi}_p-\upsilon_m^p)\sin(\hat{\varphi}_p-\omega_n)\big)-j\ell_u\upsilon_m^p+j\ell_v\omega_n\\
&\quad-jk_gR_t\sin\hat{\theta}_p\cos(\hat{\varphi}_p-\omega_n)\Big)\\
&=\vartheta(k_g)\sum_{m=1}^{N}\sum_{n=1}^{N}\exp\left(jk_g\frac{R_tR_p}{\hat{r}_p}\left(\cos\hat{\theta}_p\cos\frac{2\pi t}{N}+\left(1-\cos\hat{\theta}_p\right)\right.\right.\\
&\quad\times\left(-\frac{1}{2}\left(\cos\left(2\hat{\varphi}_p-\frac{2\pi\left(t'-2\right)}{N}\right)\right.\right.\\
&\quad\left.\left.\left.\left.-\cos\frac{2\pi t}{N}\right)\right)\right)+j\ell_v\frac{2\pi t}{N}-j\frac{2\pi\,(m-1)}{N}s\right)\\
&=\vartheta(k_g)\sum_{t=1}^{N}\exp\left(jk_g\frac{R_tR_p}{2\hat{r}_p}\left(1+\cos\hat{\theta}_p\right)\cos\frac{2\pi t}{N}+j\ell_v\frac{2\pi t}{N}\right)\\
&\quad\times\sum_{t'=1}^{2N}\exp\left(-jk_g\frac{R_tR_p}{2\hat{r}_p}\left(1-\cos\hat{\theta}_p\right)\cos\left(2\hat{\varphi}_p\right.\right.\\
&\quad\left.\left.-\frac{2\pi\left(t'-2\right)}{N}\right)\right)\sum_{m=1}^{N}\exp\left(-j\frac{2\pi\,(m-1)}{N}s\right),
\end{aligned}\tag{6.47}$$

where $\vartheta(k_g)=\beta(k_g)\frac{\exp(-jk_g\hat{r}_p)}{2k_g\hat{r}_pN}$, $t=n-m\in\mathbb{Z}$, $t'=n+m\in\mathbb{Z}$, $s=\ell_u-\ell_v\in\mathbb{Z}$, $u,v=1,2,\cdots,U$, $p=1,2,\cdots,P$. From (6.47), when $s\neq 0$,

$\sum_{m=1}^{N} \exp\left(-j\frac{2\pi(m-1)}{N}s\right) = \frac{1-\exp(-j2\pi s)}{1-\exp\left(-j\frac{2\pi}{N}s\right)} = 0$, and the value of $\hat{\Lambda}^{p}_{i,k_g}(u,v)$ equals to zero. So we only need to consider the behavior of diagonal elements $\hat{\Lambda}^{p}_{i,k_g}(u,u)$, and it takes the form

$$
\begin{aligned}
&\hat{\Lambda}^{p}_{i,k_g}(u,u)\\
&= N\vartheta\left(k_g\right)\sum_{t=1}^{N}\exp\left(jk_g\frac{R_tR_p}{2\hat{r}_p}\left(1-3\cos\hat{\theta}_p\right)\cos\frac{2\pi t}{N}+j\ell_u\frac{2\pi t}{N}\right)\\
&\quad\times\sum_{t'=1}^{2N}\exp\left(-jk_g\frac{R_tR_p}{2\hat{r}_p}\left(1-\cos\hat{\theta}_p\right)\right.\\
&\quad\left.\times\cos\left(2\hat{\varphi}_p-\frac{2\pi\left(t'-2\right)}{N}\right)\right)\\
&= N\vartheta\left(k_g\right)\sum_{t=1}^{N}\exp\left(jk_g\frac{R_tR_p}{2\hat{r}_p}\left(1-3\cos\hat{\theta}_p\right)\cos\frac{2\pi t}{N}+jl_u\frac{2\pi t}{N}\right)\\
&\quad\times\sum_{t'=1}^{2N}\exp\left(-jk_g\frac{R_tR_p}{2\hat{r}_p}\left(1-\cos\hat{\theta}_p\right)\cos 2\hat{\varphi}_p\right.\\
&\quad\left.\times\cos\frac{2\pi\left(t'-2\right)}{N}\right)\exp\left(-jk_g\frac{R_tR_p}{2\hat{r}_p}\left(1-\cos\hat{\theta}_p\right)\sin 2\hat{\varphi}_p\,\sin\frac{2\pi\left(t'-2\right)}{N}\right)\\
&= N\vartheta\left(k_g\right)\sum_{t=1}^{N}\exp\left(j\mathfrak{I}_{k_g}\cos\frac{2\pi t}{N}\right)\exp\left(jl_u\frac{2\pi t}{N}\right)\\
&\quad\times\sum_{t'=1}^{2N}\exp\left(-j\mathfrak{R}_{k_g}\cos\frac{2\pi\left(t'-2\right)}{N}\right)\exp\left(-j\aleph_{k_g}\sin\frac{2\pi\left(t'-2\right)}{N}\right),
\end{aligned}
\tag{6.48}
$$

where

$$
\begin{aligned}
\mathfrak{I}_{k_g} &= k_g\frac{R_tR_p}{2\hat{r}_p}\left(1-3\cos\hat{\theta}_p\right),\\
\mathfrak{R}_{k_g} &= k_g\frac{R_tR_p}{2\hat{r}_p}\left(1-\cos\hat{\theta}_p\right)\cos 2\hat{\varphi}_p,\\
\aleph_{k_g} &= k_g\frac{R_tR_p}{2\hat{r}_p}\left(1-\cos\hat{\theta}_p\right)\sin 2\hat{\varphi}_p.
\end{aligned}
\tag{6.49}
$$

Table 6.1 The number of floating point operations of the proposed THz MU-OAM system, the MU-OAM system [15], and the traditional MU-MIMO system [13]

Scheme		Complexity
The THz MU-OAM	Pre/Post-processing	$O(GPU^3)$
	Amplitude detection	$O(GPU^3)$
The MU-OAM [15]	Pre/Post-processing	$O(GP^4U^3)$
	Amplitude detection	$O(GPU^3)$
The MU-MIMO [13]	Pre/Post-processing	$O(GP^4N^3)$
	Amplitude detection	$O(GP^2N^3)$

Using the Taylor series expansion of the exponential function $e^x = \sum_{\alpha=0}^{+\infty} x^\alpha/\alpha!$, the value of $\hat{\Lambda}^p_{i,k_g}(u,u)$ can be approximately expressed as a function of OAM mode number ℓ and wave number k_g [5]. Therefore, for $\hat{r}_p \gg R_t, R_p$, (6.48) can be expressed as

$$\hat{\Lambda}^p_{i,k_g}(u,u) \approx -2\vartheta\left(k_g\right)\frac{N^3}{2^\tau \tau!}\frac{j^{(\overline{\tau}+\tau)}}{\left(2^{\overline{\tau}}\overline{\tau}!\right)^2}\mathfrak{R}^{\overline{\tau}}_{k_g}\aleph^{\overline{\tau}}_{k_g}\mathfrak{I}^{\tau}_{k_g}, \tag{6.50}$$

where $\tau = \min\{|\ell_u|, N-|\ell_u|\}, \overline{\tau} \in \{N, 2l\}, 0 \leq l < \frac{N}{2}$. The detailed derivation process of (6.50) is given in Appendix B. Thus, the uth diagonal element of the estimated effective MU-OAM channel matrix $\hat{\mathbf{\Lambda}}^p_G(k_g)$ can be expressed as

$$\hat{\Lambda}^p_{G,k_g}(u,u) = \frac{2\sqrt{P}N^3}{2^\tau \tau!\left(2^{\overline{\tau}}\overline{\tau}!\right)^2}\left|\vartheta\left(k_g\right)\mathfrak{R}^{\overline{\tau}}_{k_g}\aleph^{\overline{\tau}}_{k_g}\mathfrak{I}^{\tau}_{k_g}\right|. \tag{6.51}$$

Remark 3 It follows that the effective channel gain $\hat{\Lambda}^p_{G,k_g}(u,u)$ of each OAM mode is only the function of the position information of the pth user UCA when k_g, R_t, R_p, N, and ℓ_u are given to the transmitter, which leads to very simple signal detection. Then, $\hat{\mathbf{V}}^p(k_g)$ can be easily obtained by substituting (6.51) into (6.22); further, we can obtain the preprocessing matrix $\hat{\mathbf{P}}(k_g)$ by substituting (6.22) into (6.29). That is, the preprocessing matrix $\hat{\mathbf{P}}(k_g)$ can be designed only according to the position information of P user UCAs instead of all the CSI.

In the proposed THz MU-OAM communication system, the complexity of AD is mainly decided by the computation of $(\hat{\mathbf{\Lambda}}^p_G(k_g))^{-1}$; the complexities of the preprocessing and the post-processing are, respectively, decided by the computation of $\hat{\mathbf{P}}(k_g)$ and $\hat{\mathbf{T}}(k_g)$. The specific computational complexities of the proposed THz MU-OAM system, the MU-OAM system [15], and the traditional MU-MIMO system [13] are shown in Table 6.1. For intuitive comparison, the computational complexities of the three systems are shown in Fig. 6.3. It can be seen from the figure that the complexities of MU-OAM are much less than that of MU-MIMO. Especially when the number of users increases, the advantages of the MU-OAM system are more obvious. Figure 6.3 also shows that the computational complexity

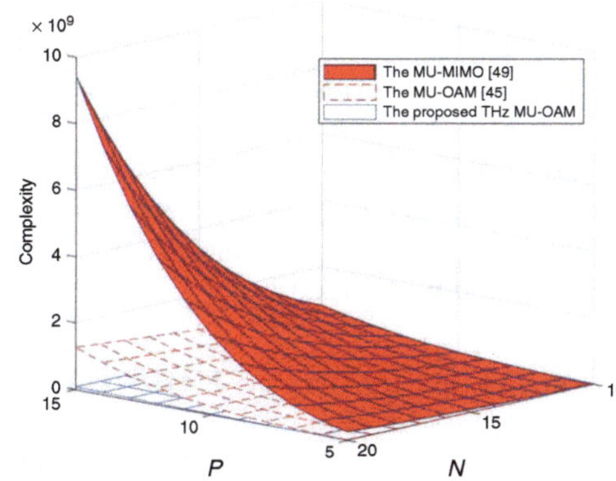

Fig. 6.3 The complexities of the proposed THz MU-OAM system, the MU-OAM system [15], and the traditional MU-MIMO system [13] vs. N and P at $G = 10$, $U = 5$

of the proposed THz MU-OAM system is much lower than that of [15] when P and U are large. The reduction of the complexity is mainly due to circumventing the SVD for $\check{\mathbf{H}}^p$.

6.4 Performance Analysis

In this section, we investigate the performances of the proposed THz MU-OAM system. First, the achievable sum rate is derived, and then, the effects of frequency on IMI and IUI are analyzed, respectively. As our proposed UL transmission scheme is a dual form of DL transmission scheme, we focus on the analysis of the DL transmission performances.

6.4.1 *The Achievable Sum Rate of THz MU-OAM System*

We assume that the total transmit power is Γ in the DL, and the average power allocated to each user is Γ/P. Considering the inevitable estimation errors of r_p, φ_p, and θ_p in practice, we have to evaluate the effect of the errors on the achievable sum rate of the THz MU-OAM system. Following (6.32), the detected OAM data symbol vector at the pth user in the DL can be written as

$$\hat{\mathbf{x}}^p(k_g) = \mathbf{x}^p(k_g) + \mathcal{L}^p(k_g) + (\hat{\mathbf{\Lambda}}_G^p(k_g))^{-1}\mathbf{F}\mathbf{n}^p(k_g), \tag{6.52}$$

where

$$\begin{aligned}\mathcal{L}^p(k_g) =& \mathcal{L}_{\mathrm{IMI}}^p(k_g) + \mathcal{L}_{\mathrm{IUI}}^p(k_g),\\ \mathcal{L}_{\mathrm{IMI}}^p(k_g) =& (\hat{\mathbf{\Lambda}}_G^p(k_g))^{-1}\mathbf{F}\Delta\mathbf{H}^p\left(k_g\right)\hat{\mathbf{P}}_p(k_g)\mathbf{x}^p(k_g),\end{aligned}$$

$$\mathcal{L}_{\rm IUI}^{p}(k_g) = \sum_{q=1,q\neq p}^{P} (\hat{\mathbf{\Lambda}}_G^{p}(k_g))^{-1}\mathbf{F}\Delta\mathbf{H}^{p}(k_g)\hat{\mathbf{P}}_q(k_g)\mathbf{x}^{q}(k_g), \tag{6.53}$$

$\mathcal{L}_{\rm IMI}^{p}(k_g)$ and $\mathcal{L}_{\rm IUI}^{p}(k_g)$ are, respectively, the U-dimensional IMI vector and IUI vector induced by the error in preprocessing matrix, and $\mathbf{n}'^{p}(k_g)$ is the noise vector. Denote $\mathbf{\Upsilon}_{\rm IMI}^{p}(k_g)$, $\mathbf{\Upsilon}_{\rm IUI}^{p}(k_g)$, and $\mathbf{\Upsilon}_{n}^{p}(k_g)$ as the $U \times U$ power matrices of $\mathcal{L}_{\rm IMI}^{p}(k_g)$, $\mathcal{L}_{\rm IUI}^{p}(k_g)$ and $\mathbf{n}'^{p}(k_g)$, i.e.,

$$\begin{aligned}
\mathbf{\Upsilon}_{\rm IMI}^{p}(k_g) &= ((\hat{\mathbf{\Lambda}}_G^{p}(k_g))^{-1}\mathbf{F}\Delta\mathbf{H}^{p}\left(k_g\right)\hat{\mathbf{P}}_p(k_g))\mathbb{E}\left(\mathbf{x}^{p}(k_g)(\mathbf{x}^{p}(k_g))^{H}\right)\\
&\quad\times ((\hat{\mathbf{\Lambda}}_G^{p}(k_g))^{-1}\mathbf{F}\Delta\mathbf{H}^{p}\left(k_g\right)\hat{\mathbf{P}}_p(k_g))^{H},\\
\mathbf{\Upsilon}_{\rm IUI}^{p}(k_g) &= \sum_{q=1,q\neq p}^{P}\left((\hat{\mathbf{\Lambda}}_G^{p}(k_g))^{-1}\mathbf{F}\Delta\mathbf{H}^{p}(k_g)\hat{\mathbf{P}}_q(k_g)\right)\mathbb{E}\left(\mathbf{x}^{q}(k_g)(\mathbf{x}^{q}(k_g))^{H}\right)\\
&\quad\times\left((\hat{\mathbf{\Lambda}}_G^{p}(k_g))^{-1}\mathbf{F}\Delta\mathbf{H}^{p}(k_g)\hat{\mathbf{P}}_q(k_g)\right)^{H},\\
\mathbf{\Upsilon}_{n}^{p}(k_g) &= \mathbb{E}\left(\mathbf{n}'^{p}(k_g)(\mathbf{n}'^{p}(k_g))^{H}\right),
\end{aligned} \tag{6.54}$$

where $\mathbf{I}_U$ is the U-dimensional unit matrix. Thus, the signal-to-interference-plus-noise ratio (SINR) at the uth mode of the pth user can be formulated as

$$\text{SINR}_p(u,k_g) = \frac{|\hat{\Lambda}_{G,k_g}^{p}(u,u)|^2\mathbb{E}(|x^{p}(u,k_g)|^2)}{[\mathbf{\Upsilon}_{\rm IMI}^{p}(k_g)]_u + [\mathbf{\Upsilon}_{\rm IUI}^{p}(k_g)]_u + [\mathbf{\Upsilon}_{n}^{p}(k_g)]_u}, \tag{6.55}$$

where $[\cdot]_u$ denotes the uth diagonal element and $[\mathbf{\Upsilon}_{\rm IMI}^{p}(k_g)]_u$, $[\mathbf{\Upsilon}_{\rm IUI}^{p}(k_g)]_u$, and $[\mathbf{\Upsilon}_{n}^{p}(k_g)]_u$, respectively, represent the power of IMI, IUI, and noise at the uth mode of the pth user. $\mathbb{E}(|x^{p}(u,k_g)|^2)$ is the average power of modulated signal transmitted on OAM beam of the uth mode of the pth user. Then, the achievable sum rate of the MU-OAM system is

$$C_{\rm MU-OAM} = \frac{1}{G}\sum_{p=1}^{P}\sum_{g=1}^{G}\sum_{u=1}^{U}\log_2\left(1+\text{SINR}_p(u,k_g)\right). \tag{6.56}$$

If the receiver has perfect CSI, the proposed communication system can obtain the same channel characteristic matrix and achievable sum rate as the MU-MIMO system in [13]. In order to have a clear comparison on the performance of the proposed system, we take the achievable sum rate of the MU-MIMO system in [13] as the upper bound of capacity, which can be expressed as

$$C_{\text{MU-MIMO}} = \frac{1}{G}\sum_{p=1}^{P}\sum_{g=1}^{G}\sum_{i=1}^{W_p}\log_2\left(1+\frac{|\Lambda'^{p}_{k_g}(i,i)|^2\mathbb{E}(|x^p(i,k_g)|^2)}{\sigma_z^2(k_g)}\right), \tag{6.57}$$

where $\Lambda'^{p}_{k_g}(i,i)$ is the ith diagonal element of $\mathbf{\Lambda}'^{p}_{k_g}$ that represents the equivalent channel matrix of pth user at the gth subcarrier and W_p is the rank of $\mathbf{\Lambda}'^{p}_{k_g}$, $\sigma_z^2(k_g)$ is the variance of AWGN $\mathbf{n}^p(k_g)$.

6.4.2 The Effects of Frequency on Interferences

In practice, IMI is prone to appear in the case short of high-precision alignment between transmit and receive beam directions. The IMI at the uth mode of the pth user comes from the interference caused by other modes of the pth user, and it can be expressed as

$$[\mathbf{\Upsilon}^{p}_{\text{IMI}}(k_g)]_u = \left|\sum_{\iota=1}^{NP}\sum_{\varpi=1}^{N}\left(\hat{\Lambda}^{p}_{G,k_g}(u,u)\right)^{-1}F(u,\varpi)\,\Delta H^{p}_{k_g}(\varpi,\iota)\,\hat{P}^{p}_{k_g}(\iota,u)\right|^2. \tag{6.58}$$

The IUI at the uth mode of the pth user comes from the interference caused by all modes used by other users, and it can be expressed as

$$[\mathbf{\Upsilon}^{p}_{\text{IUI}}(k_g)]_u = \left|\sum_{q=1,q\neq p}^{P}\sum_{\iota=1}^{NP}\sum_{\varpi=1}^{N}\left(\hat{\Lambda}^{p}_{G,k_g}(u,u)\right)^{-1}F(u,\varpi)\,\Delta H^{p}_{k_g}(\varpi,\iota)\,\hat{P}^{q}_{k_g}(\iota,u)\right|^2. \tag{6.59}$$

Proposition 4 *In the proposed THz MU-OAM system, the IMI and IUI decrease with the increase of the carrier frequency.*

Proof (6.58) can be calculated in detail as

$$\begin{aligned}[\mathbf{\Upsilon}^{p}_{\text{IMI}}(k_g)]_u &= \left|\sum_{\iota=1}^{NP}\sum_{\varpi=1}^{N}\frac{1}{\sqrt{N}}\left(\hat{\Lambda}^{p}_{G,k_g}(u,u)\right)^{-1}e^{-j\frac{2\pi\ell_u(\varpi-1)}{N}}\right.\\ &\quad\left.\times\Delta H^{p}_{k_g}(\varpi,\iota)\frac{\hat{\Lambda}^{p}_{G,k_g}(u,u)}{\hat{\Lambda}^{p}_{i,k_g}(u,u)}\rho\right|^2\\ &= \left|\frac{1}{\hat{\Lambda}^{p}_{i,k_g}(u,u)}\right|^2\left|\sum_{\iota=1}^{NP}\sum_{\varpi=1}^{N}\frac{1}{\sqrt{N}}e^{-j\frac{2\pi\ell_u(\varpi-1)}{N}}\Delta H^{p}_{k_g}(\varpi,\iota)\,\rho\right|^2\end{aligned}$$

$$= \frac{1}{{k_g}^{2(\tau+2\overline{\tau}-1)}} \frac{1}{\left(\frac{N^2}{2^\tau \tau!(2^{\overline{\tau}}\overline{\tau}!)^2} \frac{\mathfrak{R}_{k_g}^{\overline{\tau}} \aleph_{k_g}^{\overline{\tau}} \mathfrak{I}_{k_g}^{\tau}}{\widehat{r}_p {k_g}^{\tau+2\overline{\tau}}} \right)^2}$$

$$\times \left| \sum_{\iota=1}^{NP} \sum_{\varpi=1}^{N} \frac{1}{\sqrt{N}} e^{-j\frac{2\pi \ell_u(\varpi-1)}{N}} \Delta H_{k_g}^p(\varpi,\iota)\,\rho \right|^2, \quad (6.60)$$

where ρ represents the item unrelated to k_g in $\hat{P}_{k_g}^p(\iota, u)$. It can be seen from (6.60) that $[\boldsymbol{\Upsilon}_{\text{IMI}}^p(k_g)]_u$ is a function of $\hat{r}_p$, $\hat{\varphi}_p$, $\hat{\theta}_p$, and k_g and that its coefficient is $\frac{1}{{k_g}^{2(\tau+2\overline{\tau}-1)}}$. From (6.58) and (6.59), it can be seen that $[\boldsymbol{\Upsilon}_{\text{IUI}}^p(k_g)]_u$ has the same coefficient $\frac{1}{{k_g}^{2(\tau+2\overline{\tau}-1)}}$ with $[\boldsymbol{\Upsilon}_{\text{IMI}}^p(k_g)]_u$. When the values of $\hat{r}_p$, $\hat{\varphi}_p$, and $\hat{\theta}_p$ are constant, the values of $[\boldsymbol{\Upsilon}_{\text{IMI}}^p(k_g)]_u$ and $[\boldsymbol{\Upsilon}_{\text{IUI}}^p(k_g)]_u$ decrease as the carrier frequency rises from the mmWave band to the THz band. That is, the higher the frequency, the lower the corresponding interference.

Remark 4 In addition, Proposition 4 could be verified from the perspective of the beam divergence angle of OAM beams. The OAM beam divergence angle $\gamma(k_g)$ can be expressed as

$$\gamma(k_g) = \arg\left(J_\ell \left(\frac{2\pi f_g}{c} \times \frac{R_t R_p}{r_p} \right) \right), \quad (6.61)$$

where $J_\ell(\cdot)$ denotes the first-kind Bessel function of order ℓ. (6.61) shows that when R_t, R_p, and r_p remain unchanged, increasing the carrier frequency will increase the argument of the Bessel function, which makes the beam divergence angle smaller. Figure 6.4 shows the variation of beam divergence angle of OAM signals with mode $\ell = 2$ in different frequencies. Thus, the OAM beam in the THz band shows a better convergence effect than that of the mmWave band. Figures 6.5 and 6.6 illustrate the IMI and IUI in different frequencies. Since the divergence angle of THz becomes smaller, the IMI and IUI are reduced to a large extent.

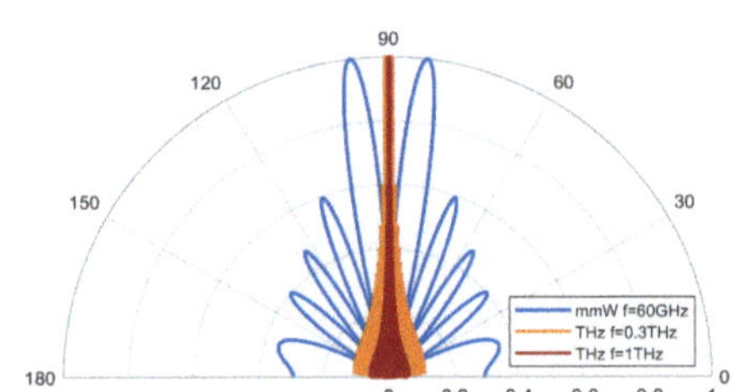

Fig. 6.4 Beam divergence angle of OAM signal with mode $\ell = 2$ in different frequencies

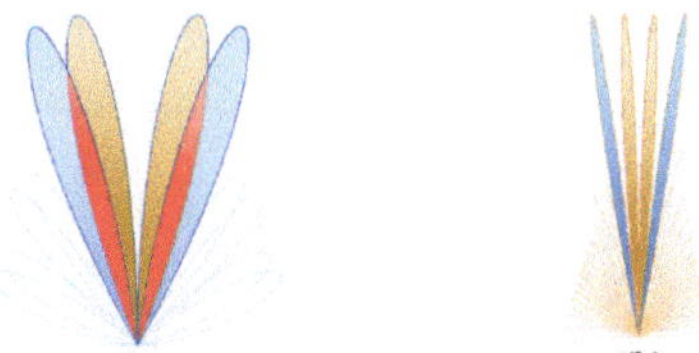

Fig. 6.5 The illustration of IMI of MU-OAM system in mmWave frequency (**a**) and THz frequency (**b**). The yellow and blue beams represent the OAM beams with modes $\ell = 1$ and $\ell = 2$, respectively. The red overlaps of beams in (**a**) represent the source of IMI when there is estimation error

Fig. 6.6 The illustration of IUI of MU-OAM system in mmWave frequency (**a**) and THz frequency (**b**). The mauve, yellow, and blue beams represent the OAM beams with mode $\ell = 1$ for users 1, 2, and 3, respectively. The red overlaps of beams in (**a**) represent the source of IUI

6.5 Numerical Simulations and Results

In this section, we show the performances of the proposed THz MU-OAM system through numerical simulations. We choose $G = 10$ subcarriers, $N = 5$ antennas in the pth user UCA, and $U = 5$ OAM modes with $\ell_1, \ell_2, \ldots, \ell_5 = -2, -1, \ldots, 2$, $\lambda = c/f$; the coordinates of the four selected user UCAs in the coordinate system of BS UCA are $(r_1, \varphi_1, \theta_1) = (900\lambda, 0°, 3°)$, $(r_2, \varphi_2, \theta_2) = (950\lambda, 90°, 6°)$, $(r_3, \varphi_3, \theta_3) = (1000\lambda, 180°, 9°)$, $(r_4, \varphi_4, \theta_4) = (1050\lambda, 270°, 12°)$. According to the OAM-based distance and AoA estimation method [5], we assume that the variances of the position parameter estimation errors σ_θ^2, σ_φ^2, and σ_r^2 are equal to σ_e^2 for convenience. Figure 6.7a shows the achievable sum rates of the proposed THz MU-OAM system vs. the user UCA radius R_p at $R_t = 10\lambda$, $f = 0.3$THz and $P = 3$. We can see from the figure that given the radius of the BS UCA is 10λ, the achievable sum rate reaches the maximum when the radius of the user UCA is equal to 25λ. Figure 6.7b shows the achievable sum rates of THz MU-OAM vs. the BS UCA radius R_t at $R_p = 25\lambda$, $f = 0.3$THz, and $P = 3$. We can also see that given the radius of the user UCA is 25λ, the achievable sum rate reaches the maximum when the radius of the BS UCA is equal to 10λ. Therefore, we assume that the radii of the BS UCA and the user UCA are 10λ and 25λ, respectively. That is, $R_t = 10\lambda$, $R_1 = R_2 = R_3 = R_4 = 25\lambda$. The radii will be adjusted appropriately according to λ when considering $f = 60$GHz and $f = 1$THz. The MU-OAM and MU-MIMO

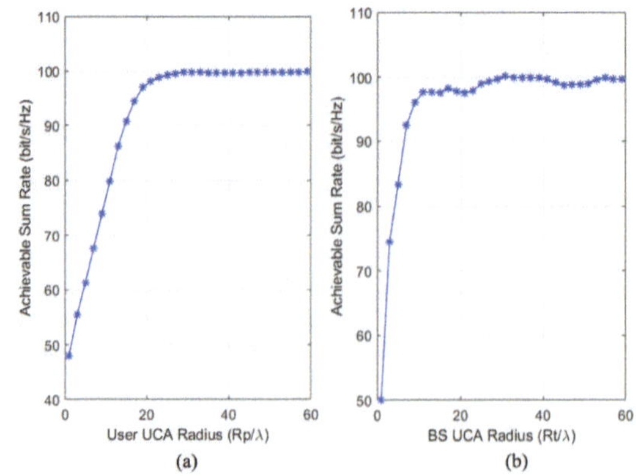

Fig. 6.7 (**a**) Achievable sum rates of THz MU-OAM vs. user UCA radius R_p at $R_t = 10\lambda$, $f = 0.3$THz, and $P = 3$. (**b**) Achievable sum rates of THz MU-OAM vs. BS UCA radius R_t at $R_p = 25\lambda$, $f = 0.3$THz, and $P = 3$

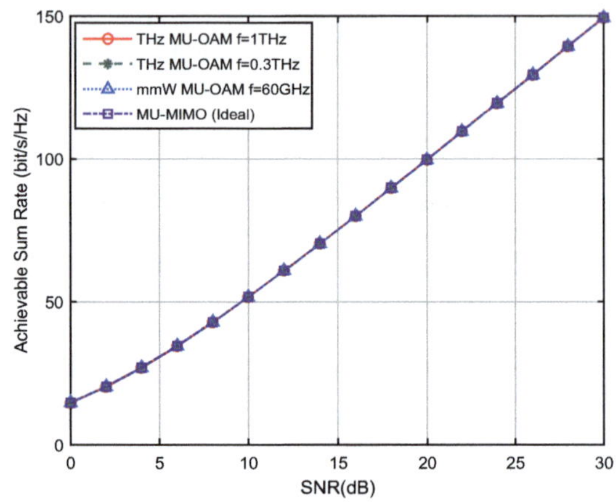

Fig. 6.8 Achievable sum rates of MU-OAM and MU-MIMO vs. carrier frequency at $\sigma_e^2 = 0$ and $P = 3$

systems with the carrier frequency $f = 1$THz and the position parameter estimation error variance $\sigma_e^2 = 0$ are the ideal baseline in the simulations.

We first simulate how the achievable sum rates of the MU-OAM system vary with the carrier frequency at the position parameter estimation error variance $\sigma_e^2 = 0$ and the number of users $P = 3$. From Fig. 6.8, we can see that the proposed MU-OAM systems which in different carrier frequencies with the perfect CSI at receiver can obtain the same achievable sum rate as the MU-MIMO system in [13]. That is, the frequency has no effect on the system performance when the receiver has the perfect CSI in the proposed MU-OAM system. In Fig. 6.9, we compare the achievable sum rates of the MU-OAM systems in different carrier frequencies at $\sigma_e^2 = 10^{-3}$ and $P = 3$, which are used to illustrate the feasibility and advantages of applying OAM in the THz frequency band. The MU-OAM channels at the frequencies of 1 THz, 0.3 THz, and 60 GHz are simulated for calculating the achievable sum rates. It can be seen from Fig. 6.9 that the OAM beam in the THz band outperforms that in the mmWave band when the receiver has the imperfect CSI in the proposed MU-OAM system. The higher the frequency is, the more pronounced is the channel gain. On

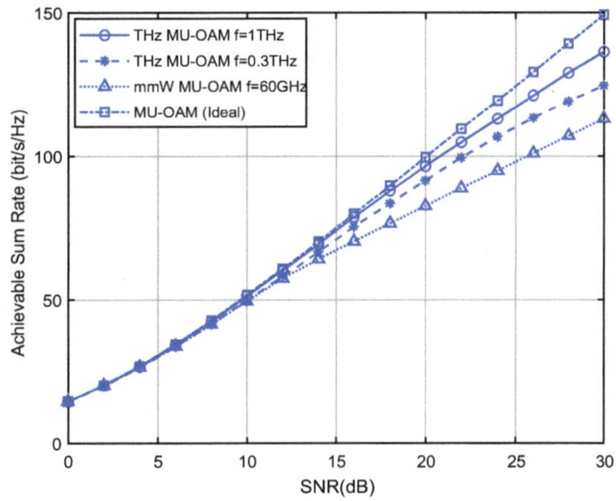

Fig. 6.9 Achievable sum rates of MU-OAM vs. carrier frequency at $\sigma_e^2 = 10^{-3}$ and $P = 3$

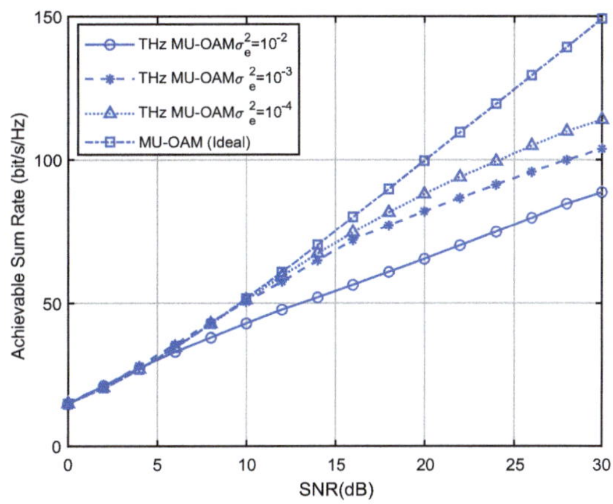

Fig. 6.10 Achievable sum rates of THz MU-OAM vs. σ_e^2 at $f = 0.3$THz and $P = 3$

the contrary, at lower carrier frequency, the channel gain is not as expected due to the increase in interference caused by the imperfect CSI at the receiver.

The achievable sum rates of the proposed THz MU-OAM system with different σ_e^2 are shown in Fig. 6.10, where $P = 3$ and the carrier frequency $f = 0.3$ THz. It can be seen that the achievable sum rate increases as σ_e^2 decreases. Figure 6.11 shows the achievable sum rates of the proposed THz MU-OAM system with different P at $f = 0.3$ THz and $\sigma_e^2 = 10^{-3}$, and we can clearly know that the achievable sum rate of the system increases with the number of users. Figure 6.12 shows the achievable sum rates of the proposed MU-OAM system with different transmission distances at $\sigma_e^2 = 10^{-3}$ and $P = 3$. It can be seen that as the transmission distance increases, the OAM beam becomes divergent, resulting in a decrease in the achievable sum rate. The proposed THz MU-OAM system has better achievable sum rate compared with the mmW MU-OAM system when the transmission distance is less than $8m$. The reason for this phenomenon is that as the carrier frequency increases, the interferences caused by the position

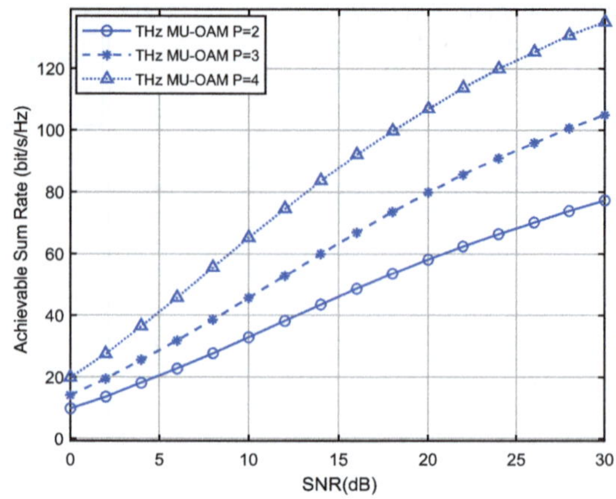

Fig. 6.11 Achievable sum rates of THz MU-OAM vs. P at $f = 0.3$THz and $\sigma_e^2 = 10^{-3}$

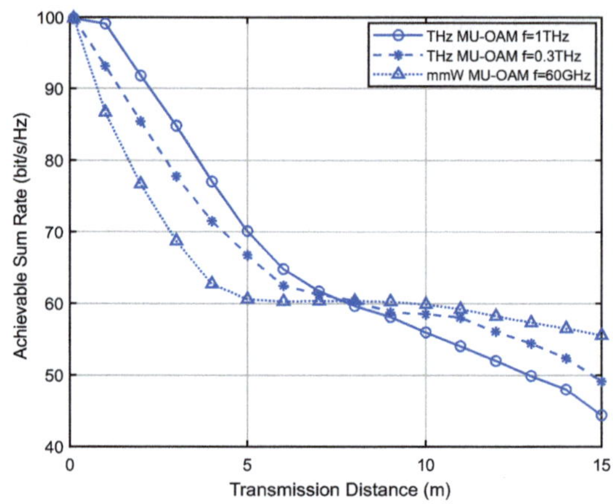

Fig. 6.12 Achievable sum rates of MU-OAM vs. transmission distance at $\sigma_e^2 = 10^{-3}$ and $P = 3$

parameter estimation errors decrease. However, the performance of the THz MU-OAM system shows the downward trend at a long transmission distance due to the rapid attenuation of THz waves. It follows that the proposed THz OAM system is suitable for short-distance large-capacity communications.

Then, we simulate how the BERs of the MU-OAM system vary with the carrier frequency at $\sigma_e^2 = 0$ and $P = 3$. From Fig. 6.13, we can see that the proposed MU-OAM systems which in different carrier frequencies with the perfect CSI at receiver can obtain the same BER performance as the MU-MIMO system. That is, the frequency has no effect on the BER performance when $\sigma_e^2 = 0$. In Fig. 6.14, we compare the BERs of the MU-OAM system in different frequencies at $\sigma_e^2 = 10^{-3}$ and $P = 3$. The MU-OAM channels at 1 THz, 0.3 THz, and 60 GHz are simulated for calculating the BERs. As can be seen from Fig. 6.14, the OAM beams in the THz band have lower BERs than the OAM beam in the mmWave band. The higher the frequency is, the lower is the BER. Conversely, at lower carrier frequency, the BER performance is not as expected due to the increase in interference.

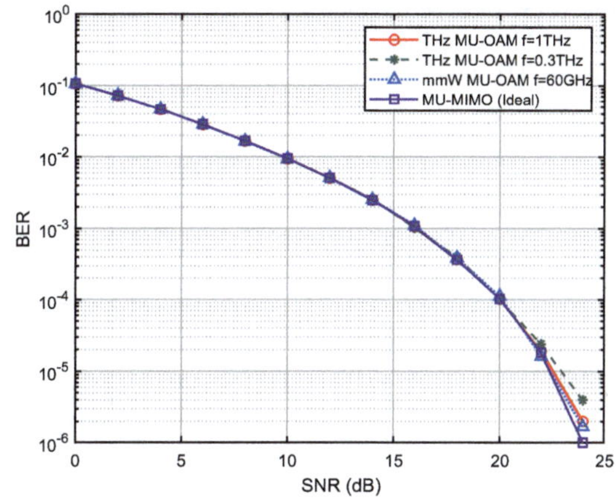

Fig. 6.13 BER performances of MU-OAM and MU-MIMO vs. carrier frequency at $\sigma_e^2 = 0$ and $P = 3$

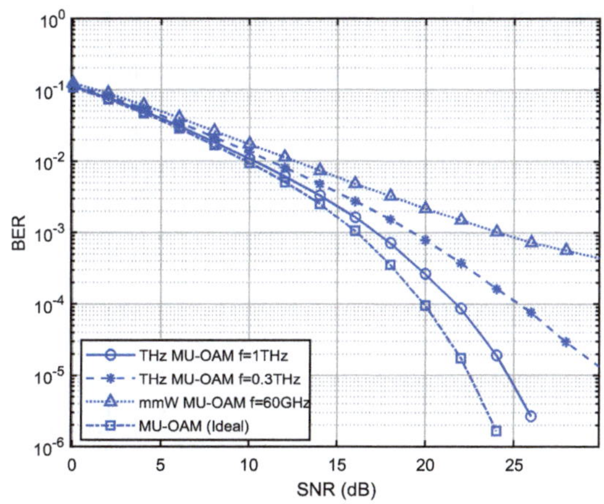

Fig. 6.14 BER performances of MU-OAM vs. carrier frequency at $\sigma_e^2 = 10^{-3}$ and $P = 3$

Figure 6.15 shows the BERs of the proposed THz MU-OAM system with different σ_e^2. It can be seen that the BER has a similar performance to achievable sum rate. The BER decreases with the decrease of σ_e^2. Considering the effect of the number of users on BER, we compare the BER performances of the proposed THz MU-OAM system with different P at $f = 0.3$ THz and $\sigma_e^2 = 10^{-3}$. It is clear from Fig. 6.16 that the BER increases as the number of users increases, because the increase in the number of users leads to an increase in IUI.

6.6 Summary

Under the background that the newly developed CMOS UCA can generate THz OAM beam, we proposed a THz MU-OAM communication system including DL and UL transmission schemes based on the THz channel characteristics modeling.

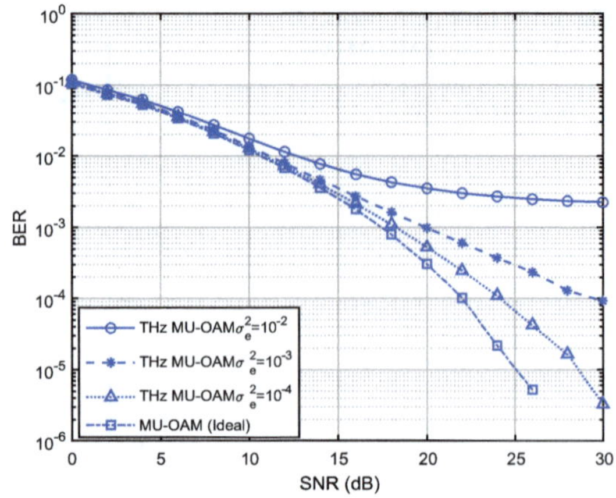

Fig. 6.15 BER performances of THz MU-OAM vs. σ_e^2 at $f = 0.3$THz and $P = 3$

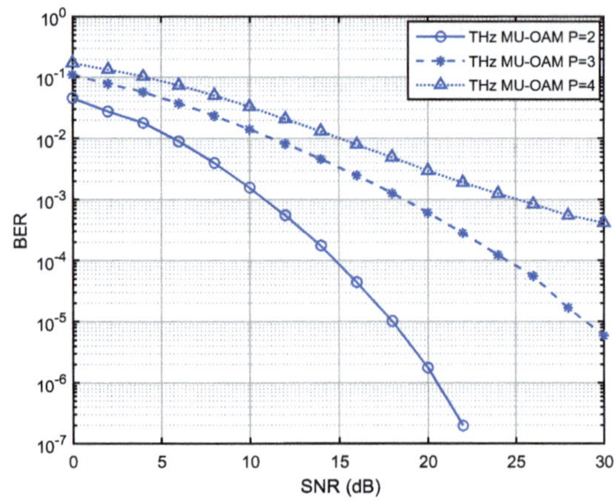

Fig. 6.16 BER performances of THz MU-OAM vs. P at $f = 0.3$THz and $\sigma_e^2 = 10^{-3}$

Since the THz channel matrices between the BS UCA and the user UCAs are completely specified by the position of all the user UCA centers, based on our proved decomposition of the THz MU-OAM channel matrix, the preprocessing matrix can be designed only according to the positions of the P user UCAs instead of the full CSI. Then, the signal detection at receiver can be simplified as an AD after despiralization. Simulation results show that the BER and the achievable sum rate of the proposed MU-OAM scheme are close to those of the ideal traditional MU-MIMO scheme at THz band but with much lower computational complexity.

Appendix A

The coordinate of the mth element of the pth user UCA is denoted as $(R_p \cos \upsilon_m^p, R_p \sin \upsilon_m^p, 0)$ in $\mathrm{Z}_p - \mathrm{X}_p\mathrm{O}_p\mathrm{Y}_p$, and coordinate of the nth element of the BS UCA is denoted as $(R_t \cos\omega_n, R_t \sin\omega_n, 0)$ in $\bar{Z} - \bar{X}\bar{O}\bar{Y}$. In order to calculate $d_{m,n}^p$, we need to express the coordinates of these two points in $Z - XOY$, where $Z - XOY$ is at the BS UCA and parallel to $Z_p - X_pO_pY_p$. According to the geometric relationship, the coordinate transformation matrix of rotating $\bar{Z} - \bar{X}\bar{O}\bar{Y}$ to $Z - XOY$ is

$$\mathbf{R}(\theta_p, \varphi_p) = \begin{bmatrix} a & b & -\sin\theta_p \cos\varphi_p \\ c & d & -\sin\theta_p \sin\varphi_p \\ \sin\theta_p \cos\varphi_p & \sin\theta_p \sin\varphi_p & \cos\theta_p \end{bmatrix}, \tag{6.62}$$

where $a = \cos\theta_p + (1 - \cos\theta_p)\sin^2\varphi_p$, $b = -(1 - \cos\theta_p)\sin\varphi_p\cos\varphi_p$, $c = -(1 - \cos\theta_p)\sin\varphi_p\cos\varphi_p$, $d = \cos\theta_p + (1 - \cos\theta_p)\cos^2\varphi_p$. Therefore, in $Z - XOY$, the coordinate of the nth element of the BS UCA can be denoted as $(x, y, z)^T = (R_t \cos\omega_n, R_t \sin\omega_n, 0)^T \mathbf{R}$. Since $Z_p - X_pO_pY_p$ and $Z - XOY$ are parallel, the coordinate of the mth element of the pth user UCA is denoted as $(R_p \cos\upsilon_m^p, R_p \sin\upsilon_m^p, r_p)$. Then, the distance between the mth element of the pth user UCA and the nth element of the BS UCA can be denoted as

$$\begin{aligned} d_{m,n}^p &= \left((x - R_p \cos\upsilon_m^p)^2 + (y - R_p \sin\upsilon_m^p)^2 + r_p^2\right)^{\frac{1}{2}} \\ &= \Big(R_t^2 + R_p^2 + r_p^2 - 2r_pR_t \sin\theta_p \cos(\varphi_p - \omega_n) \\ &\quad - 2R_tR_p\big(\cos\theta_p \cos(\upsilon_m^p - \omega_n) + (1 - \cos\theta_p) \\ &\quad \times \sin(\varphi_p - \upsilon_m^p)\sin(\varphi_p - \omega_n)\big)\Big)^{\frac{1}{2}}. \end{aligned} \tag{6.63}$$

Appendix B

From (6.48), denote

$$\wp(k_g) = N\vartheta(k_g) \sum_{t=1}^{N} \exp\left(j\Im_{k_g} \cos\frac{2\pi t}{N}\right) \exp\left(jl_u \frac{2\pi t}{N}\right), \tag{6.64}$$

$$\wp'(k_g) = \sum_{t'=1}^{2N} \exp\left(-j\Re_{k_g} \cos\frac{2\pi(t'-2)}{N}\right) \exp\left(-j\aleph_{k_g} \sin\frac{2\pi(t'-2)}{N}\right). \tag{6.65}$$

According to [5], substituting the Taylor series expansion of the exponential function $e^x = \sum_{\alpha=0}^{+\infty} x^\alpha/\alpha!$ into (6.64) and eventually have

$$\wp(k_g) \approx \vartheta\left(k_g\right) \frac{N^2}{2^\tau} \frac{j^\tau}{\tau!} \Im_{k_g}{}^{\tau}, \tag{6.66}$$

where $\tau = \min\{|\ell_u|, N - |\ell_u|\}$. Then, we substitute $e^x = \sum_{\alpha=0}^{+\infty} x^\alpha/\alpha!$ into (6.65) to get

$$\begin{aligned}
\wp'\left(k_g\right) &= \sum_{t'=1}^{2N} \exp\left(-j\Re_{k_g} \frac{\exp\left(j\frac{2\pi(t'-2)}{N}\right) + \exp\left(-j\frac{2\pi(t'-2)}{N}\right)}{2}\right) \\
&\quad \times \exp\left(-j\aleph_{k_g} \frac{\exp\left(j\frac{2\pi(t'-2)}{N}\right) - \exp\left(-j\frac{2\pi(t'-2)}{N}\right)}{2j}\right) \\
&= \sum_{t'=1}^{2N} \left(\sum_{\alpha=0}^{+\infty} \frac{(-j)^\alpha}{2^\alpha \alpha!} \sum_{l=0}^{\alpha} \binom{\alpha}{l} \exp\left(jl\frac{2\pi\left(t'-2\right)}{N}\right)\right. \\
&\quad \times \exp\left(-j\left(\alpha - l\right)\frac{2\pi\left(t'-2\right)}{N}\right) \Re_{k_g}^\alpha \\
&\quad \times \sum_{\alpha=0}^{+\infty} \frac{(-1)^{\alpha+1}}{2^\alpha \alpha!} \sum_{l=0}^{\alpha} \binom{\alpha}{l} \exp\left(jl\frac{2\pi\left(t'-2\right)}{N}\right) \\
&\quad \left. \times \exp\left(-j\left(\alpha - l\right)\frac{2\pi\left(t'-2\right)}{N}\right) \aleph_{k_g}^\alpha\right) \\
&= \sum_{t'=1}^{2N} \left(\sum_{\alpha=0}^{+\infty} \frac{(-j)^\alpha}{2^\alpha \alpha!} \sum_{l=0}^{\alpha} \binom{\alpha}{l} \exp\left(j\left(2l - \alpha\right)\frac{2\pi\left(t'-2\right)}{N}\right) \Re_{k_g}^\alpha\right. \\
&\quad \left. \times \sum_{\alpha=0}^{+\infty} \frac{(-1)^{\alpha+1}}{2^\alpha \alpha!} \sum_{l=0}^{\alpha} \binom{\alpha}{l} \exp\left(j\left(2l - \alpha\right)\frac{2\pi\left(t'-2\right)}{N}\right) \aleph_{k_g}^\alpha\right) \\
&= \sum_{\alpha=0}^{+\infty} \frac{(-j)^\alpha}{2^\alpha \alpha!} \Re_{k_g}^\alpha \sum_{\alpha=0}^{+\infty} \frac{(-1)^{\alpha+1}}{2^\alpha \alpha!} \aleph_{k_g}^\alpha \sum_{l=0}^{\alpha} \binom{\alpha}{l} \\
&\quad \times \sum_{t'=1}^{2N} \exp\left(j2\left(2l - \alpha\right)\frac{2\pi\left(t'-2\right)}{N}\right).
\end{aligned} \tag{6.67}$$

It can be seen from (6.67) that

$$\sum_{t'=1}^{2N} \exp\left(j2\left(2l - \alpha\right)\frac{2\pi\left(t'-2\right)}{N}\right) = \begin{cases} 0, & \mathrm{Mod}_N\left(2l - \alpha\right) \neq 0; \\ 2N, & \mathrm{Mod}_N\left(2l - \alpha\right) = 0. \end{cases} \tag{6.68}$$

When $\alpha > N$, the coefficient of $\mathfrak{R}_{k_g}^{\alpha}$ is much smaller due to the term $\frac{1}{2^{\alpha}\alpha!}$, and hence the term $\left\{\mathfrak{R}_{k_g}^{\alpha} \,|\alpha > N\right\}$ is counted into the higher-order infinitesimal of $\mathfrak{R}_{k_g}^{\alpha}$ denoted as $o\left\{\mathfrak{R}_{k_g}^{\alpha}\right\}$ as $\mathfrak{R}_{k_g}^{\alpha}$ goes to zero. Similarly, the term $\left\{\aleph_{k_g}^{\alpha} \,|\alpha > N\right\}$ is counted into $o\left\{\aleph_{k_g}^{\alpha}\right\}$. When $0 \le \alpha \le N$, the set that meets the condition $\mathrm{Mod}_N\,(2l - \alpha) = 0$ is (recall that $0 \le l \le \alpha$)

$$D = \{\alpha\,|N, 2l\}\,, 0 \le l < \frac{N}{2}. \tag{6.69}$$

Thus, (6.67) can be approximately obtained as

$$\wp'\left(k_g\right) \approx 2N\frac{(-j)^{\overline{\tau}}}{2^{\overline{\tau}}\overline{\tau}!}\frac{(-1)^{\overline{\tau}+1}}{2^{\overline{\tau}}\overline{\tau}!}\mathfrak{R}_{k_g}^{\overline{\tau}}\aleph_{k_g}^{\overline{\tau}} = -\frac{2N(j)^{\overline{\tau}}}{\left(2^{\overline{\tau}}\overline{\tau}!\right)^2}\mathfrak{R}_{k_g}^{\overline{\tau}}\aleph_{k_g}^{\overline{\tau}}, \tag{6.70}$$

where $\overline{\tau} \in \{N, 2l\}\,, 0 \le l < \frac{N}{2}$. Therefore, (6.50) can be approximated as

$$\Lambda_{i,k_g}^{p}(u, u) \approx -2\vartheta\left(k_g\right)\frac{N^3}{2^{\tau}\tau!}\frac{j^{(\overline{\tau}+\tau)}}{\left(2^{\overline{\tau}}\overline{\tau}!\right)^2}\mathfrak{R}_{k_g}^{\overline{\tau}}\aleph_{k_g}^{\overline{\tau}}\mathfrak{I}_{k_g}^{\tau}. \tag{6.71}$$

References

1. S.K. Agrawal, K. Sharma, 5G millimeter wave (mmWave) communications, in *Proc. 3rd Int. Conf. Comput. Sustain. Global Develop. (INDIACom)* (IEEE, 2016), pp. 3630–3634
2. I.F. Akyildiz, J.M. Jornet, C. Han, Terahertz band: Next frontier for wireless communications. Phys. Commun. **12**, 16–32 (2014)
3. L. Allen, M.W. Beijersbergen, R.J. Spreeuw, J.P. Woerdman, Orbital angular momentum of light and the transformation of Laguerre-Gaussian laser modes. Phys. Rev. A: At. Mol. Opt. Phys. **45**(11), 8185–8189 (1992)
4. S.A. Busari, S. Mumtaz, S. Al-Rubaye, J. Rodriguez, 5G millimeter-wave mobile broadband: performance and challenges. IEEE Commun. Mag. **56**(6), 137–143 (2018)
5. R. Chen, W.-X. Long, X. Wang, J. Li, Multi-mode OAM radio waves: generation, angle of arrival estimation and reception with UCAs. IEEE Trans. Wirel. Commun. **19**(10), 6932–6947 (2020)
6. R. Chen, Z. Tian, W.-X. Long, X. Wang, W. Zhang, Hybrid mechanical and electronic beam steering for maximizing OAM channel capacity. IEEE Trans. Wirel. Commun. **22**(1), 534–549 (2022)
7. M.Z. Chowdhury, M. Shahjalal, S. Ahmed, Y.M. Jang, 6G wireless communication systems: applications, requirements, technologies, challenges, and research directions. IEEE Open J. Commun. Soc. **1**, 957–975 (2020)
8. L. Dai, J. Tan, Z. Chen, H. Vincent Poor, Delay-phase precoding for wideband THz massive mimo. IEEE Trans. Wirel. Commun. **21**(9), 7271–7286 (2022)
9. S. Dang, O. Amin, B. Shihada, M.-S. Alouini, What should 6G be? Nat. Electron. **3**(1), 20–29 (2020)

10. K. Dovelos, M. Matthaiou, H.Q. Ngo, B. Bellalta, Channel estimation and hybrid combining for wideband terahertz massive MIMO systems. IEEE J. Sel. Areas Commun. **39**(6), 1604–1620 (2021)
11. O. Edfors, A.J. Johansson, Is orbital angular momentum (OAM) based radio communication an unexploited area? IEEE Trans. Antennas Propag. **60**(2), 1126–1131 (2012)
12. J. Huang, C.-X. Wang, R. Feng, J. Sun, W. Zhang, Y. Yang, Multi-frequency mmWave massive MIMO channel measurements and characterization for 5G wireless communication systems. IEEE J. Sel. Areas Commun. **35**(7), 1591–1605 (2017)
13. W. Liu, L.L. Yang, L. Hanzo, SVD-assisted multiuser transmitter and multiuser detector design for MIMO systems. IEEE Trans. Veh. Technol. **58**(2), 1016–1021 (2009)
14. W.-X. Long, R. Chen, M. Moretti, J. Li, AoA estimation for OAM communication systems with mode-frequency multi-time ESPRIT method. IEEE Trans. Veh. Technol. **70**(5), 5094–5098 (2021)
15. W.-X. Long, R. Chen, M. Moretti, J. Xiong, J. Li, Joint spatial division and coaxial multiplexing for downlink multi-user OAM wireless backhaul. IEEE Trans. Broadcast. **67**(4), 1–15 (2021)
16. A. Moldovan, M.A. Ruder, I.F. Akyildiz, W.H. Gerstacker, Los and NLOS channel modeling for terahertz wireless communication with scattered rays, in *Proc. IEEE Globecom Workshops (GC Wkshps)* (2014), pp. 388–392
17. X. Shao, X. Chen, C. Zhong, Z. Zhang, Exploiting simultaneous low-rank and sparsity in delay-angular domain for millimeter-wave/terahertz wideband massive access. IEEE Trans. Wirel. Commun. **21**(4), 2336–2351 (2022)
18. Y. Yan et al., High-capacity millimetre-wave communications with orbital angular momentum multiplexing. Nat. Commun. **5**(1), 4876 (2014)
19. H. Yang, S. Zheng, W. He, X. Yu, X. Zhang, Terahertz orbital angular momentum: generation, detection and communication. China Commun. **18**(5), 131–152 (2021)
20. P. Yang, Y. Xiao, M. Xiao, S. Li, 6G wireless communications: vision and potential techniques. IEEE Netw. **33**(4), 70–75 (2019)
21. X. Yu, L. Zhang, X. Zhang, S. Jia, H. Hu, T. Morioka, L.K. Oxenløwe, X. Pang, *Photonics-Aided 300–500 GHz Wireless Communications Beyond 100 Gbps* (Springer, Cham, 2022), pp. 467–472
22. B. Zhai, A. Tang, C. Peng, X. Wang, SS-OFDMA: spatial-spread orthogonal frequency division multiple access for terahertz networks. IEEE J. Sel. Areas Commun. **39**(6), 1678–1692 (2021)
23. C. Zhang, Y. Zhao, Orbital angular momentum nondegenerate index mapping for long distance transmission. IEEE Trans. Wirel. Commun. **18**(11), 5027–5036 (2019)
24. J. Zhang, P. Tang, L. Tian, Z. Hu, T. Wang, H. Wang, 6–100 GHz research progress and challenges for fifth generation (5G) and future wireless communication from channel perspective. Sci. China Inf. Sci. **60**(8), 1–16 (2017)
25. W. Zhang, S. Zheng, X. Hui, R. Dong, X. Jin, H. Chi, X. Zhang, Mode division multiplexing communication using microwave orbital angular momentum: an experimental study. IEEE Trans. Wirel. Commun. **16**(2), 1308–1318 (2017)

Chapter 7
Circular Array-Based Joint OAM Radar-Communication Systems

7.1 Introduction

Novel service requirements fuel the need for developing wireless communication networks. The rapid rise of novel applications, e.g., VR, digital twin, autopilot driving, and holographic video, leads to a surge in mobile data traffic, such that the capacity of next-generation 6G wireless communication networks will reach 100 times that of existing 5G networks [15, 29]. To match the growing demand, more and more high-frequency bands such as millimeter-wave (30–300 GHz) and terahertz (0.1–10 THz) bands are being licensed [25]. Unfortunately, this further leads to increased congestion of the frequency spectrum where existing radar and other sensing systems reside. To solve this problem, the concept of a novel joint radar-communication (RadCom) system is proposed, in which the previously competing radar sensing and communication operations can be implemented simultaneously by sharing a single hardware platform and a joint signal processing framework. Based on design priorities and the underlying signal, the existing joint RadCom schemes can be classified into the communication-centric schemes, radar-centric schemes, and joint design schemes. Due to the potential close cooperation between radar sensing and wireless communications, joint RadCom schemes are recognized as a key approach to significantly improving spectrum efficiency (SE), reducing device size, cost, and power consumption [30].

In recent years, significant progress in the research of radar-centric joint RadCom systems has been made [6, 30, 32]. In [16], the concept of the joint RadCom was proposed for the first time, providing new ideas for the research of radar and communication technologies. In [18], several joint sharing schemes, such as time-sharing, frequency-sharing, and beam-sharing, are discussed in details, laying the foundation for the design of RadCom systems in different scenarios. In [19], the continuous-phase modulation is first used to embed information into polyphase-coded frequency-modulated radar waveforms, paving the way for the radar-centered integrated waveform designs. In [21], an OFDM-based integrated

R. Chen et al., *Circular Array-Based Radio Frequency OAM Communications*, Wireless Networks, https://doi.org/10.1007/978-3-031-37589-7_7

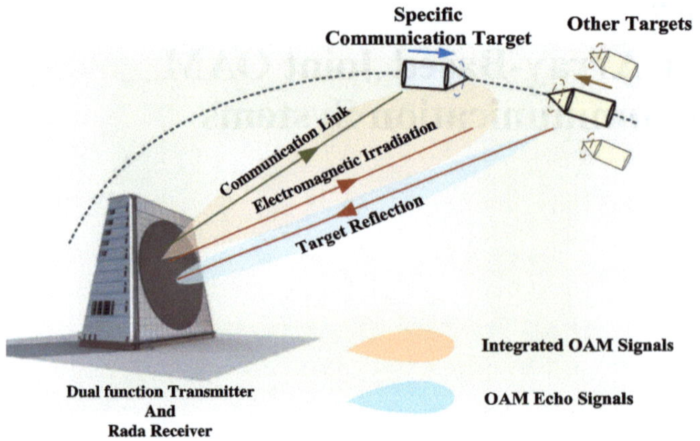

Fig. 7.1 An application scenario [4] of the proposed radar-centric joint OAM RadCom system

waveform is applied to a radar-centered MIMO RadCom system, achieving 2-D position estimation and velocity detection of targets while communicating. In [9], a novel inter-carrier interference (ICI)-aware sensing algorithm is applied to a radar-centered MIMO-OFDM RadCom system to estimate delay-Doppler-angle parameters of multiple targets in high-mobility scenarios. In [33], a phase-modulated-based integrated waveform optimization scheme is applied to a radar-centered MIMO RadCom system to find the best trade-off between radar and communication performances.

However, there are still several technical challenges for the practical application of existing radar-centered joint RadCom schemes. A specific problem is that in the radar-centered RadCom scenarios, especially in the typical scenario shown in Fig. 7.1, the existing joint RadCom schemes cannot realize the type classification and dynamic detection of targets without relative motion or beam scanning. The reason behind is that the plane EM waves used in the existing radar-centered joint RadCom systems cannot carry the azimuth and rotation velocity information of targets without relative motion or beam scanning. One promising approach to solve this problem is to employ vortex EM waves into the radar-centered joint RadCom systems. The phase front of the vortex EM wave carrying OAM rotates with azimuth exhibiting a helical structure $e^{j\ell\varphi}$ in space [1]. This helical phase structure provides a new degree of freedom for information transmission and radar application.

The immense potential of OAM-based wireless communications has been outlined in detail in Chaps. 2 and 3. Moreover, for radar sensing, the vortex EM wave carrying OAM can be seen as multiple-plane EM waves that simultaneously illuminate the target from continuous azimuth, achieving thus angular diversity without relative motion or beam scanning. Therefore, OAM-based radars can be regarded as a promising approach to provide the azimuthal resolution without relative motion [3, 8, 12, 14, 24, 28]. In particular, when vortex EM waves illuminate spinning targets, the rotational Doppler shift caused by spinning targets is proportional to the spinning velocity and OAM mode [10]. Based on this property, OAM-based radars can also be used to estimate the spinning velocity of targets even when there is no radial motion between the radar and targets [5, 13, 31]. In [5], the rotational Doppler shift of spinning targets is first observed, laying a good foundation for OAM-based rotating target detection. In [31], a method based on the phase structure of OAM beams is proposed to estimate the spinning velocity of the targets. In [13], both simulations and proof-of-concept experiments show the effectiveness of the OAM radar for imaging of spinning targets. To sum up, OAM imaging technology has been deeply studied. However, how to modulate the communication information on the existing OAM radar waveform to apply to the typical radar-centered RadCom scenario shown in Fig. 7.1 is still an open problem.

In the past decade, OAM-based communications and radar sensing have been investigated in parallel without much overlapping. In practice, they have quite a few commonalities in terms of signal processing algorithms, devices, and, to a certain extent, system architecture. Meanwhile, they can also potentially share the spectrum. Considering the problems faced by the existing radar-centered joint RadCom schemes and the significant advantages of OAM radar sensing, we propose a first radar-centric joint OAM RadCom scheme including the 3-D super-resolution position estimation, rotation velocity detection, and specific target communication. The novelty and major contributions of this chapter are summarized as follows:

- We present a first radar-centric joint OAM RadCom scheme based on UCAs, which modulates information signals on the existing OAM radar waveform by digital modulation. By taking full advantage of the phase structure of OAM beams, the proposed OAM RadCom scheme can effectively estimate the azimuth angle and rotation velocity of targets without beam scanning while at the same time communicating with a target.
- We discuss a novel OAM-based multi-target imaging method, which combines the multiple signal classification (MUSIC) and short-time Fourier transform (STFT) algorithms to achieve the 3-D super-resolution position estimation and rotation velocity detection of multiple targets. According to the unique expression of OAM echo signals, we first design a novel search vector for the MUSIC-based position estimation method, which performs a 2-D search for the amplitude and phase of OAM echo signals in the frequency and OAM mode domains, respectively, thus achieving the 3-D super-resolution imaging of multiple targets without beam scanning. Moreover, we discuss a STFT-based rotation velocity detection method, which can estimate the rotation velocity of multiple targets simultaneously by taking full advantage of the rotational Doppler characteristics of OAM echo signals, providing a basis for distinguishing the type of targets.
- The imaging and communication performances of the proposed OAM RadCom system are analyzed by deriving for the first time the posterior Cramér-Rao bound (PCRB) of the OAM-based target imaging. Thereafter, the transmitted integrated OAM beams are optimized by minimizing the PCRB under the constraints dictated by the requirements of the communication part.

The remainder of this chapter is organized as follows. Based on the feasibility of generating and receiving OAM beams by UCAs, we model the UCA-based radar-centric joint OAM RadCom system in Sect. 4.2. In Sect. 4.3, the OAM-based 3-D position estimation and rotation velocity detection methods are proposed. After that, we analyze the PCRB and SE of the joint OAM RadCom system and optimize the transmitted integrated OAM beams in Sect. 4.4. Simulation results are shown in Sect. 4.5.

7.2 Joint OAM Radar-Communication System

7.2.1 System Model

Considering the necessity of UCA to generate and receive radio OAM beams, we propose a UCA-based radar-centric joint OAM RadCom system as shown in Fig. 7.1, which consists of one dual-function transmitter with an N-element UCA, one radar receiver with an M-element UCA, and P targets each with a single communication antenna. To simplify the mathematical descriptions that follow, we assume that the dual-function transmit and radar receive UCAs are co-located and have the same radius so that a target in the far field is at the same spatial angle with respect to both the transmitter and receiver.

In the proposed joint OAM RadCom system, the dual-function transmitter periodically transmits integrated OAM beams modulated with information. For easier analysis, we suppose that the transmitter sends data only to a specific target in each period. Then, the specific target would receive integrated OAM signals. Moreover, the radar receiver would receive OAM echo signals from all targets. Therefore, by transmitting integrated OAM signals, information can be transmitted to the specific target, and the state parameters of all the targets can be estimated by OAM echo signals at the radar receiver. Moreover, we assume that the wireless communication between the dual-function transmitter and the specific target is performed through a LoS transmission, and the communication channel can be accurately estimated by the channel estimation method mentioned earlier.

7.2.2 Signal Model

To provide the range-angle-dependent beam pattern for distinguishing multiple targets, we consider an OAM-OFDM-based signal scheme as shown in Fig. 7.2. In the proposed radar-centric joint OAM RadCom system, we assume that the dual-function transmitter sends integrated signals by U OAM modes ($0 \leq U \leq M$) and W subcarriers. To avoid inter-mode interference, we suppose that the U integrated OAM signals are transmitted sequentially in a specific sequence during each period

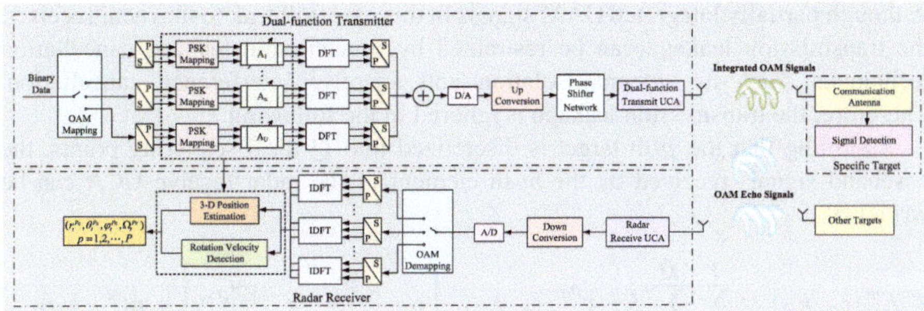

Fig. 7.2 The block diagram of the radar-centric joint OAM RadCom system

[3, 12, 14]. Meanwhile, to embed communication information into integrated OAM beams, we introduce M_p-ary PSK in each transmitted OAM waveform. Moreover, considering the different gains of integrated OAM beams with different modes, we design a weight factor A_u for the power distribution among different mode OAM beams to optimize the performances of the system. Then, the u-th signals transmitted by the n-th transmit element at the w-th subcarrier can be expressed as

$$x_n(\ell_u, k_w) = A_u e^{i\ell_u\phi} s(\ell_u, k_w), n = 1, 2, \cdots, N, \tag{7.1}$$

where $\phi = 2\pi(n-1)/N$ is the azimuthal angle of the n-th transmit element, $s(\ell_u, k_w) = e^{i\psi_p}$ is the u-th integrated OAM symbol at the w-th subcarrier, $\psi_p \in [0, \frac{2\pi}{M_p}, \cdots, \frac{2\pi(m_p-1)}{M_p}]$, and A_u is the weight of the u-th integrated OAM symbol with $\sum_{u=1}^{U} |A_u|^2 = 1$.

Based on (2.1), for an arbitrary point $\mathbb{P}(r_t, \theta_t, \varphi_t)$ in the far field at arbitrary time t, the electric field intensity $E_T(\boldsymbol{r}_t, \ell_u, k_w)$ generated by the dual-function transmitter can be written as

$$\begin{aligned} E_T(\boldsymbol{r}_t, \ell_u, k_w) &\approx \frac{\beta_t}{4\pi} \frac{e^{ik_w r_t}}{r_t} \sum_{n=1}^{N} e^{-i(\boldsymbol{k}_{w,t}\cdot\boldsymbol{r}_n - \ell_u\phi)} s(\ell_u, k_w) \\ &\approx \frac{\beta_t N e^{ik_w r_t} e^{i\ell_u\varphi_t}}{4\pi r_t} i^{-\ell_u} J_{\ell_u}(k_w R \sin\theta_t) s(\ell_u, k_w), \end{aligned} \tag{7.2}$$

where $\boldsymbol{r}_t$ is the position vector of $\mathbb{P}(r_t, \theta_t, \varphi_t)$, R is the radius of UCAs, $\boldsymbol{r}_n = R\left(\mathbf{x}'\cos\phi + \mathbf{y}'\sin\phi\right)$, and $\mathbf{x}'$ and $\mathbf{y}'$ are the unit vectors of X-axis and Y-axis of the coordinate system at the transmit/receive UCA, respectively.

7.2.2.1 OAM Radar Echo Signal Model

We assume that the radar receiver continually receives OAM echo signals, even when the dual-function transmitter is transmitting integrated OAM signals.

Although partially integrated OAM signals in this case will leak to the radar receiver, the transmission leakage can be restrained by transmission leakage cancellation techniques, such as antenna isolation and adaptive interference cancellation. Therefore, the transmission leakage is ignored in the following.

Assuming that the p-th target is discretized into Q ideal scattering points, the baseband signals received by the m-th element of the radar receive UCA can be written as

$$E_R^m(\ell_u, k_w) \approx \sum_{p=1}^{P} \sum_{q=1}^{Q} E_T(\mathbf{r}_t^{p_q}, \ell_u, k_w) \cdot \left[\sigma_{p_q} \frac{e^{ik_w r_t^{p_q}}}{r_t^{p_q}} e^{-i\boldsymbol{k}_{w,t}^{p_q} \boldsymbol{r}_m} \right] + n(\ell_u, k_w), \tag{7.3}$$

where $\boldsymbol{r}_t^{p_q}$ and σ_{p_q} are the position vector and the radar cross section (RCS) of the q-th scattering point in the p-th target, $\boldsymbol{r}_m = R\left(\mathbf{x}' \cos\alpha + \mathbf{y}' \sin\alpha\right)$, $\alpha = 2\pi(m-1)/M$ is the azimuthal angle of the m-th receive element, and $n(\ell_u, k_w)$ is the additive noise with zero mean and variance ξ^2.

After that, all the signals received by the M elements of the radar receiver are combined, and the OAM echo signal on the u-th mode at the w-th subcarrier can be expressed as

$$\begin{aligned} E_R(\ell_u, k_w) &= \sum_{m=1}^{M} E_R^m(\ell_u, k_w) + n(\ell_u, k_w) \\ &\approx \frac{\beta M N}{4\pi} i^{-\ell_u} \sum_{p=1}^{P} \sum_{q=1}^{Q} \sigma_{p_q} \frac{e^{i2k_w r_t^{p_q}}}{(r_t^{p_q})^2} e^{i\ell_u \varphi_t^{p_q}} J_{\ell_u}\left(k_w R \sin\theta_t^{p_q}\right) \\ &\quad J_0\left(k_w R \sin\theta_t^{p_q}\right) \times s(\ell_u, k_w) + n(\ell_u, k_w), \end{aligned} \tag{7.4}$$

where $(r_t^{p_q}, \theta_t^{p_q}, \varphi_t^{p_q})$ is the coordinate of the q-th scattering point of the p-th target in the coordinate system of the transmit/receive UCA at the time instant t and $\beta = \beta_t \beta_r$, β_r models all the constants relative to radar receive antenna elements. Thereafter, how to perform OAM imaging at the radar receiver based on (7.4) will be detailed in Sect. 4.3.

7.2.2.2 Specific Target Communication Signal Model

For the specific target whose centroid is located at $(r_t^{\bar{p}_0}, \theta_t^{\bar{p}_0}, \varphi_t^{\bar{p}_0})$, the received equivalent baseband signal on the u-th OAM mode at the w-th subcarrier can be expressed as

$$
\begin{aligned}
y_{\bar{p}}(\ell_u, k_w) &= \sum_{n=1}^{N} h_n^{\bar{p}}(k_w) x_n(\ell_u, k_w) + z_{\bar{p}}(\ell_u, k_w) \\
&\approx \frac{A_u \beta'}{2k_w r_t^{\bar{p}_0}} \exp\left(-ik_w r_t^{\bar{p}_0}\right) \sum_{n=1}^{N} \exp\left[i\ell_u \phi - ik_w R \sin\theta_t^{\bar{p}_0} \cos\left(\varphi_t^{\bar{p}_0} + \phi\right)\right] \\
&\quad \times s(\ell_u, k_w) + z_{\bar{p}}(\ell_u, k_w), \qquad (7.5)
\end{aligned}
$$

where

$$
h_n^{\bar{p}}(k_w) \approx \frac{\beta'}{2k_w r_t^{\bar{p}_0}} \exp\left[-ik_w r_t^{\bar{p}_0} - ik_w R \sin\theta_t^{\bar{p}_0} \cos(\varphi_t^{\bar{p}_0} + \phi)\right]
$$

is the channel coefficient from the n-th element of the transmit UCA to the receive antenna of the specific target, $\beta' = \beta_t \beta_r'$, β_r' models all constants relative to the antenna element of the specific target, and $z_{\bar{p}}(\ell_u, k_w)$ is the additive noise with zero mean and variance ξ^2. Thereafter, how to detect the received signals at the specific target to recover communication data will be specified in Sect. 4.4.

7.3 OAM-Based Rotating Target Imaging

In this section, we propose a novel OAM-based multi-target imaging method combining MUSIC and STFT algorithms, so that it can detect the rotation velocity of multiple targets while it performs 3-D super-resolution imaging, providing a basis for distinguishing the type of target.

7.3.1 OAM-Based 3-D Position Estimation

7.3.1.1 Problem Formulation

After the radar receiver is bit synchronized, the OAM echo signal on the u-th mode at the w-th subcarrier in (7.4) can be simplified as

$$
\begin{aligned}
E_R'(\ell_u, k_w) &= -\frac{E_R(\ell_u, k_w)}{|\eta(\ell_u, k_w)|} \cdot \frac{s(\ell_u, k_w)^*}{|s(\ell_u, k_w)|} \cdot i^{\ell_u} \\
&= \sum_{p=1}^{P} \sum_{q=1}^{Q} \sigma_{p_q} \frac{e^{i2k_w r_t^{p_q}}}{(r_t^{p_q})^2} e^{i\ell_u \varphi_t^{p_q}} J_{\ell_u}\left(k_w R \sin\theta_t^{p_q}\right) J_0\left(k_w R \sin\theta_t^{p_q}\right) \\
&\quad + n'(\ell_u, k_w), \qquad (7.6)
\end{aligned}
$$

where $\eta(\ell_u, k_w) = \frac{\beta MN}{4\pi} s(\ell_u, k_w)$ consists of the known parameters of the radar receiver and $n'(\ell_u, k_w)$ is the noise. Then, all the signals received on the U OAM modes at the W subcarriers can be collected in the matrix

$$\mathbf{E}'_R = \begin{bmatrix} E'_R(\ell_1, k_1) & E'_R(\ell_1, k_2) & \cdots & E'_R(\ell_1, k_W) \\ E'_R(\ell_2, k_1) & E'_R(\ell_2, k_2) & \cdots & E'_R(\ell_2, k_W) \\ \vdots & \vdots & \ddots & \vdots \\ E'_R(\ell_U, k_1) & E'_R(\ell_U, k_2) & \cdots & E'_R(\ell_U, k_W) \end{bmatrix}. \tag{7.7}$$

For easier analysis, the number of targets is assumed to be known to the radar receiver a priori [26, 27].

The aim of 3-D position estimation is to obtain the distance r_t^{pq}, the azimuthal angle φ_t^{pq}, and the elevation angle θ_t^{pq} of each scattering point of the targets from the elements of the matrix (7.7). From (7.6), we observe that the azimuthal angle φ_t^{pq} is coupled with the OAM mode number ℓ_u, and so is r_t^{pq} with k_w. The elevation angle θ_t^{pq} is associated with both ℓ_u and k_w. Therefore, we propose to estimate $\{(\theta_t^{pq}, \varphi_t^{pq})|p = 1, 2, \cdots, P, q = 1, 2, \cdots, Q\}$ by processing the columns of the matrix (7.7), whose elements depend on ℓ_u, and to estimate $\{(r_t^{pq}, \theta_t^{pq})|p = 1, 2, \cdots, P, q = 1, 2, \cdots, Q\}$ by processing the rows of the matrix (7.7), whose elements depend on k_w. To ensure the accuracy of OAM-based target imaging, we assume that $U \geq PQ$ and $W \geq PQ$. Eventually, the estimates of $(\theta_t^{pq}, \varphi_t^{pq})$ in the OAM mode domain can be combined with the estimates of $(r_t^{pq}, \theta_t^{pq})$ in the frequency domain to obtain the 3-D positions of all the scattering points.

7.3.1.2 Estimation in OAM Mode Domain

The MUSIC algorithm, a subspace-based super-resolution algorithm, provides an elegant means for estimating the parameters of complex sinusoidal signals embedded in white Gaussian noise [20]. In the estimation of θ_t^{pq} and φ_t^{pq}, we first denote the w-th column of $\mathbf{E}'_R$ as a column vector $\mathbf{e}_w$, i.e.,

$$\mathbf{e}_w = \mathbf{E}'_R(:, w) = \left[E'_R(\ell_1, k_w), E'_R(\ell_2, k_w), \cdots, E'_R(\ell_U, k_w)\right]^T,$$

so that $\mathbf{e}_w$ can be expressed in compact form as

$$\mathbf{e}_w = \mathbf{A}_w \mathbf{s}_w + \mathbf{n}_w, \tag{7.8}$$

where $\mathbf{A}_w = [e^{i\ell_u \varphi_t^{pq}} J_{\ell_u}(k_w R \sin\theta_t^{pq})]_{U \times PQ}$ is the direction matrix containing angle information of all the scattering points, $\mathbf{s}_w = \boldsymbol{\Sigma}_w \boldsymbol{\sigma}_s$, $\boldsymbol{\Sigma}_w = \mathrm{diag}\{J_0(k_w R \sin\theta_t^{11}) \frac{e^{i2k_w r_t^{11}}}{(r_t^{11})^2}, \cdots, J_0(k_w R \sin\theta_t^{PQ}) \frac{e^{i2k_w r_t^{PQ}}}{(r_t^{PQ})^2}\}$, $\boldsymbol{\sigma}_s = [\sigma_{11}, \cdots, \sigma_{PQ}]^{\mathrm{T}}$, and $\mathbf{n}_w$ is the noise vector. In practice, the OAM echo signals from different scattering points are usually incoherent, i.e.,

$$\mathbb{E}\left\{\sigma_{p_q}\sigma^*_{p'_{\tilde{q}}}\right\} = \begin{cases} \mathbb{E}\left\{\sigma_{p_q}\sigma^*_{p'_{\tilde{q}}}\right\}, & p_q = p'_{\tilde{q}}, \\ 0, & p_q \neq p'_{\tilde{q}}. \end{cases} \tag{7.9}$$

Then, the covariance matrix of $\mathbf{e}_w$ can be written as

$$\mathbf{R}_{\mathbf{e}_w} = \mathbb{E}\left\{\mathbf{e}_w\mathbf{e}_w^H\right\} = \mathbf{A}_w\mathbf{R}_{\mathbf{s}_w}\mathbf{A}_w^H + \mathbf{R}_{\mathbf{n}_w}, \tag{7.10}$$

where $\mathbf{R}_{\mathbf{s}_w} = \mathbf{\Sigma}_w\mathbb{E}\left\{\boldsymbol{\sigma}_s\boldsymbol{\sigma}_s^H\right\}\mathbf{\Sigma}_w^H$, $\mathbf{R}_{\mathbf{n}_w} = \mathbb{E}\left\{\mathbf{n}_w\mathbf{n}_w^H\right\}$. The EVD of $\mathbf{R}_{\mathbf{e}_w}$ is

$$\mathbf{R}_{\mathbf{e}_w} = \sum_{u=1}^{U}\lambda_{u,w}\mathbf{q}_{u,w}\mathbf{q}_{u,w}^H = \mathbf{Q}_w\mathbf{\Lambda}_w\mathbf{Q}_w^H, \tag{7.11}$$

where $\{\lambda_{u,w}|u = 1, 2, \cdots, U\}$ are the eigenvalues of $\mathbf{R}_{\mathbf{e}_w}$, $\mathbf{\Lambda}_w = \text{diag}\{\lambda_{1,w}, \lambda_{2,w}, \cdots, \lambda_{U,w}\}$, $\mathbf{q}_{u,w}$ is the eigenvector corresponding to the eigenvalue $\lambda_{u,w}$, and $\mathbf{Q}_w = [\mathbf{q}_{1,w}, \cdots, \mathbf{q}_{U,w}]$.

In practice, the echo signals and noise are independent of each other; thus, the covariance matrix $\mathbf{R}_{\mathbf{e}_w}$ can be decomposed into two mutually orthogonal parts:

$$\mathbf{R}_{\mathbf{e}_w} = \mathbf{Q}_w^s\mathbf{\Lambda}_w^s\mathbf{Q}_w^{s\,H} + \mathbf{Q}_w^n\mathbf{\Lambda}_w^n\mathbf{Q}_w^{n\,H}, \tag{7.12}$$

where $\mathbf{\Lambda}_w^s$ is the PQ-dimensional diagonal matrix containing the larger PQ eigenvalues of $\mathbf{R}_{\mathbf{e}_w}$, $\mathbf{Q}_w^s$ is the $U \times PQ$ signal subspace composed of the eigenvectors corresponding to the larger PQ eigenvalues, $\mathbf{\Lambda}_w^n$ is the $(U - PQ)$-dimensional diagonal matrix containing the smaller $U - PQ$ eigenvalues of $\mathbf{R}_{\mathbf{e}_w}$, and $\mathbf{Q}_w^n$ is the $U \times (U - PQ)$ noise subspace composed of the eigenvectors corresponding to the smaller $U - PQ$ eigenvalues.

Based on (7.10) and (7.12), we can see that

$$\mathbf{R}_{\mathbf{e}_w}\mathbf{Q}_w^n = \mathbf{A}_w\mathbf{R}_{\mathbf{s}_w}\mathbf{A}_w^H\mathbf{Q}_w^n + \mathbf{R}_{\mathbf{n}_w}\mathbf{Q}_w^n = \mathbf{R}_{\mathbf{n}_w}\mathbf{Q}_w^n, \tag{7.13}$$

i.e., $\mathbf{A}_w\mathbf{R}_{\mathbf{s}_w}\mathbf{A}_w^H\mathbf{Q}_w^n = 0$. Since $\mathbf{R}_{\mathbf{s}_w} = \mathbf{\Sigma}_w\mathbb{E}\left\{\boldsymbol{\sigma}_s\boldsymbol{\sigma}_s^H\right\}\mathbf{\Sigma}_w^H$ is a full rank matrix, which must be reversible, $\mathbf{A}_w^H\mathbf{Q}_w^n = 0$, that is, the direction matrix $\mathbf{A}_w$ is orthogonal to the noise subspace $\mathbf{Q}_w^n$.

Based on the orthogonality between $\mathbf{A}_w$ and $\mathbf{Q}_w^n$, the spatial-spectral function can be constructed as

$$P_w(\theta_t, \varphi_t) = \frac{1}{\mathbf{a}_w^H(\theta_t, \varphi_t)\mathbf{Q}_w^n\mathbf{Q}_w^{n\,H}\mathbf{a}_w(\theta_t, \varphi_t)}, \tag{7.14}$$

where $\mathbf{a}_w(\theta_t, \varphi_t) = [e^{i\ell_1\varphi_t} J_{\ell_1}(k_w R \sin\theta_t), e^{i\ell_2\varphi_t} J_{\ell_2}(k_w R \sin\theta_t), \cdots, e^{i\ell_U\varphi_t} J_{\ell_U}(k_w R \sin\theta_t)]^{\mathrm{T}}$, $\theta_t \in [0, \pi]$, $\varphi_t \in [0, 2\pi]$. Then, the estimates $\{(\hat{\theta}_{t,w}^{pq}, \hat{\varphi}_{t,w}^{pq})|p = 1, 2, \cdots, P, q = 1, 2, \cdots, Q\}$ can be obtained through the 2-D spectrum peak searching in (7.14).

Hence, there are W estimates of $\{(\theta_t^{pq}, \varphi_t^{pq})|p = 1, 2, \cdots, P, q = 1, 2, \cdots, Q\}$ in total, which can be expressed as

$$\hat{\theta}_{t,w}^{pq} = \theta_t^{pq} + \varepsilon_{\theta,w}^{pq}, \ \hat{\varphi}_{t,w}^{pq} = \varphi_t^{pq} + \varepsilon_{\varphi,w}^{pq}, \tag{7.15}$$

where $\varepsilon_{\theta,w}^{pq}$ and $\varepsilon_{\varphi,w}^{pq}$ represent the estimation errors, $p = 1, 2, \cdots, P$, $q = 1, 2, \cdots, Q$, $w = 1, 2, \cdots, W$. Suppose that $\{\varepsilon_{\theta,w}^{pq}|w = 1, 2, \cdots, W\}$ and $\{\varepsilon_{\varphi,w}^{pq}|w = 1, 2, \cdots, W\}$ have the same average variance $\mathrm{Var}(\varepsilon_{\theta}^{pq})$ and $\mathrm{Var}(\varepsilon_{\varphi}^{pq})$, respectively. Accordingly, since it is $\mathrm{Var}\left(\frac{1}{W}\sum_{w=1}^{W}\hat{\varphi}_{t,w}^{pq}\right) = \frac{\mathrm{Var}\left(\varepsilon_{\varphi}^{pq}\right)}{W}$, we can adopt

$$\hat{\varphi}_t^{pq} = \frac{1}{W}\sum_{w=1}^{W}\hat{\varphi}_{t,w}^{pq} \tag{7.16}$$

as the estimate of φ_t^{pq}.

7.3.1.3 Estimation in Frequency Domain

In the estimation of r_t^{pq} and θ_t^{pq}, we similarly denote the u-th row of $\mathbf{E}'_R$ as a column vector $\mathbf{e}_u$, i.e.,

$$\mathbf{e}_u = \mathbf{E}'_R(u, :) = \left[E'_R(\ell_u, k_1), E'_R(\ell_u, k_2), \cdots, E'_R(\ell_u, k_W)\right]^T,$$

so that $\mathbf{e}_u$ can be expressed in compact form as

$$\mathbf{e}_u = \mathbf{B}_u\mathbf{s}_u + \mathbf{n}_u, \tag{7.17}$$

where $\mathbf{B}_u = [e^{i2k_w r_t^{pq}} J_{\ell_u}(k_w R \sin\theta_t^{pq}) J_0(k_w R \sin\theta_t^{pq})]_{W\times PQ}$ is the position matrix containing 2-D position information of all the scattering points, $\mathbf{s}_u = \mathbf{\Sigma}_u\boldsymbol{\sigma}_s$, $\mathbf{\Sigma}_u = \mathrm{diag}\{\frac{e^{i\ell_u\varphi_t^{11}}}{(r_t^{11})^2}, \cdots, \frac{e^{i\ell_u\varphi_t^{PQ}}}{(r_t^{PQ})^2}\}$, and $\mathbf{n}_u$ is the noise vector. Then, by computing the covariance matrix $\mathbf{R}_{\mathbf{e}_u} = \mathbb{E}\left\{\mathbf{e}_u\mathbf{e}_u^H\right\}$ and following the same method outlined earlier, we can obtain the estimates $\{(\hat{r}_{t,u}^{pq}, \hat{\theta}_{t,u}^{pq})|p = 1, 2, \cdots, P, q = 1, 2, \cdots, Q, u = 1, 2, \cdots, U\}$, so that eventually it is

$$\hat{r}_t^{pq} = \frac{1}{U}\sum_{u=1}^{U}\hat{r}_{t,u}^{pq}, \tag{7.18}$$

Algorithm 1 OAM-based 3-D Position Estimation

Input: $\mathbf{E}'_R$
Output: $\{(\hat{r}_t^{p_q}, \hat{\theta}_t^{p_q}, \hat{\varphi}_t^{p_q})|p = 1, 2, \cdots, P, q = 1, 2, \cdots, Q\}$
procedure
$\mathbf{e}_w \leftarrow \mathbf{E}'_R(:, w), w = 1, 2, \cdots, W$
$\mathbf{R}_{\mathbf{e}_w} \leftarrow \mathbb{E}\left\{\mathbf{e}_w \mathbf{e}_w^H\right\}, w = 1, 2, \cdots, W$
For $w = 1 \to W$
$\mathbf{Q}_w, \mathbf{\Lambda}_w \leftarrow$ decompose $\mathbf{R}_{\mathbf{e}_w}$ such that $\mathbf{Q}_w \mathbf{\Lambda}_w \mathbf{Q}_w^H$
$\mathbf{\Lambda}_w \leftarrow \mathrm{diag}\{\lambda_{1,w}, \cdots, \lambda_{U,w}\}, \lambda_{1,w} \geq \cdots \geq \lambda_{PQ,w} \geq \cdots \geq \lambda_{U,w}$
$\mathbf{\Lambda}_w^n \leftarrow \mathrm{diag}\{\lambda_{PQ+1,w}, \cdots, \lambda_{U,w}\}$
$\mathbf{Q}_w^n \leftarrow$ the column of $\mathbf{Q}_w$ corresponding to $\mathbf{\Lambda}_w^n$
$\mathbf{a}_w(\theta_t, \varphi_t) \leftarrow [e^{i\ell_u\varphi_t} J_{\ell_u}(k_w R \sin\theta_t)]_{U\times 1}$
$P_w(\theta_t, \varphi_t) \leftarrow 1/(\mathbf{a}_w^H(\theta_t, \varphi_t)\mathbf{Q}_w^n \mathbf{Q}_w^{n\,H} \mathbf{a}_w(\theta_t, \varphi_t))$
$(\hat{\theta}_{t,w}^{p_q}, \hat{\varphi}_{t,w}^{p_q}) \leftarrow$ 2-D spectrum peak searching in $P_w(\theta_t, \varphi_t)$
end for
$\hat{\varphi}_t^{p_q} \leftarrow \frac{1}{W}\sum_{w=1}^{W} \hat{\varphi}_{t,w}^{p_q}, p = 1, \cdots, P, q = 1, \cdots, Q$
$\mathbf{e}_u \leftarrow \mathbf{E}'_R(u, :), u = 1, 2, \cdots, U$
$\mathbf{R}_{\mathbf{e}_u} \leftarrow \mathbb{E}\left\{\mathbf{e}_u \mathbf{e}_u^H\right\}, u = 1, 2, \cdots, U$
For $u = 1 \to U$
$\mathbf{Q}_u, \mathbf{\Lambda}_u \leftarrow$ decompose $\mathbf{R}_{\mathbf{e}_u}$ such that $\mathbf{Q}_u \mathbf{\Lambda}_u \mathbf{U}_u^H$
$\mathbf{\Lambda}_u \leftarrow \mathrm{diag}\{\lambda_{u,1}, \cdots, \lambda_{u,W}\}, \lambda_{u,1} \geq \cdots \geq \lambda_{u,PQ} \geq \cdots \geq \lambda_{u,W}$
$\mathbf{\Lambda}_u^n \leftarrow \mathrm{diag}\{\lambda_{u,PQ+1}, \cdots, \lambda_{u,W}\}$
$\mathbf{Q}_u^n \leftarrow$ the column of $\mathbf{Q}_u$ corresponding to $\mathbf{\Lambda}_u^n$
$\mathbf{b}_u(r_t, \theta_t) \leftarrow [e^{i2k_w r_t} J_{\ell_u}(k_w R \sin\theta_t) J_0(k_w R \sin\theta_t)]_{W\times 1}$
$P_u(r_t, \theta_t) \leftarrow 1/(\mathbf{b}_u^H(r_t, \theta_t)\mathbf{Q}_u^n \mathbf{Q}_u^{n\,H} \mathbf{b}_u(r_t, \theta_t))$
$(\hat{r}_{t,u}^{p_q}, \hat{\theta}_{t,u}^{p_q}) \leftarrow$ 2-D spectrum peak searching in $P_u(r_t, \theta_t)$
end for
$\hat{r}_t^{p_q} \leftarrow \frac{1}{U}\sum_{u=1}^{U} \hat{r}_{t,u}^{p_q}, p = 1, \cdots, P, q = 1, \cdots, Q$
$\hat{\theta}_t^{p_q} \leftarrow \frac{1}{U+W}\left(\sum_{w=1}^{W} \hat{\theta}_{t,w}^{p_q} + \sum_{u=1}^{U} \hat{\theta}_{t,u}^{p_q}\right), p = 1, \cdots, P, q = 1, \cdots, Q$
end procedure

and

$$\hat{\theta}_t^{p_q} = \frac{1}{U + W}\left(\sum_{w=1}^{W} \hat{\theta}_{t,w}^{p_q} + \sum_{u=1}^{U} \hat{\theta}_{t,u}^{p_q}\right). \tag{7.19}$$

The detailed procedure to obtain the 3-D estimated position $(\hat{r}_t^{p_q}, \hat{\theta}_t^{p_q}, \hat{\varphi}_t^{p_q})$ of the p_q scattering points is summarized in Algorithm 1. The proposed MUSIC-based imaging algorithm performs a 2-D search for the amplitude and phase of OAM echo signals in the frequency and OAM mode domains, respectively, thus realizing the 3-D super-resolution position estimation of multiple targets without beam scanning.

7.3.2 OAM-Based Rotation Velocity Detection

7.3.2.1 Problem Formulation

The precession characteristic of targets is the key basis for distinguishing the type of targets [7, 11]. In the scenario shown in Fig. 7.1, the targets are moving along the trajectory while rotating around their axis, and the rotation velocity is an important precession parameter. In this part, we will focus on the OAM-based rotation velocity detection to lay a foundation for distinguishing the type of targets.

For easier understanding, we take the vertex V_p of p-th target as an example to derive the theoretical expression of the rotation velocity. The motion state of the p-th target is shown in Fig. 7.3, whose centroid at time instant t has coordinates $(r_t^{p\mathrm{o}}, \theta_t^{p\mathrm{o}}, \varphi_t^{p\mathrm{o}})$. The p-th target is moving in the direction ρ_t^p at a velocity v_t^p while rotating around the OP_p-axis at an angular velocity Ω_t^p. Suppose the rotation center of V_p is O_p and the rotation radius is $\gamma_t^{p_\mathrm{V}}$, the vector $\overrightarrow{\mathrm{O}_p\mathrm{V}_p}$ can be written as

$$\overrightarrow{\mathrm{O}_p\mathrm{V}_p} = \begin{pmatrix} x_t^V \\ y_t^V \\ z_t^V \end{pmatrix} = \mathbf{R}_{\mathrm{ro}} \cdot \begin{pmatrix} \gamma_t^{p_\mathrm{V}} \cos(\Omega_t^p t + \psi_0^{p_\mathrm{V}}) \\ \gamma_t^{p_\mathrm{V}} \sin(\Omega_t^p t + \psi_0^{p_\mathrm{V}}) \\ 0 \end{pmatrix}, \tag{7.20}$$

where

$$\mathbf{R}_{\mathrm{ro}} = \begin{bmatrix} \cos\varphi_t^{p\mathrm{o}} & -\sin\varphi_t^{p\mathrm{o}} & 0 \\ \cos\theta_t^{p\mathrm{o}}\sin\varphi_t^{p\mathrm{o}} & \cos\theta_t^{p\mathrm{o}}\cos\varphi_t^{p\mathrm{o}} & -\sin\theta_t^{p\mathrm{o}} \\ \sin\theta_t^{p\mathrm{o}}\sin\varphi_t^{p\mathrm{o}} & \sin\theta_t^{p\mathrm{o}}\cos\varphi_t^{p\mathrm{o}} & \cos\theta_t^{p\mathrm{o}} \end{bmatrix}$$

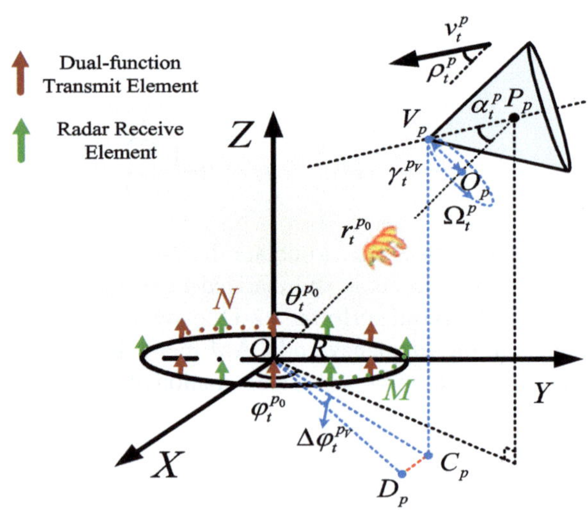

Fig. 7.3 The diagram of the OAM-based rotation velocity detection

is the rotation matrix determined by the the position of the p-th target and $\psi_0^{p_V}$ is the initial azimuth angle of the vertex V_p. When V_p rotates around OP_p-axis, the corresponding linear velocity vector $\boldsymbol{v}_V$ is perpendicular to $\overrightarrow{OP_p}$ and $\overrightarrow{O_pV_p}$, thus, the linear velocity vector $\boldsymbol{v}_V$ can be expressed as

$$\begin{aligned}\boldsymbol{v}_V &= (v_t^x, v_t^y, v_t^z) = \hat{\mathbf{r}}_t^{p_O} \times \overrightarrow{O_pV_p} \\ &= \left(y_t^{p_O}z_t^V - z_t^{p_O}y_t^V, z_t^{p_O}x_t^V - x_t^{p_O}z_t^V, x_t^{p_O}y_t^V - y_t^{p_O}x_t^V\right),\end{aligned} \tag{7.21}$$

where $\hat{\mathbf{r}}_t^{p_O} = \mathbf{r}_t^{p_O}/|\mathbf{r}_t^{p_O}| = (x_t^{p_O}, y_t^{p_O}, z_t^{p_O})^T = (\sin\theta_t^{p_O}\cos\varphi_t^{p_O}, \sin\theta_t^{p_O}\sin\varphi_t^{p_O}, \cos\theta_t^{p_O})^T$ is the unit vector of $\overrightarrow{OP_p}$. Then, within the period Δt, the moving distance of C_p that is the projection of the vertex V_p in the XOY-plane can be expressed as

$$\begin{aligned}\Delta d_t^{p_V} &= (\boldsymbol{v}_V \cdot \hat{\mathbf{r}}_\perp^{p_O})\Delta t = \frac{-y_t^{p_O}v_t^x + x_t^{p_O}v_t^y}{\sqrt{(x_t^{p_O})^2 + (y_t^{p_O})^2}}\Delta t \\ &= -\gamma_t^{p_V}\Omega_t^p\sqrt{\sin^2\varphi_t^{p_O} + \cos^2\varphi_t^{p_O}\sin^2\theta_t^{p_O}}\cos\left(\Omega_t^p t + \psi_0^{p_V} + \delta_t^{p_O}\right)\Delta t,\end{aligned} \tag{7.22}$$

where $\hat{\mathbf{r}}_\perp^{p_O} = \frac{1}{\sqrt{(x_t^{p_O})^2+(y_t^{p_O})^2}}(-y_t^{p_O}, x_t^{p_O}, 0)^T$ is the unit vector on the XOY-plane projected by the vector perpendicular to $\overrightarrow{OP_p}$, $\delta_t^{p_O} = \arctan(\cos\varphi_t^{p_O}\sin\theta_t^{p_O}/\sin\varphi_t^{p_O})$. After that, the change of the azimuth angle of the vertex V_p can be expressed as

$$\Delta\varphi_t^{p_V} = \lim_{\Delta t\to 0}\frac{\Delta d_t^{p_V}}{|\overrightarrow{OC_p}|}. \tag{7.23}$$

With (7.23), we can derive the theoretical rotational Doppler shift of the vertex V_p induced by azimuthal change.

Based on (7.6), the phase term of the OAM echo signal of the vertex V_p on the u-th mode at the w-th subcarrier can be written as

$$\Phi_t^{p_V}(\ell_u, k_w) = 2k_w r_t^{p_V} + \ell_u\varphi_t^{p_V}. \tag{7.24}$$

Thus, the Doppler shift of the vertex V_p can be expressed as

$$f_D^{p_V}(\ell_u, k_w) = \frac{1}{2\pi}\cdot\frac{\mathrm{d}\Phi_t^{p_V}(\ell_u, k_w)}{\mathrm{d}t} = f_L^{p_V}(k_w) + f_\Omega^{p_V}(\ell_u), \tag{7.25}$$

where

$$f_L^{p_V}(k_w) = \frac{1}{2\pi}k_w v_t^p\cos\rho_t^p \tag{7.26}$$

is the linear Doppler shift induced by distance variation and

$$f_{\Omega}^{p_{\mathrm{V}}}(\ell_u) = \frac{\ell_u}{2\pi} \cdot \lim_{\Delta t \to 0} \frac{\Delta \varphi_t^{p_{\mathrm{V}}}}{\Delta t} = \frac{\ell_u}{2\pi} g_t^p \Omega_t^p \cos\left(\Omega_t^p t + \psi_0^{p_{\mathrm{V}}} + \delta_t^{p_{\mathrm{O}}}\right) \tag{7.27}$$

is the rotational Doppler shift induced by azimuthal change, and $g_t^p = \frac{-\gamma_t^{p_{\mathrm{V}}}\sqrt{\sin^2\varphi_t^{p_{\mathrm{O}}} + \cos^2\varphi_t^{p_{\mathrm{O}}}\sin^2\theta_t^{p_{\mathrm{O}}}}}{(r_t^{p_{\mathrm{O}}} - \gamma_t^{p_{\mathrm{V}}}/\tan\alpha_t^p)\sin\theta_t^{p_{\mathrm{O}}}}$. From (7.27), we can observe that the rotational Doppler shift of V_p is frequency independent and determined only by the OAM mode number ℓ_u and angular velocity Ω_t^p, so that the higher the OAM mode is, the larger is the rotational Doppler shift. In the next part, we will discuss in detail how to extract the rotational Doppler shift of each scattering point from (7.6) and obtain the angular velocity Ω_t^p.

7.3.2.2 Detection of Ω_t^p

The rotation of a rigid body is essentially a kind of non-uniform motion. Thus, the echo signals are non-linear and non-stationary [23]. The core problem of extracting and analyzing the rotational Doppler shift of each scattering point from (7.6) is the processing of time-varying signals. Due to its suitability for non-stationary signal analysis, the STFT method has long been used for Doppler imaging [2, 23]. After time-frequency processing of the OAM echo signal $E_R'(\ell_u, k_w)$, we can obtain the time-frequency distribution of each scattering point superposed by the linear Doppler shift and the rotational Doppler shift. Then, the OAM echo signals $\{E_R'(0, k_w)|w = 1, 2, \cdots, W\}$ are utilized to separate the linear and rotational Doppler shifts, and the time-frequency distribution of the p_q-th scattering point only consisting of the rotational Doppler shift can be expressed as

$$\hat{f}_D^{p_q}(\ell_u) = f_D^{p_q}(\ell_u, k_w) - f_D^{p_q}(0, k_w), \ \ell_u \neq 0. \tag{7.28}$$

After that, the rotation period $T_{t,u}^{p_q}$ of p_q-th scattering point can be obtained from the time-frequency distribution of $\hat{f}_D^{p_q}(\ell_u)$, and the angular velocity can be calculated as $\hat{\Omega}_{t,u}^{p_q} = 2\pi / T_{t,u}^{p_q}$.

Hence, there are $U - 1$ estimates of $\{\Omega_t^{p_q}|p = 1, 2, \cdots, P, q = 1, 2, \cdots, Q\}$ in total, which can be expressed as

$$\hat{\Omega}_{t,u}^{p_q} = \Omega_t^{p_q} + \varepsilon_{\Omega,u}^{p_q}, \tag{7.29}$$

where $\varepsilon_{\Omega,u}^{p_q}$ represents the estimation errors, $p = 1, 2, \cdots, P$, $q = 1, 2, \cdots, Q$, $u = 1, 2, \cdots, U - 1$. Suppose that $\{\varepsilon_{\Omega,u}^{p_q}|u = 1, 2, \cdots, U - 1\}$ have the same average variance $\mathrm{Var}(\varepsilon_{\Omega}^{p_q})$; thus,

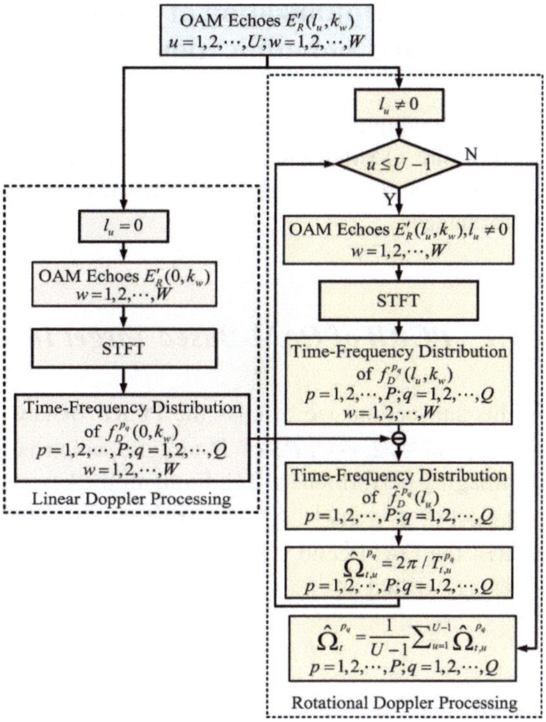

Fig. 7.4 The flowchart of the OAM-based rotation velocity detection

$$\mathrm{Var}\left(\frac{1}{U-1}\sum_{u=1}^{U-1}\hat{\Omega}_{t,u}^{p_q}\right) = \frac{\mathrm{Var}\left(\varepsilon_{\Omega}^{p_q}\right)}{U-1}. \tag{7.30}$$

Therefore, $\hat{\Omega}_t^{p_q} = \frac{1}{U-1}\sum_{u=1}^{U-1}\hat{\Omega}_{t,u}^{p_q}$ is adopted as the estimate of $\Omega_t^{p_q}$, and the detailed procedure is summarized in Fig. 7.4. It should be noted that the detected rotation velocity of the scattering points in the same target should be the same, except that at the centroid of the target, its rotation velocity is detected as zero since its rotation Doppler shift is zero.

Thus, OAM-based rotating target imaging is completed. In the numerous estimated parameters, the 3-D estimated positions $\{(\hat{r}_t^{p_O}, \hat{\theta}_t^{p_O}, \hat{\varphi}_t^{p_O})|p = 1, 2, \cdots, P\}$ of the centroid of each target and the rotation velocities $\{\hat{\Omega}_t^{p_V}|p = 1, 2, \cdots, P\}$ of the vertex of each target are regarded as the most important estimated parameters, which represent the key state characteristic of each target at time instant t.

7.4 Beam Optimization of Joint OAM Radar-Communication System

In the following part, we will first analyze the performances of the proposed radar-centric joint OAM RadCom system and then optimize the transmitted integrated OAM beams to achieve the best trade-off between the imaging and communication performances.

7.4.1 PCRB of OAM-Based Target Imaging

In this subsection, we discuss the PCRB of key state parameters $\{(r_t^{p_O}, \theta_t^{p_O}, \varphi_t^{p_O}, \Omega_t^{p_V})|p = 1, 2, \cdots, P\}$ of P targets, which is regarded as the important metric for imaging performance [22]. First, all the key state parameters to be estimated are collected in $\boldsymbol{\vartheta} = [(r_t^{1_O}, \theta_t^{1_O}, \varphi_t^{1_O}, \Omega_t^{1_V}), \cdots, (r_t^{P_O}, \theta_t^{P_O}, \varphi_t^{P_O}, \Omega_t^{P_V})]^T$. To derive the PCRB, the OAM echo signal received on the u-th mode at the w-th subcarrier in (7.4) is rewritten as

$$E_R(\ell_u, k_w) = p(\ell_u, k_w, \boldsymbol{\vartheta}) + n(\ell_u, k_w), \tag{7.31}$$

where

$$p(\ell_u, k_w, \boldsymbol{\vartheta}) = \frac{\beta M N}{4\pi} i^{-\ell_u} s(\ell_u, k_w) \times \sum_{p=1}^{P} \sum_{q=1}^{Q} \sigma_{p_q} \frac{e^{i2k_w r_t^{p_q}}}{(r_t^{p_q})^2} e^{i\ell_u \varphi_t^{p_q}} J_{\ell_u}\left(k_w R \sin\theta_t^{p_q}\right) J_0\left(k_w R \sin\theta_t^{p_q}\right).$$

Then, all the OAM echo signals received on the U modes at the W subcarriers can be expressed in compact form as

$$\mathbf{E}_R = \mathbf{P}(\boldsymbol{\vartheta}) + \mathbf{N}, \tag{7.32}$$

where $\mathbf{E}_R = [E_R(\ell_u, k_w)]_{U\times W}$, $\mathbf{P}(\boldsymbol{\vartheta}) = [p(\ell_u, k_w, \boldsymbol{\vartheta})]_{U\times W}$, and $\mathbf{N} = [n(\ell_u, k_w)]_{U\times W}$.

After that, the Fisher information matrix $\mathbf{J}$ [22] with respect to $\boldsymbol{\vartheta}$ can be expressed as (7.33).

$$\mathbf{J} = \mathbb{E}\left\{\left[\frac{\partial}{\partial\boldsymbol{\vartheta}} \ln f(\mathbf{E}_R, \boldsymbol{\vartheta})\right]\left[\frac{\partial}{\partial\boldsymbol{\vartheta}} \ln f(\mathbf{E}_R, \boldsymbol{\vartheta})\right]^T\right\} = -\mathbb{E}\left\{\frac{\partial}{\partial\boldsymbol{\vartheta}}\left[\frac{\partial}{\partial\boldsymbol{\vartheta}} \ln f(\mathbf{E}_R|\boldsymbol{\vartheta})\right]^T\right\}$$

$$= \frac{1}{\xi^2}$$

$$\begin{bmatrix} \sum_{u=1}^{U}\sum_{w=1}^{W}\mathrm{Re}\left[\frac{\partial p(\ell_u,k_w,\boldsymbol{\vartheta})}{\partial r_t^{10}}\frac{\partial p(\ell_u,k_w,\boldsymbol{\vartheta})}{\partial r_t^{10}}\right] & \sum_{u=1}^{U}\sum_{w=1}^{W}\mathrm{Re}\left[\frac{\partial p(\ell_u,k_w,\boldsymbol{\vartheta})}{\partial r_t^{10}}\frac{\partial p(\ell_u,k_w,\boldsymbol{\vartheta})}{\partial \theta_t^{10}}\right] & \cdots & \sum_{u=1}^{U}\sum_{w=1}^{W}\mathrm{Re}\left[\frac{\partial p(\ell_u,k_w,\boldsymbol{\vartheta})}{\partial r_t^{10}}\frac{\partial p(\ell_u,k_w,\boldsymbol{\vartheta})}{\partial \Omega_t^{P_V}}\right] \\ \sum_{u=1}^{U}\sum_{w=1}^{W}\mathrm{Re}\left[\frac{\partial p(\ell_u,k_w,\boldsymbol{\vartheta})}{\partial \theta_t^{10}}\frac{\partial p(\ell_u,k_w,\boldsymbol{\vartheta})}{\partial r_t^{10}}\right] & \sum_{u=1}^{U}\sum_{w=1}^{W}\mathrm{Re}\left[\frac{\partial p(\ell_u,k_w,\boldsymbol{\vartheta})}{\partial \theta_t^{10}}\frac{\partial p(\ell_u,k_w,\boldsymbol{\vartheta})}{\partial \theta_t^{10}}\right] & \cdots & \sum_{u=1}^{U}\sum_{w=1}^{W}\mathrm{Re}\left[\frac{\partial p(\ell_u,k_w,\boldsymbol{\vartheta})}{\partial \theta_t^{10}}\frac{\partial p(\ell_u,k_w,\boldsymbol{\vartheta})}{\partial \Omega_t^{P_V}}\right] \\ \vdots & \vdots & \ddots & \vdots \\ \sum_{u=1}^{U}\sum_{w=1}^{W}\mathrm{Re}\left[\frac{\partial p(\ell_u,k_w,\boldsymbol{\vartheta})}{\partial \Omega_t^{P_V}}\frac{\partial p(\ell_u,k_w,\boldsymbol{\vartheta})}{\partial r_t^{10}}\right] & \sum_{u=1}^{U}\sum_{w=1}^{W}\mathrm{Re}\left[\frac{\partial p(\ell_u,k_w,\boldsymbol{\vartheta})}{\partial \Omega_t^{P_V}}\frac{\partial p(\ell_u,k_w,\boldsymbol{\vartheta})}{\partial \theta_t^{10}}\right] & \cdots & \sum_{u=1}^{U}\sum_{w=1}^{W}\mathrm{Re}\left[\frac{\partial p(\ell_u,k_w,\boldsymbol{\vartheta})}{\partial \Omega_t^{P_V}}\frac{\partial p(\ell_u,k_w,\boldsymbol{\vartheta})}{\partial \Omega_t^{P_V}}\right] \end{bmatrix},$$

where

$$\begin{aligned}
\frac{\partial p(\ell_u, k_w, \boldsymbol{\vartheta})}{\partial r_t^{p0}} &= \frac{\beta M N}{4\pi} i^{-\ell_u} \sigma_{p0} \frac{2e^{i2k_w r_t^{p0}}}{(r_t^{p0})^2} e^{i\ell_u \varphi_t^{p0}} J_{\ell_u}\left(k_w R \sin\theta_t^{p0}\right) \\
&\quad \times J_0\left(k_w R \sin\theta_t^{p0}\right)\left(ik_w - \frac{1}{r_t^{p0}}\right) s(\ell_u, k_w), \\
\frac{\partial p(\ell_u, k_w, \boldsymbol{\vartheta})}{\partial \theta_t^{p0}} &= \frac{\beta M N}{4\pi} i^{-\ell_u} \sigma_{p0} \frac{e^{i2k_w r_t^{p0}}}{(r_t^{p0})^2} e^{i\ell_u \varphi_t^{p0}} k_w R \cos\theta_t^{p0} \cdot \left(\frac{1}{2} J_0\left(k_w R \sin\theta_t^{p0}\right)\right. \\
&\quad \times J_{\ell_u-1}(k_w R \sin\theta_t^{p0}) - \frac{1}{2} J_0(k_w R \sin\theta_t^{p0}) J_{\ell_u+1}(k_w R \sin\theta_t^{p0}) \\
&\quad \left. - J_{\ell_u}(k_w R \sin\theta_t^{p0}) J_1(k_w R \sin\theta_t^{p0})\right) s(\ell_u, k_w), \\
\frac{\partial p(\ell_u, k_w, \boldsymbol{\vartheta})}{\partial \varphi_t^{p0}} &= \frac{\beta M N}{4\pi} i^{-\ell_u+1} \ell_u \sigma_{p0} \frac{e^{i2k_w r_t^{p0}}}{(r_t^{p0})^2} e^{i\ell_u \varphi_t^{p0}} J_{\ell_u}(k_w R \sin\theta_t^{p0}) \\
&\quad \times J_0(k_w R \sin\theta_t^{p0}) s(\ell_u, k_w), \\
\frac{\partial p(\ell_u, k_w, \boldsymbol{\vartheta})}{\partial \Omega_t^{p_V}} &= \frac{\beta M N}{4\pi} i^{-\ell_u+1} \ell_u g_t^p t \sigma_{p_V} \frac{e^{i2k_w r_t^{p_V}}}{(r_t^{p_V})^2} e^{i\ell_u \varphi_t^{p_V}} J_{\ell_u}(k_w R \sin\theta_t^{p_V}) \\
&\quad J_0(k_w R \sin\theta_t^{p_V}) \\
&\quad \cos(\Omega_t^{p_V} t + \psi_0^{p_V} + \delta_t^{p0}), \qquad p = 1, 2, \cdots, P, \qquad (7.33)
\end{aligned}$$

Based on (7.33), the PCRBs of key state parameters to be estimated in $\boldsymbol{\vartheta}$ can be expressed as

$$\mathbb{E}\left\{(\vartheta_{\mathrm{i}} - \hat{\vartheta}_{\mathrm{i}})^2\right\} \geq [\mathbf{J}^{-1}]_{\mathrm{ii}}, \ \mathrm{i} = 1, 2, \cdots, 4P, \qquad (7.34)$$

where ϑ_{i} is the i-th key state parameter in $\boldsymbol{\vartheta}$, $\hat{\vartheta}_{\mathrm{i}}$ is the estimate of ϑ_{i} and $[\mathbf{J}^{-1}]_{\mathrm{ii}}$ is the i-th diagonal element of the inverse matrix of $\mathbf{J}$.

7.4.2 Spectral Efficiency of Joint OAM RadCom System

In this subsection, we discuss the SE of the radar-centric joint OAM RadCom system based on (7.5). During a period, the equivalent baseband signal vector $\mathbf{y}_{\bar{p}}(k_w)$ received by the specific target at the w-th subcarrier can be expressed as

$$\mathbf{y}_{\bar{p}}(k_w) = \mathbf{H}_{\bar{p}}(k_w)\mathbf{X}(k_w) + \mathbf{z}_{\bar{p}}(k_w) = \mathbf{H}_{\bar{p}}(k_w)\mathbf{F}_U^H\mathbf{AS}(k_w) + \mathbf{z}_{\bar{p}}(k_w), \tag{7.35}$$

where $\mathbf{y}_{\bar{p}}(k_w) = [y_{\bar{p}}(\ell_u, k_w)]_{1\times U}$ and $y_{\bar{p}}(\ell_u, k_w)$ is given by (7.5), $\mathbf{H}_{\bar{p}}(k_w) = [h_n^{\bar{p}}(k_w)]_{1\times N}$ is the channel matrix from the dual-function transmitter to the specific target, $\mathbf{X}(k_w) = [A_u e^{i\ell_u\phi} s(\ell_u, k_w)]_{N\times U}$ is the transmit signal matrix generated by the dual-function transmitter, $\mathbf{F}_U^H$ is the (partial) DFT matrix used to generate integrated OAM beams, $\mathbf{A} = \text{diag}\{A_1, \cdots, A_u\}$ is the U-dimensional weight matrix of the integrated OAM beams, $\mathbf{S}(k_w) = \text{diag}\{s(\ell_1, k_w), \cdots, s(\ell_U, k_w)\}$ is the U-dimensional integrated OAM symbol matrix at the w-th subcarrier, and $\mathbf{z}_{\bar{p}}(k_w) = [z(\ell_1, k_w), \cdots, z(\ell_U, k_w)]$ is the noise vector.

In the proposed radar-centric joint OAM RadCom system, we assume to detect the received OAM signals by employing the conventional zero-forcing detection method. Then, after obtaining the CSI by the parametric channel estimation method mentioned in previous chapters, the detected data symbol matrix at the specific target can be expressed as

$$\begin{aligned}\hat{\mathbf{S}}_{\bar{p}}(k_w) &= (\hat{\mathbf{H}}_{\bar{p}}(k_w)\mathbf{F}_U^H)^{\dagger}\left(\mathbf{H}_{\bar{p}}(k_w)\mathbf{F}_U^H\mathbf{AS}(k_w) + \mathbf{z}_{\bar{p}}(k_w)\right)\\ &= \mathbf{AS}(k_w) + \mathcal{I}_{\bar{p}}(k_w) + \mathbf{Z}'_{\bar{p}}(k_w),\end{aligned} \tag{7.36}$$

where $\hat{\mathbf{H}}_{\bar{p}}(k_w)$ is the estimated channel matrix, $\mathcal{I}_{\bar{p}}(k_w) = \big((\hat{\mathbf{H}}_{\bar{p}}(k_w)\mathbf{F}_U^H)^{\dagger}\mathbf{H}_{\bar{p}}(k_w)\mathbf{F}_U^H - \mathbf{I}_U\big)\mathbf{AS}(k_w)$ is the remaining interferences induced by the error in $\hat{\mathbf{H}}_{\bar{p}}(k_w)$, $\mathbf{I}_U$ is the U-dimensional unit matrix, and $\mathbf{Z}'_{\bar{p}}(k_w) = (\hat{\mathbf{H}}_{\bar{p}}(k_w)\mathbf{F}_U^H)^{\dagger}\mathbf{z}_{\bar{p}}(k_w)$.

Define $\mathbf{R}_{\mathbf{AS}}^{\bar{p}}(k_w)$, $\mathbf{R}_{\mathcal{I}}^{\bar{p}}(k_w)$ and $\mathbf{R}_{\mathbf{Z}'}^{\bar{p}}(k_w)$ as the $U \times U$ covariance matrices of $\mathbf{AS}(k_w)$, $\mathcal{I}_{\bar{p}}(k_w)$ and $\mathbf{Z}'_{\bar{p}}(k_w)$, i.e.,

$$\begin{aligned}\mathbf{R}_{\mathbf{AS}}^{\bar{p}}(k_w) &= \mathbf{A}\mathbb{E}\left\{\mathbf{S}(k_w)\mathbf{S}^H(k_w)\right\}\mathbf{A}^H,\\ \mathbf{R}_{\mathcal{I}}^{\bar{p}}(k_w) &= \left((\hat{\mathbf{H}}_{\bar{p}}(k_w)\mathbf{F}_U^H)^{\dagger}\mathbf{H}_{\bar{p}}(k_w)\mathbf{F}_U^H - \mathbf{I}_U\right)\mathbf{A}\mathbb{E}\left\{\mathbf{S}(k_w)\mathbf{S}^H(k_w)\right\}\mathbf{A}^H\\ &\quad \left((\hat{\mathbf{H}}_{\bar{p}}(k_w)\mathbf{F}_U^H)^{\dagger}\mathbf{H}_{\bar{p}}(k_w)\mathbf{F}_U^H - \mathbf{I}_U\right)^H,\\ \mathbf{R}_{\mathbf{Z}'}^{\bar{p}}(k_w) &= \left(\hat{\mathbf{H}}_{\bar{p}}(k_w)\mathbf{F}_U^H\right)^{\dagger}\mathbb{E}\left\{\mathbf{z}_{\bar{p}}(k_w)\mathbf{z}_{\bar{p}}^H(k_w)\right\}\left((\hat{\mathbf{H}}_{\bar{p}}(k_w)\mathbf{F}_U^H)^{\dagger}\right)^H,\end{aligned} \tag{7.37}$$

then, the SINR on the u-th OAM mode at the w-th subcarrier can be formulated as

$$\mathrm{SINR}_{\bar{p}}(\ell_u, k_w) = \frac{\left[\mathbf{R}_{\mathbf{AS}}^{\bar{p}}(k_w)\right]_{uu}}{\left[\mathbf{R}_{I}^{\bar{p}}(k_w)\right]_{uu} + \left[\mathbf{R}_{\mathbf{Z}'}^{\bar{p}}(k_w)\right]_{uu}}, \tag{7.38}$$

where $[\cdot]_{uu}$ is the u-th diagonal element of the matrix. Therefore, the average SE at the specific communication target during each period can be written as

$$C_{\mathrm{av}}^{\bar{p}} = \frac{1}{U}\sum_{w=1}^{W}\sum_{u=1}^{U}\log_2\left(1 + \mathrm{SINR}_{\bar{p}}(\ell_u, k_w)\right). \tag{7.39}$$

7.4.3 Optimization of Integrated OAM Beams

From (7.34) and (7.39), we can see that both the imaging and communication performances of the joint OAM RadCom system are closely associated with the weights $\{A_u | u = 1, 2, \cdots, U\}$ of integrated OAM beams. Different power distribution schemes will directly affect the performances of the system. To achieve the best trade-off between the imaging and communication performances, we formulate the optimization problem to minimize the PCRB, subject to the SE constraint for the specific target as well as a transmit power budget, i.e.,

$$\begin{aligned} &\min_{A_1,\cdots,A_U} \sum_{\mathfrak{i}=1}^{4P} [\mathbf{J}^{-1}]_{\mathfrak{i}\mathfrak{i}} \\ &\text{s.t. } C_{\mathrm{av}}^{\bar{p}} \geq C_{\mathrm{av}}^{\min}, \ \sum_{u=1}^{U} |A_u|^2 = 1, \end{aligned} \tag{7.40}$$

where $C_{\mathrm{av}}^{\min}$ is the minimum average SE required by the specific target.

Since $\sum_{\mathfrak{i}=1}^{4P} [\mathbf{J}^{-1}]_{\mathfrak{i}\mathfrak{i}}$ can not be expressed as a function of $\boldsymbol{\vartheta}$ in the closed form, we propose to solve the optimization problem (7.40) by the exhaustive search method, which is the simplest optimization method [17]. In the exhaustive search method, the optimal value of the problem (7.40) can be obtained by calculating the function values at several equally spaced points. We first select a search step size $\Delta = 1/\mathfrak{N}$ for the weight $A_u \in (0, 1)$ and calculate the function values of all search points $\{A_u^{\mathfrak{n}_u} | \mathfrak{n}_u = 1, 2, \cdots, \mathfrak{N}, u = 1, 2, \cdots, U\}$ based on the problem (7.40), where $\mathfrak{N}$ is the number of intermediate points and $A_u^{\mathfrak{n}_u}$ is the $\mathfrak{n}_u$-th intermediate point of the weight A_u. According to the calculation result, the search point $\{A_u^{\bar{\mathfrak{n}}_u} | u = 1, 2, \cdots, U\}$ that minimizes the PCRB and meets the constraints is adopted as the optimal value of the problem (7.40), and the detailed procedure is summarized in Fig. 7.5. Based on the optimal weights, the dual-function transmitter distributes

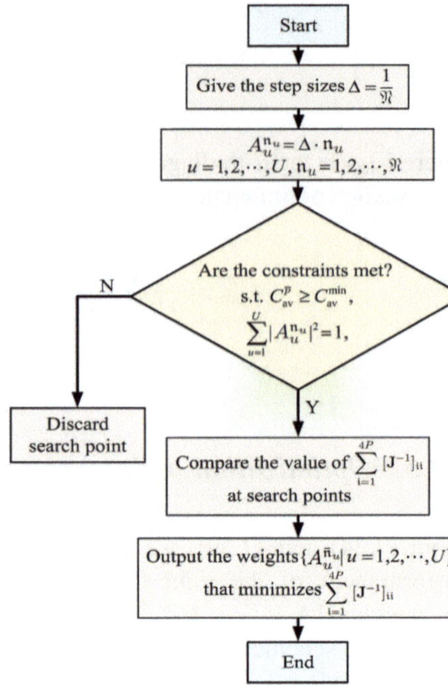

Fig. 7.5 The flowchart of the weight optimization of integrated OAM beams with the exhaustive search method

power for U integrated OAM beams in a period to achieve the best performance trade-off of the radar-centric joint OAM RadCom system.

7.5 Summary

In this section, we show the performances of the proposed scheme by numerical simulations. We first verify the proposed OAM-based 3-D position estimation and rotation velocity detection methods at different SNRs and compare the MSE of the proposed methods with the PCRB of the OAM-based imaging. Then, we plot the PCRB and SE of the radar-centric joint OAM RadCom system vs. SNR. Finally, we show the performance trade-off under the different SE constraints. Unless otherwise stated, the SNRs in all the figures are defined as the ratio of the transmitted signal power versus the noise power.

We choose $W = 16$ subcarriers from 9.979 GHz to 10.695 GHz corresponding to the wave numbers $k_1, k_2, \cdots, k_{16} = 209, 210, \ldots, 224$, $M = N = 17$, $U = 16$ OAM modes with $\ell_1, \ell_2, \cdots, \ell_{16} = -8, -7, \cdots, +7$, $R = 30\lambda_1$, $\lambda_1 = 2\pi/k_1$, $P = 3$ with the key state parameters of the specific communication target $(r_t^{1o}, \theta_t^{1o}, \varphi_t^{1o}, \Omega_t^{1v}) = (82.5\,\text{m}, 20°, 70°, 8\pi)$ and other targets $(r_t^{2o}, \theta_t^{2o}, \varphi_t^{2o}, \Omega_t^{2v}) = (170\,\text{m}, 80°, 20°, 10\pi)$, $(r_t^{3o}, \theta_t^{3o}, \varphi_t^{3o}, , \Omega_t^{3v}) = (165\,\text{m}, 75°, 25°, 11.5\pi)$ and $Q = 3$ scattering points per target.

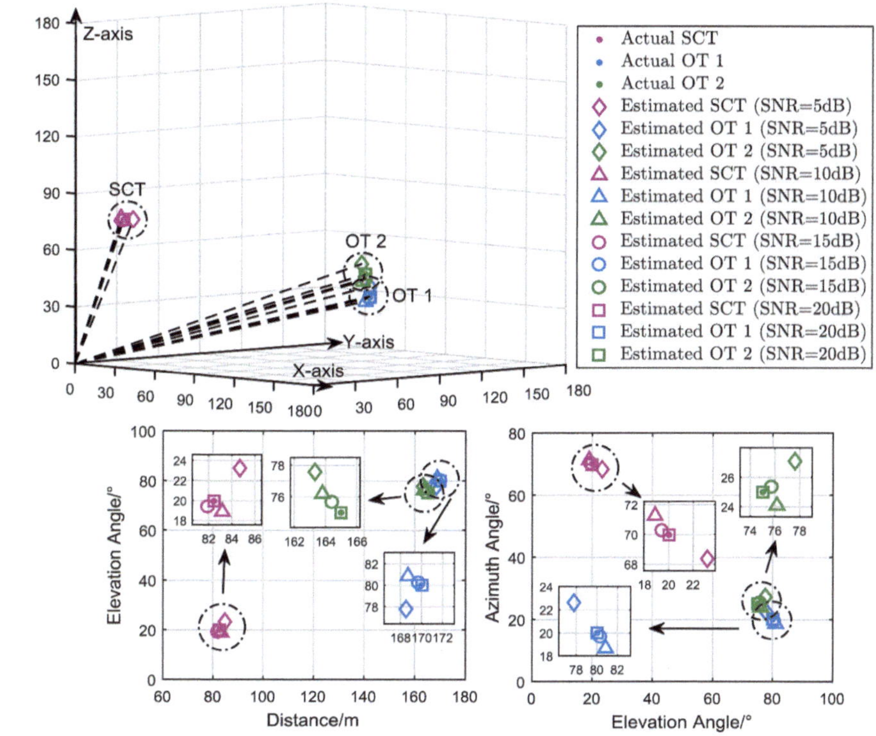

Fig. 7.6 The 3-D position estimation results of the proposed method. SCT specific communication target, OT other target

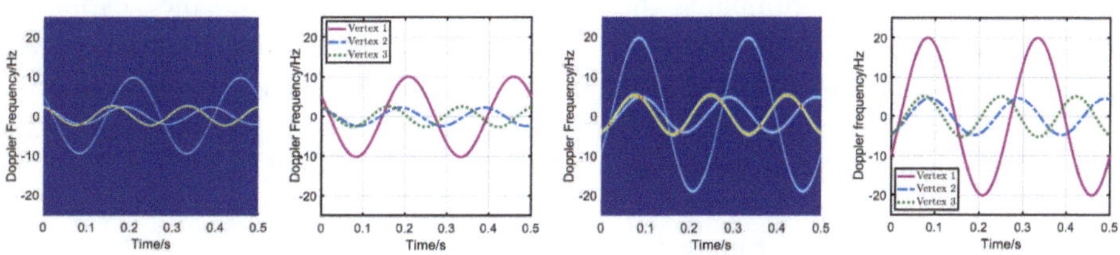

Fig. 7.7 The time-frequency simulation results at 15 dB. (**a**) STFT image with $\ell_u = -1$. (**b**) Theoretical Doppler curve with $\ell_u = -1$. (**c**) STFT image with $\ell_u = +2$. (**d**) Theoretical Doppler curve with $\ell_u = +2$

Then, by using the proposed method, the estimated positions of P targets are shown in Fig. 7.6. As we can see from the figure, the estimated positions of P targets approach the actual positions as the SNR increases, e.g., when SNR reaches 20 dB, it is $(\hat{r}_t^{1_0}, \hat{\theta}_t^{1_0}, \hat{\varphi}_t^{1_0}) = (82.502\,\text{m}, 19.997°, 69.995°)$, $(\hat{r}_t^{2_0}, \hat{\theta}_t^{2_0}, \hat{\varphi}_t^{2_0}) = (169.996\,\text{m}, 80.002°, 19.998°)$, $(\hat{r}_t^{3_0}, \hat{\theta}_t^{3_0}, \hat{\varphi}_t^{3_0}) = (165.000\,\text{m}, 74.996°, 24.994°)$, which are very close to the actual positions.

Figure 7.7 shows the time-frequency distributions of the vertices of P targets. It can be seen that the theoretical curves almost coincide with the time-frequency

Table 7.1 The estimation results of Ω_t^p

	$\hat{\Omega}_t^{1v}$	$\hat{\Omega}_t^{2v}$	$\hat{\Omega}_t^{3v}$
5 dB	8.281π	10.225π	11.961π
10 dB	7.861π	10.163π	11.720π
15 dB	8.045π	9.901π	11.628π
20 dB	8.000π	10.004π	11.495π

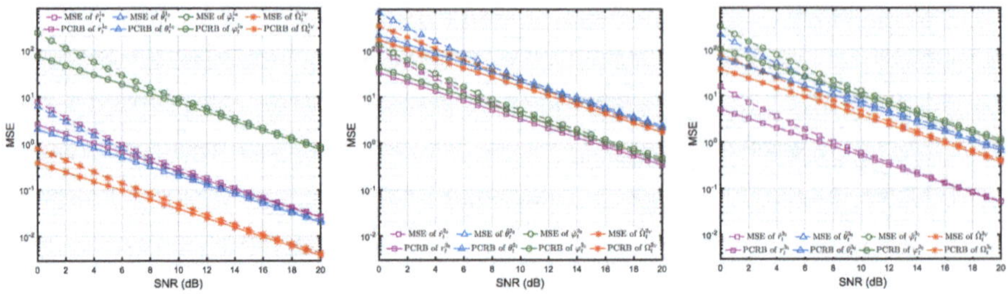

(a) (b) (c)

Fig. 7.8 The comparison of the MSEs and PCRBs. (**a**) Specific communication target $(\hat{r}_t^{1o}, \hat{\theta}_t^{1o}, \hat{\varphi}_t^{1o}, \hat{\Omega}_t^{1v})$. (**b**) Other target 1 $(\hat{r}_t^{2o}, \hat{\theta}_t^{2o}, \hat{\varphi}_t^{2o}, \hat{\Omega}_t^{2v})$. (**c**) Other target 2 $(\hat{r}_t^{3o}, \hat{\theta}_t^{3o}, \hat{\varphi}_t^{3o}, \hat{\Omega}_t^{3v})$

simulation results, proving the effectiveness of the rotational Doppler shift model derived in Sect. 4.3. Thereafter, the estimated rotation velocities are shown in Table 7.1. The estimated measures approach the actual velocities as the SNR increases. Meanwhile, noise has little influence on the rotation velocity detection when using the STFT method.

In Fig. 7.8, we compare the MSEs and PCRBs of the estimates $\{(\hat{r}_t^{po}, \hat{\theta}_t^{po}, \hat{\varphi}_t^{po}, \hat{\Omega}_t^{pv})|p = 1, 2, 3\}$ under the condition of average power distribution. The MSE is defined as $\mathbb{E}\{(\hat{x} - x)^2\}$, where $\hat{x}$ denotes the estimate of x. As the SNR increases, the MSEs of the estimates decrease and approach the PCRBs gradually. At high SNRs, the MSEs of the estimates $\{(\hat{r}_t^{po}, \hat{\theta}_t^{po}, \hat{\varphi}_t^{po}, \hat{\Omega}_t^{pv})|p = 1, 2, 3\}$ are very close to their PCRBs, proving the effectiveness of the proposed OAM-based rotating target imaging method.

After that, the PCRBs and SE of the radar-centric joint OAM RadCom system are simulated under QPSK modulation. An example of the system performance under a random power allocation with $A_1 = 0, A_2 = 0.071, A_3 = 0.354, A_4 = 0.354, A_5 = 0.283, A_6 = 0.141, A_7 = 0.141, A_8 = 0.283, A_9 = 0.3, A_{10} = 0.283, A_{11} = 0.141, A_{12} = 0.141, A_{13} = 0.283, A_{14} = 0.354, A_{15} = 0.354, A_{16} = 0.071$ is shown in Fig. 7.9b, whose PCRBs are superior to those under the average power distribution shown in Fig. 7.9a, verifying that the PCRBs of the system can be improved by designing the weights of integrated OAM beams.

Based on the simulation results of SE in Fig. 7.9, we assume that the SEs required by the specific target are 6 bit/s/Hz in in Fig. 7.10a and 7 bit/s/Hz in Fig. 7.10b, the transmit SNR is 15 dB, and the number of intermediate points is $\mathfrak{N} = 10$. It can be seen from Fig. 7.10 that under different SE constraints, the PCRBs of the system have different optimization results. More importantly, comparing Figs. 7.9 and 7.10,

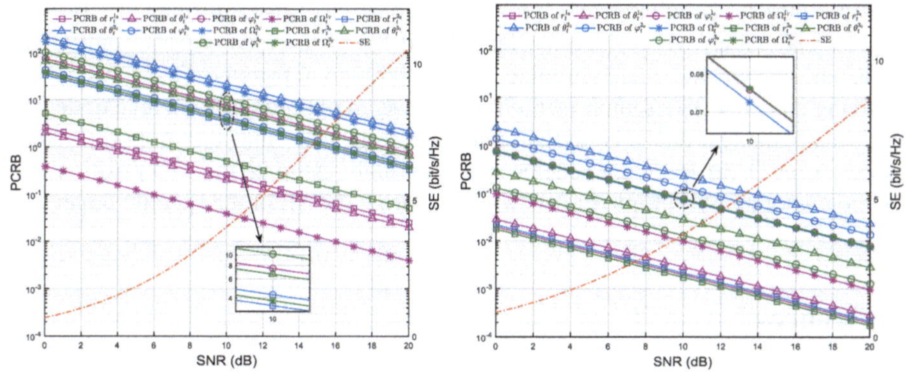

(a) (b)

Fig. 7.9 The PCRBs and SE of the radar-centric joint OAM RadCom system vs SNR. (**a**) Average power distribution. (**b**) An example of superior performance under random power allocation

Fig. 7.10 Minimum PCRBs under the different SE constraints. (**a**) The required SE = 6 bit/s/Hz. (**b**) The required SE = 7 bit/s/Hz

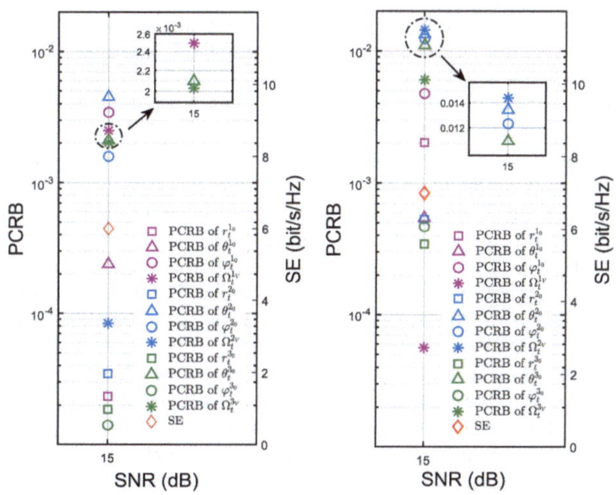

(a) (b)

we can see that by using the proposed optimization method, the PCRBs of the joint OAM RadCom system are greatly improved, while the SEs of the specific target are guaranteed, realizing the best performance trade-off of the system.

In this chapter, we propose a novel UCA-based radar-centric joint OAM RadCom scheme including the OAM-based 3-D position estimation, rotation velocity, detection and specific target communication. In terms of the OAM-based position estimation, we propose a 3-D super-resolution position estimation method based on the MUSIC algorithm. By performing the 2-D search for the amplitude and phase of OAM echo signals in the frequency and OAM mode domains, respectively, the proposed imaging method realizes the 3-D super-resolution imaging of multiple targets without beam scanning. Moreover, we discuss a rotation velocity detection method based on the STFT algorithm, which takes full advantage of the rotational Doppler characteristics of OAM echo signals to estimate the rotation velocities of multiple targets, simultaneously providing a basis for distinguishing the type of targets. Thereafter, we analyze the PCRB and SE of the radar-centric joint OAM RadCom system and then optimize the transmitted integrated OAM beams by applying an exhaustive search method to achieve an optimal performance trade-off. Both mathematical analysis and simulation results show that the proposed radar-centric joint OAM RadCom scheme can accurately estimate the 3-D position and rotation velocity of multiple targets while ensuring the SE of the specific target.

References

1. L. Allen, M.W. Beijersbergen, R.J. Spreeuw, J.P. Woerdman, Orbital angular momentum of light and the transformation of Laguerre-Gaussian laser modes. Phys. Rev. A: At. Mol. Opt. Phys. **45**(11), 8185–8189 (1992)
2. V.C. Chen, S. Qian, Joint time-frequency transform for radar range-Doppler imaging. IEEE Trans. Aerosp. Electron. Syst. **34**(2), 486–499 (1998)
3. R. Chen, W.-X. Long, Y. Gao, J. Li, Orbital angular momentum-based two-dimensional super-resolution targets imaging, in *Proceedings of the IEEE Global Conference on Signal and Information Processing (GlobalSIP)* (2018), pp. 1–4
4. C.E. Cole, Missile communication links. J. Hopkins APL Tech. Dig. **28**(4), 324–330 (2010)
5. J. Courtial, K. Dholakia, D.A. Robertson, L. Allen, M.J. Padgett, Measurement of the rotational frequency shift imparted to a rotating light beam possessing orbital angular momentum. Phys. Rev. Lett. **80**(15), 3217–3219 (1998)
6. Z. Feng, Z. Fang, Z. Wei, X. Chen, Z. Quan, D. Ji, Joint radar and communication: a survey. China Commun. **17**(1), 1–27 (2020)
7. H. Gao, L. Xie, S. Wen, Y. Kuang, Micro-Doppler signature extraction from ballistic target with micro-motions. IEEE Trans. Aerosp. Electron. Syst. **46**(4), 1969–1982 (2010)
8. G. Guo, W. Hu, X. Du, Electromagenetic vortex based radar target imaging (in chinese). J. Nat. Univ. Defense Technol. **35**(6), 71–76 (2013)
9. M.F. Keskin, H. Wymeersch, V. Koivunen, MIMO-OFDM joint radar-communications: is ICI friend or foe? IEEE J. Sel. Topics Signal Process. **15**(6), 1393–1408 (2021)
10. M.P.J. Lavery, F.C. Speirits, S.M. Barnett, M.J. Padgett, Detection of a spinning object using light's orbital angular momentum. Science **341**(6145), 537–540 (2013)
11. P. Lei, J. Sun, J. Wang, W. Hong, Micromotion parameter estimation of free rigid targets based on radar micro-doppler. IEEE Trans. Geosci. Remote Sens. **50**(10), 3776–3786 (2012)
12. M. Lin, Y. Gao, P. Liu, et al., Improved OAM-based radar targets detection using uniform concentric circular arrays. Int. J. Antenn. Propag. **2016**(6), 1–8 (2016)
13. K. Liu, X. Li, Y. Gao, H. Wang, Y. Cheng, Microwave imaging of spinning object using orbital angular momentum. J. Appl. Phys. **122**(12), 124903 (2017)
14. H. Liu, K. Liu, Y. Cheng, H. Wang, Microwave vortex imaging based on dual coupled OAM beams. IEEE Sens. J. **20**(2), 806–815 (2020)
15. W.-X. Long, R. Chen, M. Moretti, W. Zhang, J. Li, A promising technology for 6G wireless networks: intelligent reflecting surface. J. Commun. Inf. Networks **6**(1), 1–16 (2021)
16. R.M. Mealey, A method for calculating error probabilities in a radar communication system. IEEE Trans. Space Electron. Telemetry **9**(2), 37–42 (1963)
17. S. Nayak, *Fundamentals of Optimization Techniques with Algorithms* (Academic Press, New York, 2020)
18. S. Quan, W. Qian, J. Guq, V. Zhang, Radar-communication integration: an overview, in *Proceedigs of the 7th IEEE/International Conference on Advanced Infocomm Technology* (2014), pp. 98–103

19. C. Sahin, J. Jakabosky, P.M. McCormick, J.G. Metcalf, S.D. Blunt, A novel approach for embedding communication symbols into physical radar waveforms, in *2017 IEEE Radar Conference (RadarConf)* (2017), pp. 1498–1503
20. R. Schmidt, Multiple emitter location and signal parameter estimation. IEEE Trans. Antennas Propag. **34**(3), 276–280 (1986)
21. Y.L. Sit, B. Nuss, T. Zwick, On mutual interference cancellation in a MIMO OFDM multiuser radar-communication network. IEEE Trans. Veh. Technol. **67**(4), 3339–3348 (2018)
22. P. Tichavsky, C.H. Muravchik, A. Nehorai, Posterior cramér-rao bounds for discrete-time nonlinear filtering. IEEE Trans. Signal Process. **46**(5), 1386–1396 (1998)
23. Y. Wang, K. Liu, H. Liu, J. Wang, Y. Cheng, Detection of rotational object in arbitrary position using vortex electromagnetic waves. IEEE Sensors J. **21**(4), 4989–4994 (2021)
24. J. Wang, K. Liu, H. Liu, K. Cao, Y. Cheng, H. Wang, 3-D object imaging method with electromagnetic vortex. IEEE Trans. Geosci. Remote Sens. **60**, 1–12 (2022)
25. World Radiocommunication Conference (WRC) (2018). https://www.itu.int/en/ITU-R/conferences/wrc/Pages/default.aspx
26. H.-T. Wu, J.-F. Yang, F.-K. Chen, Source number estimators using transformed Gerschgorin radii. IEEE Trans. Signal Process. **43**(6), 1325–1333 (1995)
27. Y. Yang, F. Gao, C. Qian, G. Liao, Model-aided deep neural network for source number detection. IEEE Signal Process. Lett. **27**, 91–95 (2020)
28. H. Yuan, Y. Chen, Y. Luo, J. Liang, Z. Wang, A resolution-improved imaging algorithm based on uniform circular array. IEEE Antennas Wireless Propag. Lett. **21**(3), 461–465 (2021)
29. Z. Zhang, Y. Xiao, Z. Ma, M. Xiao, Z. Ding, X. Lei, G.K. Karagiannidis, P. Fan, 6G wireless networks: vision, requirements, architecture, and key technologies. IEEE Veh. Technol. Mag. **14**(3), 28–41 (2019)
30. J.A. Zhang, F. Liu, C. Masouros, R.W. Heath, Z. Feng, L. Zheng, A. Petropulu, An overview of signal processing techniques for joint communication and radar sensing. IEEE J. Sel. Topics Signal Process. **15**(6), 1295–1315 (2021)
31. M. Zhao, X. Gao, M. Xie, W. Zhai, W. Xu, S. Huang, W. Gu, Measurement of the rotational doppler frequency shift of a spinning object using a radio frequency orbital angular momentum beam. Opt. Lett. **41**(11), 2549–2552 (2016)
32. L. Zheng, M. Lops, Y.C. Eldar, X. Wang, Radar and communication coexistence: An overview: a review of recent methods. IEEE Signal Process. Mag. **36**(5), 85–99 (2019)
33. X. Zhou, L. Tang, Y. Bai, Y.-C. Liang, Performance analysis and waveform optimization of integrated FD-MIMO radar-communication systems. IEEE Trans. Wireless Commun. **20**(11), 7490–7502 (2021)

Chapter 8
Emerging Applications and Conclusions

8.1 Application of OAM in Particle Manipulation and Imaging

8.1.1 Optical and Acoustical Particle Manipulation

OAM has found applications also outside the field of optical, radio, and acoustic communications. Thanks to the work of Nobel laureate Arthur Ashkin, light has since long been used to trap small particles in what we call *optical tweezers*. Optical tweezers use a single tightly focused beam with a large-enough gradient force around the focus to overcome the linear momentum of the light, attracting the particles toward the center [2]. In 1995, [12] has shown that, by transferring the OAM carried by photons to small particles, lasers with OAM can be employed to achieve the rotation of the trapped particles, transforming the tweezers into *optical spanners* that cause the objects to spin [22]. When a circularly polarized LG beam with $\ell = 1$ is used to interact with the particles, the total angular momentum is either $\sim 0.06\,\hbar$ or $\sim 2.06\,\hbar$ per photon, depending on the spin direction of the beam, resulting in the stopping or the rotation of the particles. Similarly, using the same method but with a different beam with $\ell = 3$, two different rotational velocities of the particles can be observed [7]. In the above work, the size of the particles is larger than the beam, and the particles are considered to be trapped on or near the axis. When the size is smaller than the ring beam, the particles are generally trapped in the off-axis area. In this case, the spin of the trapped particles depends on the SAM of the ring beam, and the ring beam with the helical phase structure associated with OAM imposes an azimuthal scattering force on the particle to rotate the particle around the beam axis [10, 21].

When interacting with the absorbing particles, OAM carried by acoustic vortices can also be transferred to the particles, exerting a torque on them. This mechanical effect can be applied in the form of *acoustic tweezers* and *acoustic spanners*. Acoustic vortices can manipulate larger-size particles than optical vortices due to

R. Chen et al., *Circular Array-Based Radio Frequency OAM Communications*, Wireless Networks, https://doi.org/10.1007/978-3-031-37589-7_8

larger wavelength and have great application prospects in the field of ultrasonic medicine. Courtney et al. implemented acoustic tweezers using first-order and superimposed high-order Bessel acoustic vortices [5, 6]. Their experiment uses a transducer array to generate the required Bessel acoustic vortex field and can control the movement of the vortex center by adjusting the drive signal of the array element, achieving particle capture at different locations. Unlike optical vortices, acoustic vortices cannot carry SAM, and when vortex beams act on particles, only OAM transfer occurs. In [25], it is experimentally demonstrated that acoustic vortices in free space can transmit acoustic OAM to an object and make it rotate. The experiment uses a torsion pendulum to measure the angular momentum of acoustic vortices and compares the effective acoustic torque obtained with different topological charges. Moreover, in [1], the amount of OAM transmitted by an acoustic vortex to an absorbing disk in a viscous liquid has been quantitatively measured.

From the perspective of simplicity and flexibility, optical OAM for particle manipulation is usually generated using SPPs and SLMs, while acoustic OAM for particle manipulation are generated using UCTAs.

8.1.2 Optical and Radar Imaging

OAM applied to an optical or radar imaging system can break the limits of resolution or sensitivity. The stimulated emission depletion (STED) microscope uses a ring beam with a helical phase distribution to suppress the fluorescence of particles around the scanning point, achieving a resolution beyond the diffraction limit [13]. Applying OAM to diffraction tomography can also make a significant breakthrough with respect to the diffraction limits associated with traditional techniques [14].

Inserting a spiral phase mask into the Fourier plane of an optical imaging system can effectively convert the point spread function of the system into an annular intensity cross section with an $\exp(i\ell\theta)$ phase distribution [20]. This can be used to observe the light around bright objects [23] or to achieve spiral phase contrast imaging that is often used to observe bright edges of phase objects [8, 26]. In [24], this method is used to suppress the brightness of stars and make orbital planets more visible. In addition, when a helical phase beam is used as reference wave of an interference system, the interference pattern is a chiral spiral stripe, so that the protrusions and depressions of the sample surface can be distinguished by observing the direction of rotation of the spiral stripe [9].

In 2013, [11] applied OAM to the field of radar imaging and proposed the idea that vortex electromagnetic waves have the potential for radar target imaging. The paper developed an echo model of the ideal point scattering target under vortex electromagnetic wave illumination and implemented radar imaging using the back projection algorithm and the filtered FFT based algorithm. Afterward, [17] established echo signal models for multiple-input multiple-output and multiple-input single-output systems. The OAM-based radar imaging system model is shown

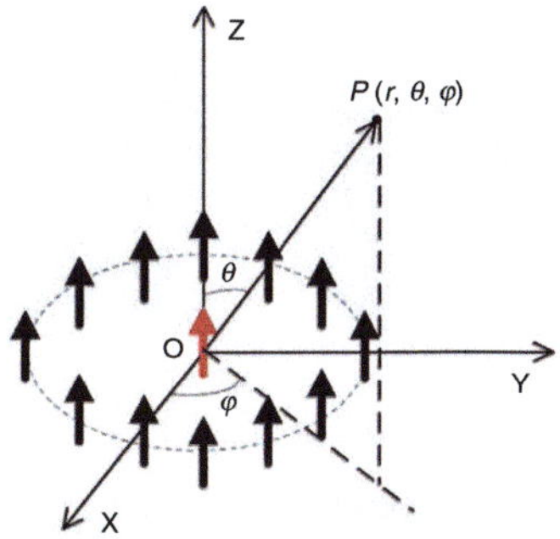

Fig. 8.1 Schematic diagram of radar target detection based on vortex electromagnetic waves. The target position P is denoted by (r, θ, φ). In the multiple-input multiple-output mode, all antenna in the UCAA (black arrows) are used to receive the echo signal, while only a single antenna (red arrow) is used in the multiple-input single-output mode [17]

in Fig. 8.1. Assume that the target is made up of M ideal scattering points, denoted by $P_m(r_m, \theta_m, \varphi_m)$ and corresponding radar cross-section σ_m, $m = 1, \cdots, M$. In the multiple-input single-output mode, a single antenna is set at the original point to receive the echo signal. The normalized echo signal is expressed as

$$s(k, \ell) = \sum_{m=1}^{M} \frac{\sigma_m}{r_m^2} e^{i2kr_m} e^{i\ell\varphi_m} J_\ell(ka \sin\theta_m), \tag{8.1}$$

where $k = 2\pi f/c$ is the wave number, $J_\ell(ka \sin\theta)$ is the Bessel function of the first kind, and a is the radius of the UCAA. As can be seen from (8.1), the OAM mode number ℓ and the azimuth φ satisfy the dual relationship, and so do the frequency k and target range r. Thus, the 2-D FFT can be used in estimating the range and azimuth of targets. This shows that the OAM-based radar imaging system has the prospect of acquiring the azimuth information of target.

OAM-based radar detection provides a new set of ideas and solutions for the development of accurate target imaging. Due to its helical phase structure, a vortex electromagnetic wave carrying OAM can be regarded as a traditional plane wave illuminating a target from multiple consecutive angles, which is equivalent to realizing continuous sampling in a two-dimensional space in a short time, obtaining a good degree of spatial diversity. Therefore, applying multimodal OAM to radar imaging enables azimuthal imaging without relative motion [27]. Moreover, when vortex electromagnetic waves with different helical phase structures hit the target in space, the phase differences associated with the various OAM modes make the backscattering characteristics different and improve the radar cross-section diversity gain [30]. In conventional radar imaging, the azimuth resolution is generally enhanced by increasing the aperture size, while vortex electromagnetic waves can provide a high resolution that is not limited by the array aperture, as it has been verified by a proof-of-concept experiment in [19]. The comparison of the results of

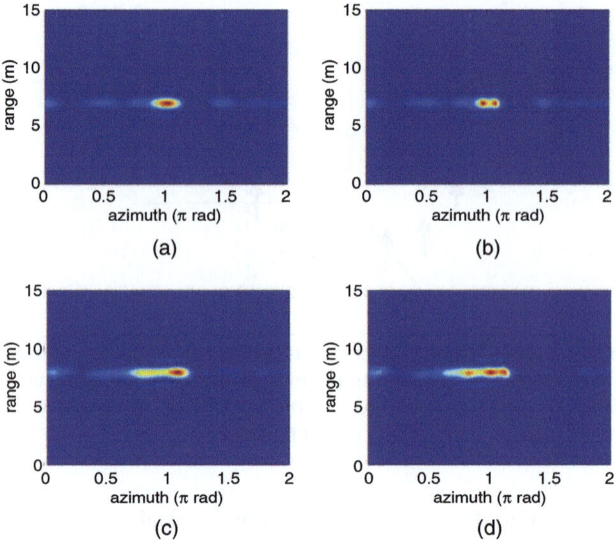

Fig. 8.2 Comparisons of imaging results between the traditional array imaging and the electromagnetic vortex imaging: (**a**) traditional array imaging of two reflectors, (**b**) electromagnetic vortex imaging of two reflectors, (**c**) traditional array imaging of three reflectors, and (**d**) electromagnetic vortex imaging of three reflectors [19]

traditional array imaging and electromagnetic vortex imaging is shown in Fig. 8.2, where it is shown that the imaging resolution is clearly improved by using vortex electromagnetic waves. Similarly, it has been demonstrated that applying vortex electromagnetic waves to synthetic aperture radar (SAR) imaging can also achieve higher azimuthal resolution than planar electromagnetic waves [3].

Recently, research on OAM-based radar imaging has mainly focused on azimuth imaging algorithms. Due to the existence of the Rayleigh limit, traditional radar imaging algorithms can only provide limited azimuth resolution. Some efficient algorithms have been proposed for OAM-based super-resolution radar imaging, such as UCAA-based autoregressive models and power spectral density (PSD) estimation algorithm [18], UCAA-based echo models, and multiple signal classification (MUSIC) algorithm [15, 16], estimating signal parameter via rotational invariance techniques (ESPRIT) algorithm [4] and least squares algorithms [28, 29].

There are also many challenges in OAM-based radar imaging. Because the center of vortex electromagnetic beams is null and the angular direction of the main lobe varies with the OAM modes, it is difficult to simultaneously illuminate the target with vortex electromagnetic beams of different modes, which results in limited echo energy. The difference of echo intensity will lead to amplitude modulation of the echo signal, resulting in a deterioration of the imaging capability. These effects can be mitigated by carefully designing the radius and excitation of the UCAA to adjust the main lobes of the vortex electromagnetic waves with different modes [18, 27,

29]. In addition, detecting dynamic targets is also challenging, which requires more in-depth research.

Optical OAM beams for imaging is generated using SPPs and SLMs, while radio OAM beams for radar imaging are generated using UCAAs. OAM beam divergence and beam steering must be considered in order to make OAM waves with multiple modes illuminate the same target. More in-depth research on topics such as high-precision azimuth imaging and dynamic target detection are the object of future research.

8.2 Conclusions

In recent years, OAM has been extensively studied in many fields of application, especially in the field of communications. A large number of studies have shown that it is possible to multiplex a set of orthogonal OAM modes on the same frequency channel so to achieve a high spectral efficiency. In this book, we provide a comprehensive overview of the generation and detection methods of optical, radio, and acoustic OAM, together with their applications in communications, particle manipulation, and imaging. Particular attention has been devoted to the issues that still obstruct or limit the application of OAM in practice.

OAM technology is on the brink of been deployed in many communication systems, but there are still some problems that need to be solved. For example, many of the experiments related to OAM multiplexing still use bulky and expensive components, which are not suitable for practical implementation and large-scale use. How to generate and detect high-order and multi-mode OAM beams with a finite-size aperture is challenging. Future success of OAM heavily relies on the development of system components such as transmitters, multiplexers, demultiplexers, and receivers. These components are required to provide reductions in cost and size and to be compatible with existing technologies. Moreover, the state of the propagation channel has a large impact on OAM-based links, and the problems arising in some complex and severe channel conditions need to be addressed and solved. In spite of the many challenges in applying OAM to communication systems, on the base of the research progress surveyed in this book, we remain optimistic that these challenges will be gradually resolved in the future and that OAM will help make breakthroughs in future communications and radar target detection systems.

References

1. A. Anhäuser, R. Wunenburger, E. Brasselet, Acoustic rotational manipulation using orbital angular momentum transfer. Phys. Rev. Lett. **109**(3), 034301 (2012)

2. A. Ashkin, J.M. Dziedzic, J.E. Bjorkholm, S. Chu, Observation of a single-beam gradient force optical trap for dielectric particles. Opt. Lett. **11**(5), 288–290 (1986)
3. X. Bu, Z. Zhang, L. Chen, X. Liang, H. Tang, X. Wang, Implementation of vortex electromagnetic waves high-resolution synthetic aperture radar imaging. IEEE Antennas Wireless Propag. Lett. **17**(5), 764–767 (2018)
4. R. Chen, W.-X. Long, Y. Gao, J. Li, Orbital angular momentum-based two-dimensional super-resolution targets imaging, in *Proceedings of the IEEE Global Conference Signal Information Processing* (2018)
5. C.R.P. Courtney, B.W. Drinkwater, C.E.M. Demore, S. Cochran, A. Grinenko, P.D. Wilcox, Dexterous manipulation of microparticles using Bessel-function acoustic pressure fields. Appl. Phys. Lett. **102**(12), 123508 (2013)
6. C.R.P. Courtney, C.E.M. Demore, H. Wu, A. Grinenko, P.D. Wilcox, S. Cochran, B.W. Drinkwater, Independent trapping and manipulation of microparticles using dexterous acoustic tweezers. Appl. Phys. Lett. **104**(15), 154103 (2014)
7. M.E.J. Friese, J. Enger, H. Rubinsztein-Dunlop, N.R. Heckenberg, Optical angular-momentum transfer to trapped absorbing particles. Phys. Rev. A **54**(2), 1593–1596 (1996)
8. S. Fürhapter, A. Jesacher, S. Bernet, M. Ritsch-Marte, Spiral phase contrast imaging in microscopy. Opt. Express **13**(3), 689–694 (2005)
9. S. Fürhapter, A. Jesacher, S. Bernet, M. Ritsch-Marte, Spiral interferometry. Opt. Lett. **30**(15), 1953–1955 (2005)
10. V. Garcés-Chávez, D. Mcgloin, M.J. Padgett, W. Dultz, H. Schmitzer, K. Dholakia, Observation of the transfer of the local angular momentum density of a multiringed light beam to an optically trapped particle. Phys. Rev. Lett. **91**(9), 093602 (2003)
11. G. Guo, W. Hu, X. Du, Electromagnetic vortex based radar target imaging. J. Nat. Univ. Defense Technol. **35**(6), 71–76 (2013)
12. H. He, M.E.J. Friese, N.R. Heckenberg, H. Rubinsztein-Dunlop, Direct observation of transfer of angular momentum to absorptive particles from a laser beam with a phase singularity. Phys. Rev. Lett. **75**(5), 826–829 (1995)
13. S.W. Hell, J. Wichmann, Breaking the diffraction resolution limit by stimulated emission: stimulated-emission-depletion fluorescence microscopy. Opt. Lett. **19**(11), 780–782 (1994)
14. L. Li, F. Li, Beating the rayleigh limit: Orbital-angular-momentum-based super-resolution diffraction tomography. Phys. Rev. E **88**(3), 033205 (2013)
15. M. Lin, Y. Gao, P. Liu, J. Liu, Super-resolution orbital angular momentum based radar targets detection. Electron. Lett. **52**(13), 1168–1170 (2016)
16. M. Lin, Y. Gao, P. Liu, J. Liu, Improved OAM-based radar targets detection using uniform concentric circular arrays. Int. J. Antennas Propag. **2016**, 1–8 (2016)
17. K. Liu, Y. Cheng, Z. Yang, H. Wang, Y. Qin, X. Li, Orbital-angular-momentum-based electromagnetic vortex imaging. IEEE Antennas Wireless Propag. Lett. **14**, 711–714 (2015)
18. K. Liu, Y. Cheng, X. Li, Y. Qin, H. Wang, Y. Jiang, Generation of orbital angular momentum beams for electromagnetic vortex imaging. IEEE Antennas Wireless Propag. Lett. **15**, 1873–1876 (2016)
19. K. Liu, Y. Cheng, Y. Gao, X. Li, Y. Qin, H. Wang, Super-resolution radar imaging based on experimental OAM beams. Appl. Phys. Lett. **110**(16), 164102 (2017)
20. C. Maurer, A. Jesacher, S. Bernet, M. Ritsch-Marte, What spatial light modulators can do for optical microscopy. Laser Photonics Rev. **5**(1), 81–101 (2011)
21. A.T. O'Neil, I. Macvicar, L. Allen, M.J. Padgett, Intrinsic and extrinsic nature of the orbital angular momentum of a light beam. Phys. Rev. Lett. **88**(5), 053601 (2002)
22. N.B. Simpson, K. Dholakia, L. Allen, M.J. Padgett, Mechanical equivalence of spin and orbital angular momentum of light: an optical spanner. Opt. Lett. **22**(1), 52–54 (1997)
23. G.A. Swartzlander, Peering into darkness with a vortex spatial filter. Opt. Lett. **26**(8), 497–499 (2001)
24. G.A. Swartzlander, E.L. Ford, R.S. Abdul-Malik, L.M. Close, M.A. Peters, D.M. Palacios, D.W. Wilson, Astronomical demonstration of an optical vortex coronagraph. Opt. Express **16**(14), 10 200–10 207 (2008)

25. K. Volke-Sepúlveda, A.O. Santillán, R.R. Boullosa, Transfer of angular momentum to matter from acoustical vortices in free space. Phys. Rev. Lett. **100**(2), 024302 (2008)
26. J. Wang, W. Zhang, Q. Qi, S. Zheng, L. Chen, Gradual edge enhancement in spiral phase contrast imaging with fractional vortex filters. Sci. Rep. **5**, 15826 (2015)
27. T. Yuan, H. Wang, Y. Qin, Y. Cheng, Electromagnetic vortex imaging using uniform concentric circular arrays. IEEE Antennas Wireless Propag. Lett. **15**, 1024–1027 (2016)
28. T. Yuan, H. Liu, Y. Cheng, Y. Qin, H. Wang, Orbital-angular-momentum-based electromagnetic vortex imaging by least-squares method, in *Proceedings of the IEEE International Geoscience Remote Sensing Symposium* (2016), pp. 6645–6648
29. T. Yuan, Y. Cheng, H. Wang, Y. Qin, Beam steering for electromagnetic vortex imaging using uniform circular arrays. IEEE Antennas Wireless Propag. Lett. **16**, 704–707 (2017)
30. C. Zhang, D. Chen, Large-scale orbital angular momentum radar pulse generation with rotational antenna. IEEE Antennas Wireless Propag. Lett. **16**, 2316–2319 (2017)

Index

R. Chen et al., *Circular Array-Based Radio Frequency OAM Communications*, Wireless Networks, https://doi.org/10.1007/978-3-031-37589-7

MIX
Papier aus verantwortungsvollen Quellen
Paper from responsible sources
FSC® C105338

If you have any concerns about our products,
you can contact us on
ProductSafety@springernature.com

In case Publisher is established outside the EU,
the EU authorized representative is:
Springer Nature Customer Service Center GmbH
Europaplatz 3, 69115 Heidelberg, Germany

Printed by Libri Plureos GmbH
in Hamburg, Germany